Campylobacter Infection in Man and Animals

Editor

Jean-Paul Butzler, M.D., Ph.D.

Director, Infectious Disease Unit
Vrije Universiteit Brussel, and
Department of Clinical Microbiology
University St. Pierre Hospital
Brussels, Belgium

Director, WHO Collaborating Center
for *Campylobacter jejuni*

CRC Press
Taylor & Francis Group
Boca Raton London New York

CRC Press is an imprint of the
Taylor & Francis Group, an **informa** business

First published 1984 by CRC Press
Taylor & Francis Group
6000 Broken Sound Parkway NW, Suite 300
Boca Raton, FL 33487-2742

Reissued 2018 by CRC Press

Library of Congress Cataloging in Publication Data
Main entry under title:

Campylobacter infection in man and animals.

Bibliography: p.
Includes index.
1. Campylobacter infections. 2. Campylobacter.
I. Butzler. Jean-Paul, 1941- [DNLM: 1. Campylobacter infections. 2. Campylobacter infections — Veterinary. QW 154 C199]
QR201.C25C35 1984 616.9'355 83-7363
ISBN 0-8493-5446-3

A Library of Congress record exists under LC control number: 83007363

ISBN 13: 978-1-315-89129-3 (hbk)
ISBN 13: 978-1-351-07039-3 (ebk)

Visit the Taylor & Francis Web site at http://www.taylorandfrancis.com and the
CRC Press Web site at http://www.crcpress.com

INTRODUCTION

Frequent in animals, particularly in bovines and ovines, campylobacter has been known for more than 40 years exclusively as a veterinary disease. In 1909, two veterinary surgeons, McFadyean and Stockman, discovered the animal disease during a survey on epizootic abortion in ewes.[1] They observed an unknown bacterium, frequently present in abortion, isolated from aborted fetuses and resembling a vibrion.

In 1913, the same authors demonstrated that their organism could be observed in infectious abortions, but World War I prevented diffusion of their work.[2]

In 1919, Theobald Smith, while investigating infectious abortions of bovines in the U.S., isolated (apart from the "Bacillus of Bang") another bacterium which he described as a spirillum.[3] Finishing his study, he became acquainted with McFadyean and Stockman's work and assumed they had been studying the same bacteria. He confirmed this, together with Taylor, and proposed "*Vibrio fetus*" as a name.[3]

In 1931, Jones, Orcutt, and Little[4] attributed winter dysentery in calves to infection with a "vibrio" they called *Vibrio jejuni* and in 1944 Doyle[5] described a similar organism associated with swine dysentery which in man was a milkborne outbreak of acute diarrheal illness reported by Levy in 1946;[6] organisms resembling *Vibrio jejuni* were seen in blood cultures from several of the victims; but they could not be isolated on solid media and thus positively identified. In 1947 Vinzent isolated *Vibrio fetus* from the blood of three pregnant women, admitted because of fever of unknown origin.[7] The illness lasted about 4 weeks, and two of the three women aborted. On examination, the placenta revealed large necrotic and inflammatory areas.

In 1949, Stegenga and Terpstra demonstrated the pathogenic role of *C. fetus venerealis* in enzootic sterility in cows.[8] In 1959, Florent was able to distinguish two types of *C. fetus* by their biochemical characteristics and pathogenic powers: *C. fetus venerealis* and *C. fetus intestinalis*.[9]

In 1957, E. King described a *Vibrio* that presented several features in common with the agents described by Vinzent, but had different biochemical and antigenic characteristics. She called it *related Vibrio*.[10] This condition was for a long time unrecognized. Indeed, until 1972 only 12 cases of related *Vibrio* infections were known: 7 infants, 2 children, and 3 adults.[11-14] The reason for the paucity of reports was that the selective culture techniques necessary for the isolation of the *Vibrio*, later renamed by Sebald and Veron as *Campylobacter*,[15] from feces were not known at that time. Consequently, the infection could be diagnosed only from the blood of bacteremic patients. However, King believed that the infection was not so rare as these few reports suggested and she emphasized the need to devise a method for culturing the organisms from feces. It is sad that such a method was not developed in her lifetime.

It was not until 1972 in Brussels that the application of veterinary techniques to the culture of human material provided the necessary breakthrough.[16] The Belgian team isolated *Campylobacter jejuni* from five percent of children with diarrhea[17] and later Skirrow[18] confirmed and extended this observation. Since then other workers have reported similar findings[19-24] and in some laboratories *Campylobacter* isolations outnumbered those of *Salmonella* and *Shigella* together.

In human pathology we distinguish *C. fetus intestinalis*, now called *C. fetus* subsp. *fetus* and *C. jejuni*. *C. fetus* subsp. *fetus* is an opportunist, chiefly attacking debilitating patients with impaired defences against infections. *C. jejuni* is one of the most common etiologic agents of bacterial diarrhea.

Veterinary medicine has to investigate the etiologic role of different new subspecies of *Campylobacter*. There is a lot of evidence to consider human campylobacteriosis as a zoonosis. This book reflects the different efforts made by veterinary and medical doctors for a better knowledge of the disease. It shows how much we depend upon each other to understand better the clinical features, pathogenesis, and epidemiology of campylobacter infections and other diseases.

I wish to thank all my colleagues and friends for their contributions.

J. P. Butzler

REFERENCES

1. **McFadyean, J. and Stockman, S.,** Report of the Departmental Committee Appointed by the Board of Agriculture and Fisheries to Inquire into Epizootic Abortion, Appendix, D., Her Majesty's Stationery Office, London, I, 156, 1909.
2. **McFadyean, J. and Stockman, S.,** Report of the Departmental Committee Appointed by the Board of Agriculture and Fisheries to Inquire into Epizootic Abortion, Appendix to Part III, Her Majesty's Stationery Office, London, 1913.
3. **Smith, T. and Taylor, M. S.,** Some morphological and biological characteristics of the spirilla associated with disease of fetal membranes in cattle, *J. Exp. Med.*, 30, 299, 1919.
4. **Jones, F. S., Little, R. B., and Orcutt, M.,** A continuation of the study of the aetiology of infectious diarrhoea (winter scours) in cattle, *J. Am. Vet. Med. Ass.*, 81, 610, 1932.
5. **Doyle, L. P.,** A Vibrio associated with swine dysentery, *Am. J. Vet. Res.*, 5, 3, 1944.
6. **Levy, A. J.,** A gastroenteritis outbreak probably due to a bovine strain of vibrio, *Yale Jour. Biol. Med.*, 18, 243, 1946.
7. **Vinzent, R., Dumas, J., and Picard, N.,** Septicémie grave au cours de la grossesse due à un Vibrion. Avortement consécutif., *Bull. Acad. Nat. Méd. Paris*, 131, 90, 1947.
8. **Stegenga, Th. and Terpstra, J. I.,** Over *Vibrio foetus* infekties bij het rund en enzoötische steriliteit, *Tijdschr. Diergeneesk.*, 74, 293, 1945.
9. **Florent, A.,** Les deux Vibrioses génitales de la bête bovine: la vibriose vénérienne, due à *V. foetus venerialis*, et la Vibriose d'origine intestinale due à *V. foetus intestinalis*, *Proc. 16th Int. Vet. Congr.*, 2, 489, 1959.
10. **King, E. O.,** Human infection with *Vibrio foetus* and a closely related *Vibrio*, *J. Inf. Dis.*, 101, 119, 1957.
11. **Wheeler, W. E. and Borchers, J.,** Vibrionic enteritis in infants, *Am. J. Dis. Child.*, 101, 60, 1961.
12. **Middelkamp, J. M. and Wolf, H. A.,** Infections due to "related vibrio", *J. Ped.*, 59, 318, 1961.
13. **Darrell, J. H., Farrell, B. C., and Mulligan, R. A.,** Case of human vibriosis, *Brit. Med. J.*, 2, 287, 1967.
14. **White, W. D.,** Human vibriosis: indigenous cases in England, *Brit. Med. J.*, 2, 283, 1967.
15. **Veron, M. and Chatelain, R.,** Taxonomic study of the genus *Campylobacter* Sebald and Veron and designation of the neotype strain for *C. foetus* (Smith and Taylor) Sebald and Veron, *Inst. J. Syst. Bact.*, 23, 122, 1973.
16. **Dekeyser, P., Gossuin-Detrain, M., Butzler, J. P., and Sternon, J.,** Acute enteritis due to related Vibrio: first positive stool culture, *J. Inf. Dis.*, 125, 390, 1972.
17. **Butzler, J. P., Dekeyser, P., Detrain, M., and Dehaen, F.,** Related Vibrio in stools, *J. Ped.*, 82, 318, 1973.
18. **Skirrow, M. B.,** Campylobacter enteritis: a "new" disease, *Brit. Med. J.*, 2, 9, 1974.
19. **Lauwers, S., De Boeck, M., and Butzler, J. P.,** Campylobacter enteritis in Brussels, *Lancet*, 1, 604, 1978.
20. **De Mol, P. and Bosmans, E.,** Campylobacter enteritis in Central Africa, *Lancet*, 1, 604, 1978.
21. **Severin, W. P. J.,** Campylobacter enteritis, *Ned. T. Gen.*, 122, 499, 1978.
22. **Lindquist, B., Kjellander, J., and Kosunen, T.,** Campylobacter enteritis in Sweden, *Br. Med. J.*, 1, 303, 1978.
23. **Blaser, M. J., Berkowitz, I. D., Laforce, F. M., Cravens, J., Reller, L. B., and Wang, W. L.,** Campylobacter enteritis and epidemiologic features, *Ann. Int. Med.*, 91, 179, 1979.
24. **Karmali, M. A. and Fleming, P. C.,** Campylobacter enteritis in children, *J. Pedatr.*, 94, 527, 1979.

THE EDITOR

Jean-Paul Butzler, M.D., Ph.D., is Professor of Clinical Microbiology and Epidemiology and Director of the Infectious Disease Unit at the Vrije Universiteit Brussel. Since 1974 he has been Director of the Department of Clinical Microbiology of the St. Pierre University Hospital in Brussels. Dr. Butzler is also Director of the World Health Organization Collaborating Center for *Campylobacter jejuni.* He is a member of many societies, among which are the Royal Society of Medicine, the British Society for the Study of Infection, the British Society for Antimicrobial Chemotherapy, the Infectious Disease Society of the Netherlands, the New York Academy of Sciences, and the American Society for Microbiology.

Among other awards, he is a recipient of the Award of the Belgian Royal Academy of Medicine for his work in infectious diseases and of the Van Beneden and Brohée awards for his research in gastroenterology. Dr. Butzler is author of over 150 articles. He has presented his research findings at more than 70 international meetings. His current major research interests include rapid diagnosis and pathogenesis of infectious diseases and the control of diarrhoeal disease in relation to child morbidity and mortality in developing countries.

To Miss Elisabeth King (1912—1966)
whose vision and diligence
paved the way

CONTRIBUTORS

R. R. Al-Mashat, Ph.D.
University of Glasgow Veterinary School
Glasgow
United Kingdom

Martin Blaser, M.D.
Chief
Infectious Disease Section
V.A. Medical Center
Denver, Colorado

Michael P. Doyle, Ph.D.
Assistant Professor
Department of Food Microbiology and Toxicology
Food Research Institute
University of Wisconsin
Madison, Wisconsin

Marcel De Boeck, Ph.D.
Department of Microbiology
St. Pieters University Hospital
Brussels
Belgium

Joseph Dekeyser, D.V.M.
Head
Department of Large Animal Pathology
National Institute for Veterinary Research
Brussels
Belgium

Patrick de Mol, M.D.
Hôpital Universitaire de Butare
Butare
Rwanda

Roger A. Feldman, M.D.
Director
Division of Bacterial Diseases
Center for Infectious Diseases
Centers for Disease Control
Atlanta, Georgia

Herman Goossens, M.D.
Department of Microbiology
St. Pieters University Hospital
Brussels
Belgium

Dennis Jones, M.D.
Director
Public Health Laboratory
Withington Hospital
Manchester
United Kingdom

Mohamed Karmali, M.D.
Assistant Professor of Medical Microbiology
The Hospital for Sick Children and
University of Toronto, and
Ontario Ministry of Health Career Scientists
Toronto, Ontario
Canada

Bertil Kaijser, M.D.
Department of Clinical Bacteriology
Institute of Medical Microbiology
University of Göteborg
Göteborg
Sweden

Sabine Lauwers, M.D.
Director of Microbiology
Department of Microbiology
A.Z. Free University of Brussels
Brussels
Belgium

G. H. K. Lawson, B.V.M. and S.
Department of Veterinary Pathology
University of Edinburgh
Edinburgh
Scotland

Hermy Lior
Chief
National Enteric Reference Centre
Laboratory Centre for Disease Control
Ottawa, Ontario
Canada

Susan M. Logan
Department of Biochemistry and Microbiology
University of Victoria
Victoria
Canada

Bibhat K. Mandal, F.R.C.P.
Consultant Physician and Lecturer
Regional Department of Infectious Diseases
University of Manchester, School of Medicine
Monsall Hospital
Manchester
United Kingdom

Diane Newell, Ph.D.
Principal Microbiologist
Public Health Laboratory
Southampton General Hospital
Southampton
United Kingdom

John L. Penner, Ph.D.
Associate Professor
Department of Medical Microbiology
Faculty of Medicine
University of Toronto
Banting Institute
Toronto
Canada

A. C. Rowland, M.R.C.V.S.
Senior Lecturer
Department of Veterinary Pathology
University of Edinburgh Veterinary Field Station
Easter Bush, Roslin
Midlothian, Scotland

Martin B. Skirrow, M.B., Ph.D., F.R.C. Path.
Consultant Microbiologist
Worcester Royal Infirmary
Worcester
United Kingdom

Peter Speelman, M.D.
Associate Scientist
International Centre for Diarrhoeal Disease Research, Bangladesh
Dhaka
Bangladesh

Marc J. Struelens, M.D.
Scientific Fellow
International Centre for Diarrhoeal Disease Research, Bangladesh
Dhaka
Bangladesh

Å. Svedhem, M.D.
Department of Clinical Bacteriology
Institute of Medical Microbiology
University of Göteborg
Göteborg
Sweden

David J. Taylor, M.A. Vet., M.B., Ph.D.
University of Glasgow Veterinary School
Glasgow
United Kingdom

David N. Taylor, M.D.
Department of Bacteriology and Clinical Laboratory Science
US Army Medical Component
Armed Forces Research Institute of Medical Sciences
Bangkok
Thailand

Diane E. Taylor, Ph.D.
Department of Medical Microbiology
Faculty of Medicine
The University of Alberta
Edmonton
Canada

Trevor J. Trust, Ph.D.
Department of Biochemistry and Microbiology
University of Victoria
Victoria
Canada

Raymond Vanhoof, M.D.
Instituut Pasteur van Brabant
Brussels
Belgium

Herman Van Landuyt, M.D.
Department of Microbiology
A.Z. St. Jan
Ruddershove
Brugge
Belgium

TABLE OF CONTENTS

Chapter 1
Taxonomy of the Genus *Campylobacter* 1
Mohamed A. Karmali and Martin B. Skirrow

Chapter 2
Clinical Aspects of *Campylobacter* Infections in Humans 21
Bibhat K. Mandal, P. De Mol, and J. P. Butzler

Chapter 3
Campylobacter jejuni in Travellers' Diarrhea 33
Peter Speelman and Marc J. Struelens

Chapter 4
Isolation of *Campylobacter jejuni* from Human Feces 39
H. Goossens, M. De Boek, H. van Landuyt, and J. P. Butzler

Chapter 5
Serotyping *Campylobacter jejuni* and *Campylobacter coli* on the Basis of Thermostable Antigens 51
S. Lauwers and J. L. Penner

Chapter 6
Serotyping of *Campylobacter jejuni* by Slide Agglutination Based on Heat-Labile Antigenic Factors 61
Hermy Lior

Chapter 7
Susceptibility of Campylobacters to Antimicrobial Agents 77
R. Vanhoof

Chapter 8
Plasmids from *Campylobacter* 87
Diane E. Taylor

Chapter 9
Serological Responses to *Campylobacter jejuni* Infection 97
D. M. Jones

Chapter 10
Diagnostic Serology for *Campylobacter jejuni* Infections Using the DIG-ELISA Principle and a Comparison with a Complement Fixation Test 105
Å. Svedhem, S. Lauwers, and B. Kaijser

Chapter 11
Experimental Studies of Campylobacter Enteritis 113
D. G. Newell

Chapter 12
Outer Membrane and Surface Structure of *Campylobacter jejuni* 133
Trevor J. Trust and Susan M. Logan

Chapter 13
Epidemiology of *Campylobacter* Infections..........143
Martin J. Blaser, David N. Taylor, and Roger A. Feldman

Chapter 14
Campylobacter in Foods..........163
Michael P. Doyle

Chapter 15
Bovine Genital Campylobacteriosis..........181
Joseph Dekeyser

Chapter 16
Enteric Infections with Catalase-Positive Campylobacters in Cattle, Sheep, and Pigs..........193
D. J. Taylor and R. R. Al-Mashat

Chapter 17
Campylobacter sputorum Subspecies *mucosalis*..........207
G. H. K. Lawson and A. C. Rowland

Index..........227

Chapter 1

TAXONOMY OF THE GENUS *CAMPYLOBACTER*

Mohamed A. Karmali and Martin B. Skirrow

TABLE OF CONTENTS

I. Introduction 2

II. General Description 2

III. Catalase-Positive Campylobacters 3
- A. Historical Background 3
- B. Differentiation of Catalase-Positive Campylobacters from Catalase-Negative Campylobacters 8
- C. Differentiation of Catalase-Positive *Campylobacter* Species and Subspecies 8
- D. Description and Pathogenic Significance of Species and Subspecies 8
  1. *C. jejuni* and *C. coli* 8
  2. *C. fetus* subsp. *fetus* and *C. fetus* subsp. *venerealis* 11
  3. Nalidixic Acid-Resistant Thermophilic Campylobacters (NARTC) 11
  4. *C. fecalis* 12
  5. Aerotolerant *Campylobacter* Species 12
  6. Nitrogen-Fixing *Campylobacter* Species CI 13
  7. Free-Living Aspartate-Fermenting *Campylobacter* Species 13
- E. Methods of Strain Discrimination in the *C. jejuni-C. coli* Group 13
  1. Serotyping 13
  2. Biotyping 13

IV. Catalase-Negative Campylobacters 14
- A. Historical Background 14
- B. Description and Differentiation of the Catalase-Negative *Campylobacter* Group 15
- C. Principal Features and Pathogenic Significance of Catalase-Negative *Campylobacter* Species and Subspecies 15
  1. *C. sputorum* subsp. *sputorum* 15
  2. *C. sputorum* subsp. *bubulus* 15
  3. *C. sputorum* subsp. *mucosalis* 16
  4. *C. concisus* 16

V. Concluding Remarks 16

References 16

I. INTRODUCTION

Even though campylobacters were first recognized about 70 years ago,[1] their remarkable association with disease is only now becoming widely appreciated. Not only are these organisms among the most frequent causes of bacterial diarrhea in man,[2,3] but they are also leading causes of enzootic sterility in cattle and sporadic abortion in various domestic animals.[4]

The taxonomy of the genus *Campylobacter* has a confusing past. Renewed interest in this genus during the past decade has led to significant improvements in the identification, nomenclature, and classification of its members. The major impetus to these improvements has undoubtedly been the practical need to improve understanding of the epidemiology of campylobacter infections in man. The extent of current progress is such that descriptions of new and unnamed species are already appearing in the literature. Continuing modifications to the classification of campylobacters are thus to be anticipated. The purpose of this chapter is to provide a historical perspective to the taxonomy of the genus *Campylobacter*, and to outline practical approaches to the identification and classification of its species and strains. For a more detailed description of the taxonomy of this genus, the reader is referred to publications by Véron and Chatelain[5] and Smibert.[6,7]

II. GENERAL DESCRIPTION

Campylobacters were originally referred to as "micro-aerophilic vibrios". The new generic term *Campylobacter* ("curved rod" in Greek) was proposed by Sebald and Véron[8] in 1963 on the grounds that the microaerophilic vibrios differed significantly from *Vibrio cholerae* and certain other vibrios and vibrio-like organisms with respect to their biochemical and physiological properties, and their DNA base-pair ratios.

Campylobacters[5-7] are small, nonsporeforming, Gram-negative bacteria that have a characteristic curved, S-shaped, or spiral morphology. The cells may vary from 0.5 to 8 μm in length, and 0.2 to 0.5 μm in width. Virtually all members of the genus are oxygen sensitive and can grow only under conditions of reduced oxygen tension that vary from almost anaerobic to microaerobic for the different species; an aerotolerant *Campylobacter* group has recently been described.[9] The cells are highly motile with a characteristic rapid, darting, corkscrew-like motility; they have a single polar flagellum at one or both ends of the cell.

Campylobacters[5-7] use amino acids and tricarboxylic acid cycle intermediates as their principal energy sources. They are oxidase positive and reduce nitrates; some species are catalase positive, others are catalase negative. Their metabolism is strictly respiratory; carbohydrates are neither fermented nor oxidized. Campylobacters are rather inert when tested in conventional biochemical media and this has made it difficult to classify them on biochemical grounds.

The genus *Campylobacter* is classified together with the genus *Spirillum* in the family Spirillaceae.[10] The distinction between these genera is based on the number of polar flagella, the ability of cells to accumulate intracellular granules of polyhydroxybutyric acid (PHB), and the DNA base composition. Campylobacters usually have a single polar flagellum and are unable to accumulate PHB, whereas the spirilla have tufts of flagella at their poles and readily accumulate intracellular granules of PHB.

The guanine-plus-cytosine (G + C) content of the genus *Campylobacter* ranges from 29 to 38 mol %,[7,9,11,12] which, as pointed out by Neill et al.,[9] is among the lowest known for bacteria; the G + C ratio of the genus *Spirillum* ranges from 38 to 65 mol %.[10]

There are now about a dozen species and subspecies recognized in the genus *Campylobacter*. These may be conveniently separated into two broad groups: the catalase-positive campylobacters, and the catalase-negative campylobacters. A list of species and subspecies

Table 1
***CAMPYLOBACTER* SPECIES AND SUBSPECIES**

Catalase positive	Catalase negative
C. fetus subsp. *fetus*[35]	*C. sputorum* subsp. *sputorum*[35]
C. fetus subsp. *venerealis*[35]	*C. sputorum* subsp. *bubulus*[35]
C. jejuni[35]	*C. sputorum* subsp. *mucosalis*[95]
C. coli[35]	*C. concisus*[96]
NARTC group[38,67]	
C. fecalis[69]	
Aerotolerant *Campylobacter* sp.[9]	
Nitrogen-fixing *Campylobacter* sp.[75]	
Free-living *Campylobacter* sp.[76]	

belonging to each of these groups is shown in Table 1. The use of the catalase test, which has proved to be a very useful taxonomic tool in classifying campylobacters, was first reported in this context in 1955 by Bryner and Frank.[13]

III. CATALASE-POSITIVE CAMPYLOBACTERS

A. Historical Background

Microaerophilic vibrios, now considered to be campylobacters, were first described in 1913 in England by McFadyean and Stockman,[1] who implicated these organisms as causal agents of abortion in sheep. A few years later, in 1918, Smith[14] in the U.S. reported on the association of similar organisms with bovine abortion. The latter organisms, which were isolated from aborted bovine fetuses, were named *Vibrio fetus* by Smith and Taylor[15] and the disease became known as vibrionic abortion. In 1927, Smith and Orcutt[16] described the isolation of microaerophilic vibrios from cultures of livers and spleens obtained from calves with diarrhea. They noted that the calf diarrhea strains differed serologically from *Vibrio fetus* and speculated that the former might be causally linked to bovine enteritis.

Investigations into the role of microaerophilic vibrios in bovine enteritis were subsequently conducted by Jones and colleagues[17-19] who, in 1931, published evidence showing a causal relationship between these organisms and winter dysentery in cattle. They noted[19] that the calf enteritis strains "while presenting certain slight morphological differences such as length, the number of coils, and to some extent the depth of coils," resembled *Vibrio fetus* sufficiently to be regarded as a closely related group. Jones et al. named the calf enteritis strains *Vibrio jejuni*.[19]

Microaerophilic vibrios became linked with yet another disease when Doyle,[20] in 1944, suggested that these organisms caused swine dysentery. Doyle named the swine organism *Vibrio coli*.[21]

A significant development during the 1940s was the association of microaerophilic vibrios with human disease. In 1946, Levy[22] described a large institutional outbreak of gastroenteritis that affected about 350 people. Microaerophilic vibrios were isolated from blood cultures of 13 of 39 patients and were seen in fecal smears from about 20% of the cases. Levy suggested that these organisms were identical to *V. jejuni* described earlier by Jones et al.[19] in association with winter dysentery. Vinzent et al.,[23] in 1947, reported the isolation of a microaerophilic vibrio that they considered to be *V. fetus* from the blood culture of a pregnant woman who aborted during the course of a febrile illness. Vinzent[24] went on to describe two further cases of a similar nature.

V. fetus, V. jejuni, and *V. coli* owe their names more to their association with specific diseases in animals rather than to any observed taxonomic differences between them. The position was somewhat clarified in 1957 when King[25] showed that catalase-positive microaerophilic vibrios could be distinguished by their ability to grow at different temperatures.

Table 2
NOMENCLATURES USED FOR *CAMPYLOBACTER FETUS*

Original description	*Vibrio fetus* (Smith and Taylor, 1919)[15]	
Florent (1959)[27]	*Vibrio fetus* var. *intestinalis*	*Vibrio fetus* var. *venerealis*
Véron and Chatelain (1973)[5,a]	*Campylobacter fetus* subsp. *fetus*	*Campylobacter fetus* subsp. *venerealis*
Smibert (1974)[6]	*Campylobacter fetus* subsp. *intestinalis*	*Campylobacter fetus* subsp. *fetus*

[a] Officially recognized nomenclature in the "Approved Lists" of bacterial names.[35]

She noted that strains of *V. fetus* were able to grow at 25 and 37°C, but not at 42°C; by contrast, a closely related group of organisms which she termed "related vibrios" failed to grow at 25°C, but grew at 37°C, and even better at 42°C. Notably all four human isolates of "related vibrios" that she studied were obtained from blood cultures of infants or young children with diarrhea. In a subsequent report,[26] King suggested that her "related vibrios" were identical to the *V. jejuni* of Jones et al.[19] and *V. coli* of Doyle.[21] King[26] also hinted at possible morphological differences between *V. fetus* and the "related vibrios" when she observed that in cells of the latter the undulations tended to be closer together than in *V. fetus*.

During the late 1940s and early 1950s the clinical spectrum of abortion in cattle associated with *V. fetus* was becoming better understood. It became clear that there were two distinct clinical entities of bovine vibrionic abortion. The first was the syndrome of sporadic bovine abortion occurring among pregnant cows that were members of an otherwise fertile herd. The infection that resulted in abortion was acquired not venereally, but probably following a bacteremia by *V. fetus* that had established itself in the intestinal tract of the cow,[27,28] i.e., in a manner similar to that proposed for ovine vibrionic abortion.[29] The second clinical entity, of much greater economic importance, was that of sporadic abortion occurring in a herd with strikingly reduced conception rates.[30] In this syndrome of enzootic sterility or infectious infertility,[31] *V. fetus* was transmitted from the bull to the cow during coitus. The bull was a symptomless carrier of the organism on the prepuce, and transmitted the organism venereally to an entire herd. Infertility was the major consequence, although conception and subsequent abortion also sometimes occurred.

The reasons for the involvement of *V. fetus* in two apparently different abortion syndromes was clarified when Florent[27] showed in 1959 that there were in fact two different varieties of *V. fetus*. One variety, which he termed *V. fetus* var. *intestinalis,* originated in the intestine and caused sporadic abortion among members of a fertile herd; the second, which he called *V. fetus* var. *venerealis*, was the one transmitted venereally and implicated in the infertility syndrome.

Florent[27] distinguished *V. fetus* var. *intestinalis* from *V. fetus* var. *venerealis* on the basis of two tests: the ability of the strains to produce hydrogen sulfide (H_2S) in cysteine-containing media (using a lead acetate strip as indicator), and their ability to grow in the presence of 1% glycine, a test originally described by Lecce[32] for distinguishing ovine from bovine strains of *V. fetus*. *V. fetus* var. *intestinalis* grew in the presence of 1% glycine, and produced H_2S in a cysteine-containing medium, whereas *V. fetus* var. *venerealis* gave the opposite reactions.

Following the work of Sebald and Véron[8] in 1963, in which microaerophilic vibrios were assigned to the new genus *Campylobacter*, Véron and Chatelain,[5] in 1973, published the

Table 3
NOMENCLATURES USED FOR *C. JEJUNI* AND *C. COLI*

Original description	*Vibrio jejuni* (Jones et al., 1931)[19]	*Vibrio coli* (Doyle, 1948)[21]
King (1957)[25]	"Related vibrios"	
Véron and Chatelain (1973)[5a]	*Campylobacter jejuni*	*Campylobacter coli*
Smibert (1974)[6]	*Campylobacter fetus* subsp. *jejuni*	

[a] Officially recognized nomenclature in the "Approved Lists" of bacterial names.[35]

first comprehensive account of the taxonomy of this genus. In accordance with the *International Code of Nomenclature of Bacteria*,[33] Véron and Chatelain had to establish one of the species, preferably that which conformed most closely to the original description of *V. fetus* by Smith and Taylor,[15] as the type species of the genus *Campylobacter*. Since Smith and Taylor's original strains were no longer available for study, Véron and Chatelain[5] argued for selecting *V. fetus* var. *intestinalis* (Florent) as the type species and, according to the *International Code*,[33] renamed that species *C. fetus* subsp. *fetus* (Table 2). The choice of *V. fetus* var. *intestinalis* (Florent) rather than *V. fetus* var. *venerealis* for the type species was subsequently supported by Karmali et al.[34] who showed that the measurements of the wavelength and amplitude of cells corresponding to *V. fetus* var. *intestinalis* (Florent) were similar to those documented for *V. fetus* by Smith.[14]

Another major taxonomic study of the genus *Campylobacter* was published by Smibert[6] in 1974 in the 8th edition of *Bergey's Manual of Determinative Bacteriology*. In contrast to Véron and Chatelain's nomenclature, Smibert speculated that the original strains of *V. fetus* were more likely to have been *V. fetus* var. *venerealis* (Florent). He therefore designated the latter variety as the type species, and renamed it *C. fetus* subsp. *fetus* (Table 2). Smibert also gave the name *C. fetus* subsp. *jejuni* to strains that Véron and Chatelain called *C. jejuni* and *C. coli* (Table 3).

Because the nomenclatures of both Smibert[6] and Véron and Chatelain[5] have been widely used in the literature, confusion has arisen over the use of names for species and subspecies. The different nomenclatures used for catalase-positive campylobacters are listed in Tables 2 and 3. The officially recognized nomenclature for campylobacters published in the "Approved List" of bacterial names[35] is that of Véron and Chatelain.[5] It should be noted that the nomenclature to be used in the 9th edition of *Bergey's Manual of Determinative Bacteriology* (in press) is in accordance with the List of Approved Names.[101]

A dilemma until recently was whether *C. jejuni* and *C. coli* were different species or merely biotypes or variants within the same species; laboratory tests for distinguishing between the two were not available. A major breakthrough occurred in this area in 1980 when Harvey[36] showed that strains in the *C. jejuni-C. coli* group could be separated on the basis of their ability to hydrolyze hippurate.[37] Skirrow and Benjamin[38] separated the *C. jejuni-C. coli* group by cultural characteristics. Using Harvey's hippurate test,[36] they went on to show[39] that hippurate positivity was a feature of strains they considered to be *C. jejuni*, while hippurate negativity was linked with characteristics of strains they recognized as *C. coli*. It has now become established on the basis of DNA hybridization studies that the hippurate-positive strains *(C. jejuni)*, and hippurate-negative strains *(C. coli)* represent two different species.[11,12,40] There is reason to question the validity of the term *C. coli* because whereas strains now considered to be *C. coli* by Véron and Chatelain[5] and Skirrow and Benjamin[38,39] are nitrate positive, the strains originally described by Doyle[21] as *C. coli* were reported by him to be nitrate negative. This controversy is now of academic interest only, because the listing of "*C. coli* (Doyle 1948) Véron and Chatelain 1973" in the Approved

Lists of Bacterial names[35] gives official recognition to this species and assumes that the organisms studied by Doyle and Véron and Chatelain were the same. The type strain of *C. coli* (CIP 7080)[35] is nitrate positive. There are no reports in the recent literature of nitrate-negative strains in the *C. jejuni-C. coli* group. Nevertheless, in support of Doyle's nitrate-negative strains[21] there are reports in the 1950s by Kuzdas and Morse[41] and Di Liello et al.[42] which describe nitrate-negative strains of "*Vibrio jejuni*". It is thus conceivable that there may indeed exist within the *C. jejuni-C. coli* group a subgroup of nitrate-negative strains.

Apart from *C. fetus* subsp. *fetus, C. fetus* subsp. *venerealis, C. jejuni* and *C. coli*, a number of other species (Table 1) are now included in the catalase-positive *Campylobacter* group. These species are described below.

B. Differentiation of Catalase-Positive Campylobacters from Catalase-Negative Campylobacters

Catalase-negative campylobacters[7] tend to be more oxygen sensitive than catalase-positive campylobacters and thus require a lower level of oxygen tension for optimal growth than the latter. Virtually all *Campylobacter* species are able to reduce nitrates. However the reduction of nitrites is a feature only of the catalase-negative group. As a general rule, the production of hydrogen sulfide in triple sugar iron (TSI), Kligler's iron (KI), or SIM media is confined to the catalase-negative campylobacters; *C. fecalis* is the only catalase-positive *Campylobacter* species that produces hydrogen sulfide in these media.

C. Differentiation of Catalase-Positive *Campylobacter* Species and Subspecies

Practical tests for differentiating catalase-positive campylobacters are outlined in Table 4. Other differential characteristics are discussed below under the heading of individual species.

D. Description and Pathogenic Significance of Species and Subspecies

1. C. jejuni *and* C. coli

The *C. jejuni-C. coli* group of organisms has recently emerged from obscurity to rank among the leading causes of bacterial diarrhea in man.[2,3] This discovery was the result of pioneering work done by Butzler and colleagues[43] in Belgium and Skirrow[44] in England during the past decade.

The majority of human *Campylobacter* isolates from cases of gastroenteritis are *C. jejuni*[38,45,46] (Table 5). *C. jejuni* is widely distributed in the animal kingdom[4,38,45,47] and has been isolated from poultry and wild birds, domestic animals such as dogs, cats, cattle, sheep and pigs, primates, and also from various exotic species in zoos. *C. jejuni* is an established cause of sporadic abortion in sheep[4] and has, in the past, also been associated with avian infectious hepatitis,[48] bluecomb disease of turkeys,[49] and enteritis in cattle.[17-19]

C. coli accounts for only about 3 to 5% of the *Campylobacter* isolates from cases of human gastroenteritis[38,45,46] (Table 5); it appears to cause an illness in man indistinguishable from that caused by *C. jejuni*.[38,46] Although the pig is its major reservoir,[38,45] *C. coli* is also sometimes found in other domestic animals, but with a frequency considerably less than that for *C. jejuni*[45] (Table 5).

The principal features and differential characters of *C. jejuni* and *C. coli* are shown in Table 4. Both these species have a higher than usual optimal growth temperature, about 42°C.[25,26] It should be noted that the mere presence of growth at 42°C is not in itself sufficient for assigning an organism to the *C. jejuni-C. coli* group; as originally pointed out by King,[25] it must be shown that the organism grows better at 42°C than at 37°C to qualify for *C. jejuni-C. coli* status. This factor should help to resolve the identity of those strains of *C. fetus* subsp. *fetus* that reportedly also grow at 42°C.[50] Skirrow and Benjamin[38] observed that

Table 4
KEY DIFFERENTIAL CHARACTERISTICS OF CATALASE-POSITIVE CAMPYLOBACTERS[a,b]

	C. jejuni	*C. coli*	*C. fetus* subsp. *fetus*	*C. fetus* subsp. *venerealis*	NARTC	*C. fecalis*	Aerotolerant *Campylobacter* spp.
Cell size							
Average wavelength λ (μm)	1.12		1.80	2.43			
Average amplitude α (μm)	0.48		0.55	0.73			
Relative dimensions of λ and α	"Short"	"Short"	"Medium"	"Long"	"Short"	"Medium"	"Long"
Rapid coccal transformation	+	w/ –	–	–	+	–	–
Swarming on moist media	+	w/ –	–	–	–	–	–
Growth at							
25.0°C	–	–	+	+	–	w	+
30.5°C	–	d(78%)	+	+	+	+	+
37.0°C	+	+	+	+	+	+	w
43.0°C	+ +	+ +	–	–	+ +	w	–
45.5°C	d	d	–	–	+	–	–
Susceptibility to							
Nalidixic acid ¶	S	S	R	R	R	R	S
Cephalothin §	R	R	S	S	R	S	R
Hippurate hydrolysis	+	–	–	–	–	–	–
Growth in 1% glycine	+	+	+	–	+	+	
H_2S production:							
TSI or SIM medium	–	–	–	–	–	+	–
Iron/metabisulfite (FBP) medium	d	–	–	–	+/w	+	
Cysteine medium with PbAc strip	+	+	+	–	+	+	
Growth anaerobically in presence of:							
Fumarate	–	–	+	+	d(70%)		
Trimethylamine *N*-oxide	–	–	–	–	+		

Table 4 (continued)
KEY DIFFERENTIAL CHARACTERISTICS OF CATALASE-POSITIVE CAMPYLOBACTERS[a,b]

	C. jejuni	*C. coli*	*C. fetus* subsp. *fetus*	*C. fetus* subsp. *venerealis*	NARTC	*C. fecalis*	Aerotolerant *Campylobacter* spp.
Growth on triphenyl tetrazolium chloride agar (0.4 g/ℓ)	+	+	−	−	−		
Presence of C-19 fatty acid in cellular lipids	+	+	−	−	−		

Note:
+ = 90—100% strains positive
+ + = optimal reaction
− = 90—100% strains negative
d = 10—90% strains positive
w = weak positive or intermediate reaction
S = sensitive
R = resistant
¶ = 30 μg disc or 40 mg/ℓ
§ = 30 μg disc or 64 mg/ℓ

[a] Excluding the nitrogen-fixing *Campylobacter* species of McClung and Patriquin[75] and the free-living *Campylobacter* species of Laanbroek et al.[76]
[b] Information compiled from References 5-7, 34, 38, 45, 51, 53-59.

Table 5
DISTRIBUTION OF *C. JEJUNI*, *C. COLI*, AND NARTC IN MAN AND DOMESTIC ANIMALS

	Number of strains (%)				
	C. jejuni biotype 1	*C. jejuni* biotype 2	*C. coli*	NARTC	Total
Man[a]	157 (74)	42 (20)	12 (5.5)	1 (0.5)	212 (100)
Cattle	104 (83)	10 (8)	10 (8)	1 (1)	125 (100)
Sheep	55 (78)	8 (12)	7 (10)	0	70 (100)
Dogs	55 (68)	18 (22)	4 (5)	4 (5)	81 (100)
Chickens	29 (43)	21 (32)	15 (22)	2 (3)	67 (100)
Pigs	5 (5)	0	90 (95)	0	95 (100)

[a] Indigenous infections diagnosed at Worcester Royal Infirmary, U.K., over a period of 2 years.

a temperature of 43°C is more reliable for distinguishing the *C. jejuni-C. coli* group from the *C. fetus* group, but they placed greater relevance on growth at 25°C as a differential test.

It has been demonstrated by Karmali et al.[34] that the spiral forms of *C. jejuni, C. fetus* subsp. *fetus*, and *C. fetus* subsp. *venerealis* differ substantially with respect to wavelength and amplitude when observed by phase-contrast microscopy (Table 4); these differences can be distinguished by a trained observer on the basis of size alone, without actual measurement (Figure 1). *C. jejuni* rapidly undergoes coccal degeneration within 24 to 48 hr upon exposure to normal atmospheric air,[34] whereas the two *C. fetus* subspecies undergo coccal degeneration much less readily. *C. jejuni* has a pronounced tendency to swarm on moist agar plates, a feature that is absent in *C. fetus* subsp. *fetus* and *C. fetus* subsp. *venerealis*.

C. jejuni and *C. coli* are sensitive to nalidixic acid (40 mg/ℓ or a 30-μg nalidixic acid disc) and resistant to cephalothin (64 mg/ℓ or a 30-μg cephalothin disc) in contrast to the two *C. fetus* subspecies which give the opposite susceptibility pattern.[51,52] However, we have now recognized a few strains of *C. coli* (≃2% of strains in our collections) that are sensitive to cephalothin; zone sizes of between 20 to 40 mm around a 30-μg cephalothin disc[53] have been observed. A few strains of nalidixic-acid-resistant *C. jejuni* have also been recognized;[54] these are distinct from the NARTC group described below. *C. jejuni* and *C. coli* are unable to grow anaerobically in the presence of fumarate, aspartate, or nitrate, whereas *C. fetus* subsp. *fetus* and *C. fetus* subsp. *venerealis* are able to grow anaerobically in the presence of fumarate and also usually in the presence of aspartate and nitrate.[55,56] The ability of campylobacters to grow anaerobically in the presence of the latter substrates can be assessed readily by inoculation into semisolid agar and looking for growth throughout the medium as opposed to growth restricted to near the surface.[56]

Studies of the cellular fatty acid composition of campylobacters by gas-liquid chromatography[57-59] have also shown differences between the *C. jejuni-C. coli* group and the *C. fetus* group. Most strains of the former group contain a 19-carbon cyclopropane acid which is absent in *C. fetus* subsp. *fetus* and *C. fetus* subsp. *venerealis;* some strains in the *C. jejuni* — *C. coli* group have only trace amounts or undetectable levels of the 19-carbon acid.[60] Differences between the *C. jejuni—C. coli* and the *C. fetus* have also been evident in studies examining the total protein profiles of these organisms using polyacrylamide gel electrophoresis.[61] The *C. jejuni-C. coli* group are serologically distinct from the *C. fetus* group.[62]

C. jejuni and *C. coli* are differentiated on the basis of the hippurate hydrolysis test:[36,37,39]

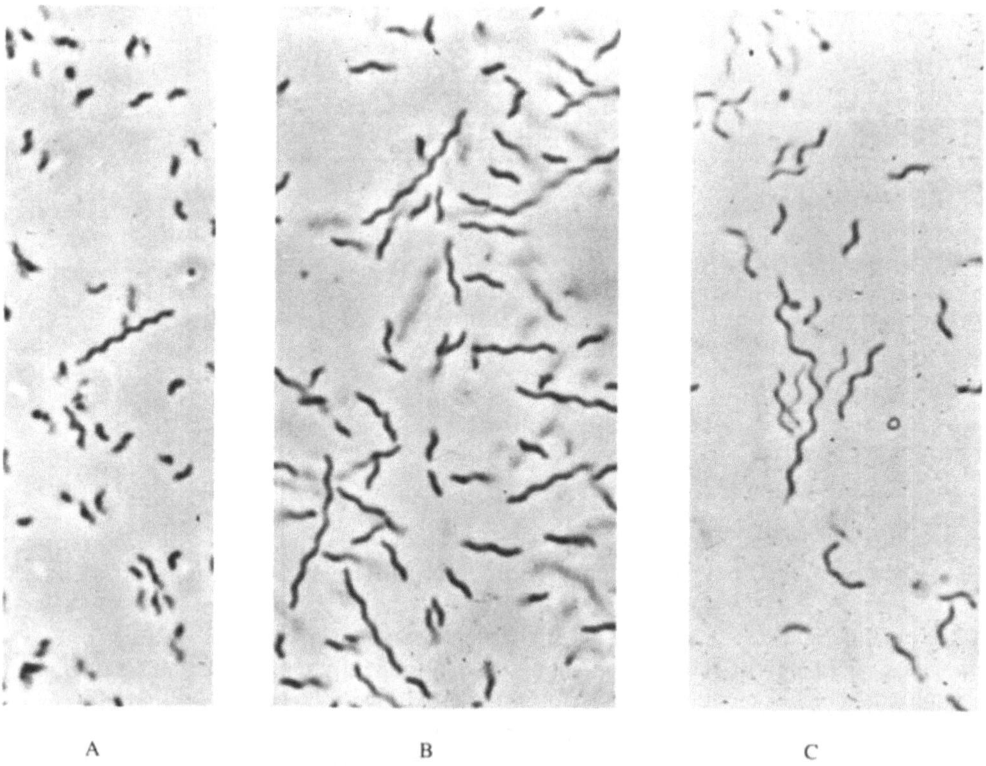

FIGURE 1. Phase-contrast photomicrographs of representative strains of *C. jejuni* (A), *C. fetus* subsp. *fetus* (B), and *C. fetus* subsp. *venerealis* (C). Magnification 1800 × for each strain.

C. jejuni is hippurate positive, and *C. coli* hippurate negative.[39] In addition, *C. coli* may be distinguished by its ability to grow at 30.5°C (78% of strains), whereas most strains (94%) of *C. jejuni* do not grow at this temperature.[38] There are other qualitative and in some cases subtle differences between the two species. *C. coli* grows somewhat more rapidly than *C. jejuni*. The tendency of *C. coli* to undergo rapid coccal transformation is less pronounced than that for *C. jejuni*, and *C. coli* shows only a limited tendency to swarm on moist agar plates. This latter difference is of practical significance because isolates of *C. coli* are likely to be missed on moist primary plates if only swarming or effuse colonies are looked for.

There have not been any studies published so far with respect to possible differences in cell dimensions between *C. jejuni* and *C. coli*. Our own observations suggest that subtle differences in cellular morphology probably do exist between the species. For example, the spiral forms of *C. coli* tend to be more "stretched out" than those of *C. jejuni*, and occasionally appear even rod-like.

Other tests that have been suggested for differentiating *C. jejuni* from *C. coli* include tolerance to brilliant green (1/100,000), triphenyl tetrazolium chloride (TTC), and 8% glucose.[5] Tolerance to brilliant green and 8% glucose have not proved to be useful tests.[38,51] Skirrow and Benjamin[38] found a wide range of tolerance to TTC which facilitated strain identification, but did not specifically separate *C. jejuni* from *C. coli*.

Strains of *C. jejuni* may be biotyped according to the H_2S test (in an iron/metabisulfite medium) described by Skirrow and Benjamin.[39] Biotype 1 strains are H_2S negative and biotype 2 strains are H_2S positive. Biotyping of *C. jejuni* is discussed in more detail later on in this chapter.

Estimates for the G+C content (mol %) of the DNA of *C. jejuni* have ranged from 29 to 34.[5,7,11,12]

2. C. fetus subsp. fetus and C. fetus subsp. venerealis

C. fetus subsp. *fetus* is the type species of the genus *Campylobacter*.[5,35] It is a recognized cause of sporadic abortion in cattle and sheep;[4] its distribution in other animal species is not known. In man, *C. fetus* subsp. *fetus* is an occasional cause of septicemic illness, particularly in patients who are immunocompromised or have other serious underlying disease.[63,64]

C. fetus subsp. *venerealis* is a major cause of enzootic sterility and abortion in cattle.[31] The isolation of this organism from other animal species has not been reported.

The differentiation of the *C. fetus* group from the *C. jejuni-C. coli* group has been discussed in the previous section and its differentiation from catalase-positive campylobacters in general is outlined in Table 4.

The number of tests for differentiating *C. fetus* subsp. *fetus* from *C. fetus* subsp. *venerealis* is limited. *C. fetus* subsp. *fetus* grows in the presence of 1% glycine and produces H_2S in cysteine-containing media (Table 4), whereas *C. fetus* subsp. *venerealis* does neither. A biotype of *C. fetus* subsp. *venerealis* termed biotype *intermedius* is recognized;[5] this produces an "intermediate" result in the latter tests in that it fails to grow in the presence of 1% glycine, but does produce H_2S in cysteine-containing media. Chang and Ogg[65] have questioned the reliability of the glycine test for separating *C. fetus* subsp. *fetus* from *C. fetus* subsp. *venerealis* because the ability to grow in the presence of 1% glycine is a transducible characteristic. The use of only the H_2S and glycine tests for separating the two *C. fetus* subspecies is clearly unsatisfactory. Karmali et al.[34] have shown that they can be distinguished on the basis of cell size with respect to the wavelength and amplitude of the spiral forms (Table 4 and Figure 1).

No distinct serological differences have been reported between *C. fetus* subsp. *fetus* (Group 3, serotype A in the classification of Berg et al.)[62] and *C. fetus* subsp. *venerealis* (Groups 1 and 2, both also serotype A).[62] The DNA relatedness between the two subspecies is close.[12] Most strains of *C. fetus* subsp. *venerealis* grow less well than *C. fetus* subsp. *fetus*.[66]

Because of the paucity of differential characters between *C. fetus* subsp. *fetus* and *C. fetus* subsp. *venerealis*, Véron and Chatelain[5] consider the latter to be a "defective mutant" of *C. fetus* subsp. *fetus* that has adapted itself to a very restricted ecological niche, the genital tract of cattle.

Estimates for the G + C content (mol %) of the DNA of *C. fetus* subsp. *fetus* and *C. fetus* subsp. *venerealis* have ranged from 33 to 36.[5,7,11,12]

3. Nalidixic Acid-Resistant Thermophilic Campylobacters (NARTC)

NARTC strains were first isolated by Skirrow and Benjamin[38] from the cloacal contents of wild seagulls of the genus *Larus*. They have been isolated from about 25% of apparently healthy seagulls, but only infrequently from other animal species[38,45] (Table 5). They have been isolated from the feces of four children, two of whom were symptomless and two who had mild recurrent diarrhea.[54] The pathogenicity of NARTC strains is not known.

The main differential characters of NARTC strains are outlined in Table 4. NARTC strains[54,67] resemble the *C. jejuni-C. coli* group in being thermotolerant and cephalothin-resistant. They resemble *C. jejuni* in having a short cellular wavelength and amplitude and in undergoing rapid coccal transformation, but differ from the latter in their decreased tendency to swarm on moist agar plates. They resemble *C. jejuni* biotype 2 strains in their ability to produce H_2S in the iron/metabisulfite medium used by Skirrow and Benjamin.[39] They resemble *C. coli* in being hippurate negative, being able to grow at 30.5°C, and showing little or no swarming on moist media. NARTC strains differ from the *C. jejuni-C. coli* group in being resistant to nalidixic acid and in having the ability to grow anaerobically in the presence of aspartate or trimethylamine *N*-oxide HCl.[56,67,68]

Recent DNA hybridization studies[40] have indicated that NARTC strains probably constitute a distinct species; the DNA relatedness between a reference NARTC strain (NCTC 11352) and representative strains of *C. jejuni* biotypes 1 and 2, and *C. coli* was found to be 17, 26, and 13%, respectively. The mean value for the G+C content (mol %) of the DNA of NARTC was found to be 32.1 ± 0.5.[11]

The new species name *C. laridis* (λαρos, Laros — a sea bird) has been proposed for the NARTC strains.[67]

4. C. fecalis

During the course of an investigation into the occurrence of *C. fetus* subsp. *fetus* in ovine feces, Firehammer[69] in 1965 isolated a hitherto unrecognized microaerophilic vibrio that was quite distinct from *C. fetus* subsp. *fetus*. This organism, which he named *Vibrio fecalis*, is now referred to as *Campylobacter fecalis*. *C. fecalis* has also been isolated from bovine semen and vagina.[7] Its pathogenic significance is not known.

C. fecalis can easily be differentiated from other catalase-positive campylobacters since it is the only member of the group that produces hydrogen sulfide in peptone iron media such as triple sugar iron, Kligler's iron, or SIM media (Table 4); this feature is more characteristic of the catalase-negative campylobacters.[7]

C. fecalis strains grow optimally at 37°C but also grow at 42°C, and poorly at 25°C; they are resistant to nalidixic acid and sensitive to cephalothin.[53]

Owen and Leaper[11] have reported the G+C content (mol %) of a single strain of *C. fecalis* to be 36.6.

5. Aerotolerant Campylobacter *Species*

Ellis et al.[70] in 1977 reported on the isolation of *Spirillum*-like organisms from the internal organs of aborted bovine fetuses. Primary isolation of these organisms was achieved only in semisolid *Leptospira* isolation media incubated at 30°C; growth was detectable as a cloudy zone slightly below the surface of the media after about 2 to 3 days of incubation. However, after one or more passes in semisolid media, visible growth could be obtained after 24 hr on blood agar under aerobic, microaerobic, and anaerobic conditions; growth occurred at 37°C, but more readily at 30°C. Ellis et al.[71] went on to isolate similar organisms from aborted porcine fetuses. Subsequent characterization of these aerotolerant organisms led to their inclusion in the genus *Campylobacter*.[9] Neill et al.[9,72] showed that these organisms (which they termed "group 2" strains) were, like the *C. jejuni-C. coli* group, susceptible to 40 mg/ℓ of nalidixic acid, but unlike the latter, grew at 25°C but not at 42°C. The isolation of aerobic vibrios conforming to the description of campylobacter from bovine genital sources has also been reported previously by Florent.[27] In addition to their proposed etiological role in bovine and porcine abortion, aerotolerant campylobacters have recently also become implicated as causes of mastitis in cows.[73]

In limited studies[53] performed on five strains of aerotolerant *Campylobacter* species supplied by Dr. S. D. Neill, we found all strains to be hippurate-negative, cephalothin-resistant (30-μg cephalothin disc), and negative for hydrogen sulfide in triple sugar iron medium; the strains did not undergo rapid coccal transformation nor did they swarm on moist agar plates. An unexpected finding was that all strains that we tested were negative for nitrate reduction using the plate nitrate method[74] (with *C. fetus* subsp. *fetus* as a positive control). Confirmation of our results would make aerotolerant *Campylobacter* species the only campylobacters with available strains known to be nitrate-negative.

Substantial pleomorphism in cellular morphology is seen when 48-hr blood agar cultures (grown at 30°C in an airtight polyethylene bag containing a culture of *S. aureus* to reduce the oxygen tension) are examined by phase-contrast microscopy;[53] single cells appear slightly curved or rod-like, while the morphology of chains of cells varies from spiral to rod-like with intermediate forms showing variable degrees of flattening of the wave forms. The size range of the wavelength and amplitude of the spiral forms resembles that of *C. fetus* subsp. *venerealis*.

Using the buoyant density method, Neill et al.[9] found the G+C content (mol %) of the aerotolerant "group 2" *Campylobacter* strains to range from 29 to 34.

6. Nitrogen-Fixing Campylobacter *Species CI*

McClung and Patriquin[75] in 1980 reported the isolation of a microaerophilic nitrogen-fixing bacterium from the roots (which had been surface sterilized) of the plant *Spartina alterniflora* Loisel growing in a Nova Scotian salt marsh. Further study of this organism, referred to as CI, supported its inclusion in the genus *Campylobacter*. It is the only plant-associated *Campylobacter* species that has been described. It is distinguished from other campylobacters by various biochemical and physiological characteristics, but particularly by its unique ability to fix nitrogen, hydrolyze urea, produce pigment from tryptophan, and grow in the presence of up to 7% sodium chloride. Its association with plant roots further distinguishes it from the other campylobacters which usually inhabit animals. The G+C content (mol %) of CI was reported to be 32.1 ± 1.0.[75]

7. "Free-Living" Aspartate-Fermenting Campylobacter *Species*

While studying the microflora of an anaerobic digester continuously fed with waste water from a potato-flour factory, Laanbroek et al.[76] isolated a free-living aspartate-fermenting microaerophilic vibrio. This was subsequently identified as a *Campylobacter* species. It could be differentiated from other campylobacters on biochemical and physiological grounds; it was found to have an unusually high (for the genus *Campylobacter*) mean G+C content (mol %) of 41.6.

E. Methods of Strain Discrimination in the *C. jejuni-C. coli* Group

The need for studying the epidemiology of *C. jejuni-C. coli* infections has led to significant progress in the development of methods for discriminating among strains of these two species by means of serotyping and biotyping. Phages have been described in campylobacters,[77] and phage-typing methods for the *C. jejuni-C. coli* group are being developed.[78] The occurrence of bacteriocins in campylobacters has not been reported. However, there have been reports of the inhibitory action of bacteriocins from *Pseudomonas aeruginosa* on some *Campylobacter* strains.[79]

1. Serotyping

Penner and Hennessy[80] in Canada and Lauwers[81] in Belgium have developed a serotyping system based on thermostable soluble antigens. The Penner and Hennessy system now recognizes over 50 different serogroups. Lior et al.,[82] also in Canada, have developed a typing system based on thermolabile antigens containing 21 different serogroups. Serotyping systems are discussed in separate chapters in this volume.

2. Biotyping

Skirrow and Benjamin[39] made the observation that some strains of *C. jejuni* caused blackening in a medium supplemented with the FBP (ferrous sulfate, sodium metabisulfite, sodium pyruvate) supplement described by Hoffman et al.[83] A positive reaction was evident within 4 hr of inoculation and clearly indicates the presence of a preformed enzyme, presumably a sulfatase, leading to the production of hydrogen sulfide. Strains that were H_2S negative in the latter test were referred to as *C. jejuni* biotype 1, and those that were H_2S positive, as biotype 2 by Skirrow and Benjamin.[39] The relative distribution of the two biotypes among human and nonhuman isolates is shown in Table 5. The precise nature of the enzyme(s) responsible for H_2S production in FBP-supplemented media as well as the optimal conditions for performing the test are currently under investigation. It should be emphasized that performance in the H_2S test using FBP media bears no relationship to H_2S production using conventional methods (cysteine, lead acetate, or T.S.I. tests).

Table 6
KEY DIFFERENTIAL CHARACTERISTICS OF CATALASE-NEGATIVE CAMPYLOBACTERS[a]

	Growth in the presence of			Dirty yellow colonies
	1% Glycine	1% Bile	3.5% NaCl	
C. sputorum subsp. *sputorum*	−	−	−	−
C. sputorum subsp. *bubulus*	+	−	+	−
C. sputorum subsp. *mucosalis*	−	+	−	+
C. concisus	−	+	−	−

[a] Information compiled from References 5—7, 89, 90, 94—97.

Skirrow and Benjamin[38] have also shown that the differential susceptibility of strains of *C. jejuni-C. coli* to triphenyl tetrazolium chloride and metronidazole may be a valuable tool in the investigation of outbreaks of infections due to these organisms.

Hébert et al.[84] have recently described the use of three tests (hippurate hydrolysis, DNA hydrolysis, and growth on charcoal-yeast extract agar) for separating the members of the *C. jejuni-C. coli* group into eight biotypes. The latter scheme remains to be tested under field conditions.

IV. CATALASE-NEGATIVE CAMPYLOBACTERS

The four main taxons in the catalase-negative *Campylobacter* group are listed in Tables 1 and 6.

A. Historical Background

Tunnicliff[85] in 1914 reported the isolation of an anaerobic vibrio that "was found in fair numbers in the sputum of a patient suffering with acute bronchitis". It was described as being a strictly anaerobic, Gram-negative organism showing one or two curves but sometimes more. When stained by the Zettnow method, the vibrio showed "one long fine wavy flagellum attached to its extremity." Tunnicliff's albeit limited description served to differentiate her anaerobic vibrio from a number of other anaerobic vibrio-like organisms often present in the human mouth. In 1940, Prevot[86] named the Tunnicliff anaerobic vibrio *Vibrio sputorum*. MacDonald[87] in 1953 isolated 12 strains of *V. sputorum* from the oral cavity and characterized them as being straight or curved rods possessing usually monotrichous polar flagella; all strains were strictly anaerobic and produced hydrogen sulfide in peptone-iron medium.

Florent[88] in 1953 isolated what he referred to as a saprophytic vibrio from the bovine vagina and semen. This vibrio differed from *Vibrio fetus* in that it was more anaerobic than the latter, and also produced abundant hydrogen sulfide in contrast to the latter. A detailed study on this organism, termed *Vibrio bubulus* by Florent,[88] was subsequently reported by Thouvenot and Florent.[89]

The isolation of anaerobic vibrios from bovine reproductive tracts similar to those described as *Vibrio bubulus* by Florent was also reported by Bryner and Frank[13] in 1955. They made the important observation that the anaerobic vibrios could be distinguished from *Vibrio fetus* on the basis of the catalase test; the former were catalase negative and the latter catalase positive. In 1963, Sebald and Véron[8] transferred *V. bubulus* to the genus *Campylobacter*.

In 1965 Loesche and colleagues[90] compared the characters of *Vibrio sputorum* and *Vibrio bubulus* and concluded that the differences between the two were insufficient to justify their consideration as two species. They suggested that the two organisms be referred to as *V.*

sputorum var. *sputorum* and *V. sputorum* var. *bubulus*, and showed that they could be distinguished on the basis of tolerance to 3.5% sodium chloride; var. *bubulus* was salt tolerant whereas var. *sputorum* was inhibited by 3.5% sodium chloride. Loesche et al.[90] also pointed out that both varieties of *V. sputorum* were not, as they had been considered until then, strict anaerobes, but were in fact microaerophiles with a somewhat lower optimal oxygen tension for growth than *V. fetus*. The two varieties of *V. sputorum* are now referred to as *C. sputorum* subsp. *sputorum*, and *C. sputorum* subsp. *bubulus*.[35]

In 1973, Rowland, Lawson, and Maxwell[91] demonstrated the presence of an intracellular vibrio-like organism within the lesions of porcine intestinal adenomatosis. Further studies showed that this organism was probably a new catalase-negative *Campylobacter* species.[92-94] The organism had many characters in common with *C. sputorum*, but it could be differentiated from *C. sputorum* subsp. *sputorum* and *C. sputorum* subsp. *bubulus* according to its tolerance to sodium chloride, glycine, and sodium deoxycholate.[94] Lawson et al. named the organism *C. sputorum* subsp. *mucosalis*.[94,95] The presence of a "dirty yellow" pigment in the colonies of *C. sputorum* subsp. *mucosalis* is another feature that distinguishes this subspecies from other campylobacters.[94]

A completely new catalase-negative *Campylobacter* species termed *C. concisus*[96] was described only recently, in 1981. This organism was isolated from gingival crevices of the human mouth, and when compared to *C. sputorum* subsp. *sputorum* and subsp. *bubulus* by DNA hybridization, showed less than 30% homology with each of the latter subspecies. The pathogenic significance of *C. concisus* is not known.

B. Description and Differentiation of the Catalase-Negative *Campylobacter* Group

Catalase-negative campylobacters have all the general characteristics of the genus *Campylobacter* discussed earlier. Apart from the catalase reaction, catalase-negative campylobacters are distinguished from catalase-positive campylobacters as follows: catalase-negative campylobacters have a lower optimal oxygen tension for growth than do the catalase-positive campylobacter group; they reduce nitrite, and produce hydrogen sulfide in peptone-iron media. By contrast, catalase-positive campylobacters fail to reduce nitrite and fail to produce hydrogen sulfide in peptone-iron medium (except for *C. faecalis*).

The key differential characteristics of catalase-negative *Campylobacter* species and subspecies are listed in Table 6.

C. Principal Features and Pathogenic Significance of Catalase-Negative *Campylobacter* Species and Subspecies

1. C. sputorum *subsp.* sputorum

This organism has been isolated only from the human mouth and intestine. Its pathogenic significance is not known. The frequent isolation of this organism from the healthy human oral cavity[94,97] suggests that it is a commensal. Gibbons et al.[97] found *C. sputorum* subsp. *sputorum* to account for about 5% of cultivable organisms isolated from gingival crevices, a proportion similar to that found for *Bacteroides melaninogenicus*. It has been isolated (filtration method) from 2% of fecal samples from healthy people.[54]

C. sputorum subsp. *sputorum* grows in the presence of 1% glycine and 1% bile, but not 3.5% sodium chloride.[5,7,90,94]

2. C sputorum *subsp.* bubulus

This organism has been isolated from the genital tracts of healthy male and female cattle and sheep.[7,88,89] Its main distinguishing characteristic is its ability to grow in the presence of at least 3.5% sodium chloride.[5,7,90,94] It has never been implicated in disease.

The mean G+C content (mol %) of this organism was recorded as 32.3 ± 0.05.[11]

3. C. sputorum *subsp.* mucosalis

This organism has been associated with lesions of porcine intestinal adenomatosis,[91-94] hemorrhagic enteropathy,[98] and regional ileitis.[99] Its presence has not been recorded in man or animals other than the pig. It is distinguished from other catalase-negative campylobacters by virtue of the "dirty yellow" colonies that it produces on solid culture media.[94] It shares with *C. concisus* the property of requiring hydrogen for microaerobic growth.[7,95,96] Detailed accounts of this organism have been reported by Rowland and Lawson and colleagues.[91,95] The G+C content (mol %) of a representative strain was found to be 33.9.[95]

4. C. concisus

This newly described species was isolated from human gingival crevices. Its pathogenic significance is not known. It has been described in detail by Tanner et al.[96] The G+C content (mol %) of this species was reported to range from 34 to 38. Like *C. sputorum* subsp. *mucosalis*, it requires hydrogen for microaerobic growth.

V. CONCLUDING REMARKS

Cowan and Steel[100] have likened taxonomy to a cocktail made up of "three components *(classification, nomenclature,* and *identification)* that are skillfully blended so that the outsider relishes the whole and cannot discern the individual ingredients". In the case of *Campylobacter*, the cocktail has, without doubt, been indigestible for many years. Recent advances in the taxonomy of this genus have made available fresh ingredients that we have attempted to blend into what we hope is a more palatable concoction.

It will be apparent that the greater emphasis in this chapter has been placed on the catalase-positive as opposed to the catalase-negative campylobacters. This is because there is much more information available on catalase-positive campylobacters and they include most of the *Campylobacter* species pathogenic to man and animals. The only catalase-negative *Campylobacter* taxon of known pathogenic significance, *C. sputorum* subsp. *mucosalis*, is discussed in detail in a separate chapter in this volume.

Finally, for the sake of historical completion, it is tempting to speculate on the identity of strains originally described by McFadyean and Stockman[1] in association with ovine abortion. The latter strains represent the very first time that organisms corresponding to the description of *Campylobacter* were described. The two groups now known to be associated with ovine abortion are *C. jejuni* and *C. fetus* subsp. *fetus*. McFadyean and Stockman's organism failed to grow under anaerobic conditions, but was "capable of active growth in a rarefied atmosphere". Its cultural activity was greatest at incubation temperatures of 35 to 37°C, but it also grew well, although slowly, at room temperature. These features suggest that McFadyean and Stockman's organism was most probably what is now referred to as *C. fetus* subsp. *fetus*.

REFERENCES

1. **McFadyean, J. and Stockman, S.,** Report of the Departmental Committee Appointed by the Board of Agriculture and Fisheries to Enquire into Epizootic Abortion, Part III, Her Majesty's Stationary Office, London, 1913.
2. **Butzler, J. P. and Skirrow, M. B.,** Campylobacter enteritis, *Clin. Gastroenterol.*, 8, 737, 1979.
3. **Karmali, M. A. and Fleming, P. C.,** Campylobacter enteritis, *Can. Med. Assoc. J.*, 120, 1525, 1979.
4. **Smibert, R. M.,** The genus *Campylobacter, Ann. Rev. Microbiol.*, 32, 673, 1978.

5. **Véron, M. and Chatelain, R.**, Taxonomic study of the genus *Campylobacter* Sebald and Véron and designation of the neotype strain for the type species, *Campylobacter fetus* (Smith and Taylor) Sebald and Véron, *Int. J. Syst. Bacteriol.*, 23, 122, 1973.
6. **Smibert, R. M.**, *Campylobacter*, in *Bergey's Manual of Determinative Bacteriology*, 8th ed., Williams & Wilkins, Baltimore, 1974, 207.
7. **Smibert, R. M.**, *Campylobacter*, in *Bergey's Manual of Determinative Bacteriology*, 9th ed., Williams & Wilkins, Baltimore, 1984.
8. **Sebald, M. and Véron, M.**, Teneur en bases de L'ADN et classification des vibrions, *Ann. Inst. Pasteur (Paris)*, 105, 897, 1963.
9. **Neill, S. D., Ellis, W. A., and O'Brien, J. J.**, Designation of aerotolerant *Campylobacter*-like organisms from porcine and bovine abortions to the genus *Campylobacter, Res. Vet. Sci.*, 27, 180, 1979.
10. **Kreig, N. R. and Smibert, R. M.**, *Spirillaceae*, in *Bergey's Manual of Determinative Bacteriology*, 8th ed., Buchanan, R. E. and Gibbons, N. E., Eds., Williams & Wilkins, Baltimore, 1974, 196.
11. **Owen, R. J. and Leaper, S.**, Base composition, size, and nucleotide sequence similarities of genome deoxyribonucleic acids from species of the genus *Campylobacter. FEMS Microbiol. Lett.*, 12, 395, 1981.
12. **Belland, R. J. and Trust, T. J.**, Deoxyribonucleic acid sequence relatedness between thermophilic members of the genus *Campylobacter, J. Gen. Microbiol.*, 128, 2515, 1982.
13. **Bryner, J. H. and Frank, A. H.**, A preliminary report on the identification of *Vibrio fetus, Am. J. Vet. Res.*, 16, 76, 1955.
14. **Smith, T.**, Spirilla associated with disease of the fetal membranes in cattle (infectious abortion), *J. Exp. Med.*, 28, 701, 1918.
15. **Smith, T. and Taylor, M. S.**, Some morphological and biological characters of the spirilla (*Vibrio fetus*, n. sp.) associated with disease of the fetal membranes in cattle, *J. Exp. Med.*, 30, 299, 1919.
16. **Smith, T. and Orcutt, M. L.**, Vibrios from calves and their serological relation to *Vibrio fetus, J. Exp. Med.*, 45, 391, 1927.
17. **Jones, F. S. and Little, R. B.**, The etiology of infectious diarrhea (winter scours) in cattle, *J. Exp. Med.*, 53, 835, 1931.
18. **Jones, F. S. and Little, R. B.**, Vibrionic enteritis in calves, *J. Exp. Med.*, 53, 845, 1931.
19. **Jones, F. S., Orcutt, M., and Little, R. B.**, Vibrios (*Vibrio jejuni*, n. sp.) associated with intestinal disorders of cows and calves, *J. Exp. Med.*, 53, 853, 1931.
20. **Doyle, L. P.**, A vibrio associated with swine dysentery, *Am. J. Vet. Res.*, 5, 3, 1944.
21. **Doyle, L. P.**, The etiology of swine dysentery, *Am. J. Vet. Res.*, 9, 50, 1948.
22. **Levy, A. J.**, A gastro-enteritis outbreak probably due to a bovine strain of vibrio, *Yale J. Biol. Med.*, 18, 243, 1946.
23. **Vinzent, R., Dumas, J., and Picard, N.**, Septicemie grave eu cours de la grossesse, due à un vibrion. Avortement consecutif, *C. R. Acad. Med.*, 131, 90, 1947.
24. **Vinzent, R.**, Une affection méconnue de la grossesse l'infection placentaire a *Vibrio foetus, La Presse Med.*, 81, 1230, 1949.
25. **King, E. O.**, Human infections with *Vibrio fetus* and a closely related vibrio, *J. Infect. Dis.*, 101, 119, 1957.
26. **King, E. O.**, The laboratory recognition of *Vibrio fetus* and a closely related *Vibrio* isolated from cases of human vibriosis, *Ann. N.Y. Acad. Sci.*, 98, 700, 1962.
27. **Florent, A.**, Les deux vibrioses génitales: la vibriose vénérienne due a *V. fetus venerialis* et al vibriose d'origine intestinale due a *V. fetus intestinalis, Meded. Veeartsenijsch. Rijksuniv. Gent.*, 3, 1, 1959.
28. **Bryner, J. H., O'Berry, P. A., and Frank, A. H.**, *Vibrio* infection of the digestive organs of cattle, *Am. J. Vet. Res.*, 25, 1048, 1964.
29. **Miller, V. A., Jensen, R., and Gilroy, J. J.**, Bacteremia in pregnant sheep following oral administration of *Vibrio fetus, Am. J. Vet. Res.*, 20, 677, 1959.
30. **Plastridge, W. N., Williams, L. F., and Petrie, D.**, Vibrionic abortion in cattle, *Am. J. Vet. Res.*, 8, 178, 1947.
31. **Clark, B. L.**, Review of bovine vibriosis, *Aust. Vet. J.*, 47, 103, 1971.
32. **Lecce, J. G.**, Some biochemical characteristics of *Vibrio fetus* and other related vibrios isolated from animals, *J. Bacteriol.*, 76, 312, 1958.
33. **Lapage, S. P., Sneath, P. H. A., Lessel, E. F., Skerman, V. B. D., Seeliger, H. P. R., and Clark, W. A.**, *International Code of Nomenclature of Bacteria*, American Society for Microbiology, Washington, D.C., 1975.
34. **Karmali, M. A., Allen, A. K., and Fleming, P. C.**, Differentiation of catalase-positive campylobacters with special reference to morphology, *Int. J. Syst. Bacteriol.*, 31, 64, 1981.
35. **Skerman, V. B. D., McGowan, V., and Sneath, P. H. A.**, Approved lists of bacterial names, *Int. J. Syst. Bacteriol.*, 30, 225, 1980.
36. **Harvey, S. M.**, Hippurate hydrolysis by *Campylobacter fetus, J. Clin. Microbiol.*, 11, 435, 1980.

37. **Hwang, M. and Ederer, G. M.,** Rapid hippurate hydrolysis method for presumptive identification of group B streptococci, *J. Clin. Microbiol.*, 1, 114, 1975.
38. **Skirrow, M. B. and Benjamin, J.,** '1001' campylobacters: cultural characteristics of intestinal campylobacters from man and animals, *J. Hyg. (Cambridge)*, 85, 427, 1980.
39. **Skirrow, M. B. and Benjamin, J.,** Differentiation of enteropathogenic campylobacter, *J. Clin. Pathol.*, 33, 1122, 1980.
40. **Leaper, S. and Owen, R. J.,** Differentiation between *Campylobacter jejuni* and allied thermophilic campylobacters by hybridization of deoxyribonucleic acids, *FEMS Microbiol. Lett.*, 15, 203, 1982.
41. **Kuzdas, C. D. and Morse, E. V.,** Physiological characteristics differentiating *Vibrio fetus* and other vibrios, *Am. J. Vet. Res.*, 17, 331, 1956.
42. **Di Liello, L. R., Poelma, L. J., and Faber, J. E.,** Biochemical and serological separation of some members of the genus *Vibrio*, *Am. J. Vet. Res.*, 20, 532, 1959.
43. **Butzler, J. P., Dekeyser, P., Detrain, M., and Dehaen, M.,** Related vibrio in stools, *J. Pediatr.*, 82, 493, 1973.
44. **Skirrow, M. B.,** *Campylobacter* enteritis: a "new" disease, *Br. Med. J.*, 2, 9, 1977.
45. **Skirrow, M. B. and Benjamin, J.,** The classification of 'thermophilic' campylobacters and their distribution in man and domestic animals, in *Campylobacter: Epidemiology, Pathogenesis, and Biochemistry*, Newell, D. G., Ed., MTP Press, Lancaster, U.K., 1982, 40.
46. **Karmali, M. A., Penner J. L., Fleming, P. C., Williams, A., and Hennessy, J. N.,** The serotype and biotype distribution of clinical isolates of *Campylobacter jejuni/coli* over a three-year period, *J. Infect. Dis.*, 147, 243, 1983.
47. **Luechtefeld, N. W. and Wang, W. L. L.,** Animal reservoirs of *Campylobacter jejuni*, in *Campylobacter: Epidemiology, Pathogenesis, and Biochemistry*, Newell, D. G., Ed., MTP Press, Lancaster, U.K., 1982, 249.
48. **Peckham, M. C.,** Avian vibrionic hepatitis, *Avian Dis.*, 2, 348, 1958.
49. **Truscott, R. B. and Morin, E. W.,** A bacteriological agent causing bluecomb disease in turkeys. II. Transmission and studies of the etiological agent, *Avian Dis.*, 8, 27, 1964.
50. **Smibert, R. M. and von Graevenitz, A.,** A human strain of *Campylobacter fetus* subsp. *intestinalis* grown at 42°C, *J. Clin. Pathol.*, 33, 509, 1980.
51. **Karmali, M. A., De Grandis, S. A., Allen, A. K., and Fleming, P. C.,** Identification, nomenclature, and taxonomy of catalase-positive campylobacters, in *Campylobacter: Epidemiology, Pathogenesis, and Biochemistry*, Newell, D. G., Ed., MTP Press, Lancaster, U.K., 1982, 35.
52. **Karmali, M. A., De Grandis, S. A., and Fleming, P. C.,** Antimicrobial susceptibility of *Campylobacter jejuni* and *Campylobacter fetus* subsp. *fetus* to eight cephalosporins with special reference to species differentiation, *Antimicrob. Agents Chemother.*, 18, 948, 1980.
53. **Karmali, M. A., Fleming, P. C., and Williams, A.,** unpublished data, 1982.
54. **Skirrow, M. B. and Benjamin, J.,** unpublished data, 1982.
55. **Véron, M., Lenvoisé-Furet, A., and Beaune, P.,** Anaerobic respiration of fumarate as a differential test between *Campylobacter fetus* and *Campylobacter jejuni*, *Curr. Microbiol.*, 6, 349, 1981.
56. **Razi, M. H. H., Park, R. W. A., and Skirrow, M. B.,** Two new tests for differentiating between strains of *Campylobacter*, *J. Appl. Bacteriol.*, 50, 55, 1981.
57. **Blaser, M. J., Moss, C. W., and Weaver, R. E.,** Cellular fatty acid composition of *Campylobacter fetus*, *J. Clin. Microbiol.*, 11, 448, 1980.
58. **Leaper, S. and Owen, R. J.,** Identification of catalase-producing *Campylobacter* species based on biochemical characteristics and on cellular fatty acid composition, *Curr. Microbiol.*, 6, 31, 1981.
59. **Curtis, M. A.,** Cellular fatty acid profiles of campylobacters, in *Campylobacter: Epidemiology, Pathogenesis, and Biochemistry*, Newell, D. G., Ed., MTP Press, Lancaster, U.K., 1982, 234.
60. **Luechtefeld, N. W. and Wang, W. L. L.,** Hippurate hydrolysis by and triphenyltetrazolium tolerance of *Campylobacter fetus*, *J. Clin. Microbiol.*, 15, 137, 1982.
61. **Veal, B., Newell, D. G., and Pearson, A. D.,** Total protein profiles — method of identification and classification for campylobacters, in *Campylobacter: Epidemiology, Pathogenesis, and Biochemistry*, Newell, D. G., Ed., MTP Press, Lancaster, U.K., 1982, 231.
62. **Berg, R. L., Jutila, J. W., and Firehammer, B. D.,** A revised classification of *Vibrio fetus*, *Am. J. Vet. Res.*, 32, 11, 1971.
63. **Bokkenheuser, V.,** *Vibrio fetus* infection in man. I. Ten new cases and some epidemiologic observations, *Am. J. Epidemiol.*, 91, 400, 1970.
64. **Schmidt, U., Chmel, H., Kaminski, Z., and Sen, P.,** The clinical spectrum of *Campylobacter fetus* infections: report of five cases and review of the literature, *Q. J. Med.*, 49, 431, 1980.
65. **Chang, W. and Ogg, J. E.,** Transduction and mutation to glycine tolerance in *Vibrio fetus*, *Am. J. Vet. Res.*, 32, 649, 1971.

66. **Mohanty, S. B., Plumer, G. J., and Faber, J. E.,** Biochemical and colonial characteristics of some bovine vibrios, *Am. J. Vet. Res.*, 23, 554, 1962.
67. **Benjamin, J., Leaper, S., Owen, R. J., and Skirrow, M. B.,** Description of *Campylobacter laridis*, a new species comprising the nalidixic acid resistant thermophilic campylobacter (NARTC) group, *Current Microbiol.*, 8, 209, 1983.
68. **Park, R. W. A., Razi, M. H. H., and Skirrow, M. B.,** Some new tests for differentiating between campylobacters, *J. Appl. Bacteriol.*, 49, 17, 1980.
69. **Firehammer, B. D.,** The isolation of vibrios from ovine feces, *Cornell Vet.*, 55, 482, 1965.
70. **Ellis, W. A., Neill, S. D., O'Brien, J. J., Ferguson, H. W., and Hanna, J.,** Isolation of *Spirillum/Vibrio*-like organisms from bovine fetuses, *Vet. Rec.*, 100, 451, 1977.
71. **Ellis, W. A., Neill, S. D., O'Brien, J. J., and Hanna, J.,** Isolation of spirillum-like organisms from pig fetuses, *Vet. Rec.*, 102, 106, 1978.
72. **Neill, S. D., Ellis, W. A., and O'Brien, J. J.,** The biochemical characteristics of *Campylobacter*-like organisms from cattle and pigs, *Res. Vet. Sci.*, 25, 368, 1978.
73. **Logan, E. F., Neill, S. D., and Mackie, D. P.,** Mastitis in dairy cows associated with an aerotolerant campylobacter, *Vet. Rec.*, 110, 229, 1982.
74. **Cook, G. T.,** A plate nitrate test for nitrate reduction, *J. Clin. Pathol.*, 3, 359, 1950.
75. **McClung, C. R. and Patriquin, D. G.,** Isolation of a nitrogen-fixing *Campylobacter* species from the roots of *Spartina alterniflora* Loisel, *Can. J. Microbiol.*, 26, 881, 1980.
76. **Laanbroek, H. J., Kingma, W., and Veldkamp, H.,** Isolation of an aspartate-fermenting, free-living *Campylobacter* species, *FEMS Lett.*, 1, 99, 1977.
77. **Firehammer, B. D. and Border, M.,** Isolation of temperate bacteriophages from *Vibrio fetus*, *Am. J. Vet. Res.*, 29, 2229, 1968.
78. **Bryner, J. H., Ritchie, A. E., and Foley, J. W.,** Techniques for phage typing *Campylobacter jejuni*, in *Campylobacter: Epidemiology, Pathogenesis, and Biochemistry*, Newell, D. G., Ed., MTP Press, Lancaster, U.K., 1982, 52.
79. **Blackwell, C. D., Winstanley, F. P., and Brunton, W. A. T.,** Sensitivity of thermophilic campylobacters to R-type pyocines of *Pseudomonas aeruginosa*, *J. Med. Microbiol.*, 15, 247, 1982.
80. **Penner, J. L. and Hennessy, J. N.,** Passive haemagglutination technique for serotyping *Campylobacter fetus* subsp. *jejuni* on the basis of soluble, heat-stable antigens, *J. Clin. Microbiol.*, 12, 732, 1980.
81. **Lauwers, S.,** Serotyping of *C. jejuni*: a useful tool in the epidemiology of campylobacter diarrhea, in *Campylobacter: Epidemiology, Pathogenesis, and Biochemistry*, Newell, D. G., Ed., MTP Press, Lancaster, U.K., 1982.
82. **Lior, H., Woodward, D. L., Edgar, J. A., Laroche, L. J., and Gill, P.,** Serotyping of *Campylobacter jejuni* by slide agglutination based on heat-labile antigenic factors, *J. Clin. Microbiol.*, 15, 761, 1982.
83. **Hoffman, P. S., Kreig, N. R., and Smibert, R. M.,** Studies on the microaerophilic nature of *Campylobacter fetus* subsp. *jejuni*. I. Physiological aspects of enhanced aerotolerance, *Can. J. Microbiol.*, 25, 1, 1979.
84. **Hébert, G. A., Hollis, D. G., Weaver, R. E., Lambert, M. A., Blaser, M. J., and Moss, C. W.,** Thirty years of campylobacters: biochemical characteristics and a biotyping proposal for *Campylobacter jejuni*, *J. Clin. Microbiol.*, 15, 1065, 1982.
85. **Tunnicliff, R.,** An anaerobic vibrio isolated from a case of acute bronchitis, *J. Infect. Dis.*, 15, 350, 1914.
86. **Prévot, A. R.,** Etudes de systématique bacterienne. V. Essai de classification des vibrions anaérobies, *Ann. Inst. Pasteur*, 64, 117, 1940.
87. **MacDonald, J. B.,** The Motile Non-Sporulating Anaerobic Rods of the Oral Cavity, Ph.D. thesis, University of Toronto, Toronto, Canada, 1953.
88. **Florent, A.,** Isolement d'un vibrion saprophyte du sperme du taureau et du vagin de la vache *(Vibrio bubulus)*, *C. R. Soc. Biol.*, 147, 2066, 1953.
89. **Thouvenot, H. and Florent, A.,** Etude d'un anaerobie du sperme du taureau et du vagin de la vache *Vibrio bubulus*, Florent 1953, *Ann. Inst. Pasteur*, 86, 237, 1954.
90. **Loesche, W. J., Gibbons, R. J., and Socransky, S.,** Biochemical characteristics of *Vibrio sputorum* and relationship to *Vibrio bubulus* 91 and *Vibrio fetus*, *J. Bacteriol.*, 89, 1109, 1965.
91. **Rowland, A. C., Lawson, G. H. K., and Maxwell,** Intestinal adenomatosis in the pig: occurrence of a bacterium in affected cells, *Nature (London)*, 243, 417, 1973.
92. **Rowland, A. C. and Lawson, G. H. K.,** Intestinal adenomatosis in the pig: immunofluorescent and electron microscopic studies, *Res. Vet. Sci.*, 17, 323, 1974.
93. **Lawson, G. H. K. and Rowland, A. C.,** Intestinal adenomatosis in the pig: a bacteriological study, *Res. Vet. Sci.*, 17, 331, 1974.
94. **Lawson, G. H. K., Rowland, A. C., and Wooding, P.,** The characterization of *Campylobacter sputorum* subsp. *mucosalis* isolated from pigs, *Res. Vet. Sci.*, 18, 121, 1975.

95. **Lawson, G. H. K., Leaver, J. L., Pettigrew, G. W., and Rowland, A. C.,** Some features of *Campylobacter sputorum* subsp. *mucosalis* subsp. nov., nom. rev. and their taxonomic significance, *Int. J. Syst. Bacteriol.*, 31, 385, 1981.
96. **Tanner, A. C. R., Badger, S., Lai, C. H., Listgarten, M. A., Visconti, R., and Socransky, S.,** *Wolinella* gen. nov., *Wolinella succinogenes (Vibrio succinogenes* Wolin et al.) comb. nov., and description of *Bacteroides gracilis* sp. nov., *Wolinella recta* sp. nov., *Campylobacter concisus* sp. nov., and *Eikenella corrodens* from humans with periodontal disease, *Int. J. Syst. Bacteriol.*, 31, 432, 1981.
97. **Gibbons, R. J., Socransky, S. S., Sawyer, S., Kapsimalis, B., and MacDonald, J. B.,** The microbiolota of the gingival crevice area of man. II. The predominant cultivable organisms, *Arch. Oral Biol.*, 8, 281, 1963.
98. **Rowland, A. C. and Lawson, G. H. K.,** Intestinal adenomatosis in the pig: a possible relationship with a hemorrhagic enteropathy, *Res. Vet. Sci.*, 18, 263, 1975.
99. **Jönsson, L. and Martinsson, K.,** Regional ileitis in pigs. Morphological and pathogenetical aspects, *Acta Vet. Scand.*, 17, 223, 1976.
100. **Cowan, S. T. and Steel, K. J.,** *Manual for the Identification of Medical Bacteria,* 2nd ed., Cambridge University Press, London, 1974.
101. **Smibert, R. M.,** personal communication, 1981.

Chapter 2

CLINICAL ASPECTS OF *CAMPYLOBACTER* INFECTIONS IN HUMANS

Bibhat K. Mandal, P. De Mol, and J. P. Butzler

TABLE OF CONTENTS

I. Introduction 22

II. Human Infection Caused by *C. fetus* subsp. *fetus* 22
A. Types of Infections 22
1. Purely Febrile Forms 22
2. Cardiac Forms 22
3. Meningitic Forms 22
4. Articular Forms 22
5. Forms with Suppuration in Other Sites 23
B. Prognosis and Treatment 23

III. *C. jejuni* Infection 23
A. Complications 24
B. Pathology 25
C. Treatment 27

References 30

I. INTRODUCTION

Only *Campylobacter fetus* subsp. *fetus* and *C. jejuni* are now recognized as pathogens in humans. The former, in contrast to the latter, nearly always attacks debilitated individuals with impaired defenses against infection and is a comparatively rare cause of disease. *C. jejuni*, on the other hand, can attack the healthiest of people and is one of the most common causes of acute diarrheal illness.

II. HUMAN INFECTION CAUSED BY *C. FETUS* SUBSP. *FETUS*

Since Vinzent et al.[1] described the first case of this infection in 1947, some 122 further cases have been published. An analysis of them does not, however, allow one to describe a characteristic clinical syndrome which might suggest the diagnosis. The diagnosis has in fact always been made in the laboratory by the isolation of the organism, usually from blood but occasionally from a local collection of pus. The most common presentation is a pure septicemia, with no foci of origin or of dissemination and with fever, which is a constant and often the only feature. Bejot[2] has described five clinical forms of *C. fetus* subsp. *fetus* infections.

A. Types of Infections

1. Purely Febrile Forms

These are the most common. Sometimes it is a simple bacteremia with a moderate spike of fever, passing off without any other symptoms.[3,4] More often it is a true septicemia with a long febrile course and many positive blood cultures. The fever usually progresses in an irregular way with bouts of fever going up to 39 or 40°C at the beginning, with rigors and a lot of sweating, lasting only for 3 or 4 days.

2. Cardiac Forms

These forms occur in two different ways. Sometimes the septicemia due to *C. fetus* subsp. *fetus* is superimposed on preexisting heart diseases; in other cases the picture is one of the acute primary endocarditis.[5] Some patients show evidence of valvular disease, most frequently of rheumatic order, in the mitral or aortic valves,[5] but others have normal heart sounds on admission and develop a primary endocarditis later. Although the cultures of the vegetations, when made possible, remained negative, it does seem very likely that these cases of acute primary endocarditis with vegetations were due to *C. fetus* subsp. *fetus* since the blood cultures were repeatedly positive.[6,7]

3. Meningitic Form

This appears as acute purulent meningitis[8,9] or meningoencephalitis.[10] Of the 12 cases reported, 6 were noticed in the newborn.[8] In all the cases the cerebrospinal fluid (CSF) was frankly purulent and in a third of them bloodstained. The organism was isolated from CSF alone in three cases, from the CSF and from the brain post mortem in two cases of neonatal meningoencephalitis, and from the blood in the remaining seven patients. The seriousness of the meningeal form depends primarily on the age of the subject; there were five deaths among the six newborn[9] but only one among the adults.

4. Articular Forms

These patients suffer from an acute purulent arthritis in a single joint. They are not common and only six cases have been reported.[12] In some cases, *C. fetus* was isolated only from the synovial fluid but in other cases also from the blood.

5. Forms with Suppuration in Other Sites

Exceptionally, suppuration in other sites has been reported, but it has not always been possible to demonstrate that the patient was suffering from a septicemia, too. The cases covered various clinical pictures: an acute pericarditis wherein the organism had also been isolated from the purulent pericardial fluid, apart from a blood culture which proved to be positive;[12] a purulent pleurisy in which the organism was isolated from the blood and the pleural pus;[2] a subcutaneous abcess of the thorax; a pustule on the face;[2] and a case which presented as an acute endocarditis but yielded *C. fetus* subsp. *fetus*, from an infected ovarian cyst.[2]

C. fetus subsp. *fetus* is an opportunist pathogen and the severity of the damage it causes is directly related to its site of multiplication. Most of the patients attacked already have serious underlying medical problems. A variety of such conditions has been described in the literature including alcoholism,[3,13] cirrhosis,[14] diabetes mellitus,[15,16] atherosclerotic or rheumatic heart disease,[15,16] gastrectomy,[17] aplasia of the marrow,[3,13] lymphoma,[3] leukemia or a lymphoproliferative disease, neoplasms,[3] steroid or immunodepressive therapy,[13] prior splenectomy,[13] Crohn's disease,[4] and tuberculosis.[13] In addition, pregnant women and their fetuses are particularly susceptible.[18]

B. Prognosis and Treatment

Campylobacter fetus subsp. *fetus* septicemia or meningitis was the primary cause of death in 17 of the 122 reported cases. Anecdotal information about the effectiveness of antimicrobial therapy is provided by case reports of campylobacteriosis. In general, chloramphenicol, aminoglycosides, and tetracycline have seemed to be effective. There are many reports available on temporary therapeutic failures with penicillins, followed by clinical and bacteriological cures with alternate antibiotics. If the patient dies, it is generally due to aggravation of the underlying pathology. The mortality rate is higher in patients with meningitis or endocarditis than in those with nonlocalized sepsis.

III. *C. JEJUNI* INFECTION

Campylobacter is now recognized as one of the most common causes of bacterial diarrhea. The organism has an international reputation and has been found in virtually every country where it has been looked for.[19-28]

The symptoms and signs of a *C. jejuni* infection are not so distinctive that the physician can differentiate it from illnesses caused by other pathogens. At the mild end of the spectrum, symptoms may last for only 24 hr and be indistinguishable from those seen in viral gastroenteritis.

Our knowledge of the clinical features of campylobacter infections is based largely on studies of patients whose symptoms were sufficiently severe to prompt their physicians to obtain a specimen of feces for culture. Initially, a patient may happen to have a temperature of over 40°C, in which case confusion and delirium may appear; rigors have occasionally been noted.[29] This prodromal state lasts for up to 2 days and is followed by nausea and abdominal cramps which are typically periumbilical. These are rapidly followed by diarrhea, which may be profuse; stools are watery or slimy and foul smelling. Fresh blood may appear in the stools after 2 or 3 days. Most fecal samples examined microscopically show an inflammatory exudate containing polymorphs and it is usually possible to see numerous *Campylobacters*, owing to their characteristic morphology.[25]

Vomiting may occur, but it is rarely a marked feature. The acute diarrhea lasts for about 2 or 3 days, by which time the patient feels weak and exhausted. Profuse watery diarrhea is most frequently described in developing countries, whereas in the industrialized parts of the world this feature seems to be exceptional. Sometimes nausea and vomiting cause sufficient dehydration and electrolyte imbalance to make admission to the hospital necessary. The illness usually does not last for longer than 1 week, but 25% of the patients will subsequently have a recurrence of the symptoms, most commonly abdominal pain. The clinical features can vary from a brief insignificant gastroenteritis to an enterocolitis with abdominal pain and bloody diarrhea which may last for several weeks.[19,20]

The term "enteritis", frequently used to describe the clinical illness caused by campylobacter infection, presupposes that the site of infection can be found in the small intestine. Certainly the occurrence of copious watery diarrhea (the hallmark of small-gut diarrhea) and the observation of ileitis in occasional patients who have undergone a laparotomy or were examined postmortem, supported this assumption.[29]

However, in a significant proportion of patients the stools contain fresh blood, pus, or mucus and this suggests that colorectal inflammation is not uncommon in campylobacter infection. Lambert et al.[30] studied 11 patients and a sigmoidoscopy and rectal biopsy found evidence of colonic inflammation in 8 of them. Colgan et al.[31] reported that 10 of the 11 patients they studied in Toronto had colitis; Blaser et al.[32] found evidence of colitis in all four of their patients.

Evidence of colonic inflammation is not uncommon even when the stools are devoid of macroscopic blood; microscopy, however, usually shows the presence of an inflammatory exudate. Sigmoidoscopy usually reveals some degree of abnormality ranging from mucosal edema and hyperemia, either with or without petechial hemorrhage, to mucosal friability. Gross alterations and slough formation are rare. The picture is that of a proctitis or proctocolitis, but in one patient reported by Loss et al.[33] colonoscopic examination revealed segmental colitis with large, shaggy ulcers and patchy areas of hyperemia, cobblestoning, and edema which resembled granulomatous colitis.

The endoscopic findings in campylobacter colitis are not distinctive as similar features can be seen, either in other forms of infective colitis such as those caused by shigella, salmonella, amoeba, and *C. difficile*, or in ulcerative colitis and Crohn's disease. A barium enema should provide confirmation of the colitis in severely affected cases, the films showing the loss of haustrations, a fine irregularity of the bowel outline, and a disturbance of the mucosal pattern.

A patient who presents with bloody diarrhea and with *Campylobacter* in his stool, is possibly suffering from primary *Campylobacter* infection, but it could be that he is experiencing the first episode of an inflammatory bowel disease with a concurrent *Campylobacter* infection. Even the histological changes and the findings at the sigmoidoscopy and after a barium enema can be identical in the two conditions.

In general, diarrhea in primary *Campylobacter* colitis is of short duration and subsides rapidly, whether spontaneously or following erythromycin therapy; but in severe and more persistent cases steroids as well as erythromycin will have to be given to cover both the possibilities. Definitive diagnosis in such cases can only be made by further observations, including rectal biopsies and a prolonged follow-up. After primary infective colitis, rectal histology almost invariably returns to normal after a month or so,[54] whereas in inflammatory bowel disease this is most uncommon.

Campylobacter infection can also be responsible for a flare-up in established cases of inflammatory bowel disease[34,35] and adequate stool bacteriology is a must in all acute relapses of this disease. Most puzzling of all: in occasional patients a seemingly primary *Campylobacter* colitis has evolved into inflammatory bowel disease.[55] We have encountered similar occurrences with salmonella and shigella infections as well, and the existence of such patients highlights the need for additional studies of the relationship between these infections and the subsequent development of inflammatory bowel disease.

A. Complications

Few patients, most of whom are teenagers or young adults, may have such severe abdominal pain that they are admitted to hospital as cases of acute appendicitis,[20] cholecystitis,[36] or peritonitis from acute appendicitis, but most of those who undergo laparatomy show inflammation of some part of the ileum and jejunum coupled with mesenteric adenitis, which more than once has been mistaken for the lesions of typhoid fever. Another potential surgical complication may arise in young infants — diarrhea is commonly mild at this age, yet blood may appear in the stools and mislead the clinician into thinking there is an intususception.[27,29] *C. jejuni* has been shown to cause infectious proctitis in homosexual men;[37] whether this manifestation of infection occurs disproportionately in this population is unknown. Some patients develop erythema nodosum[38] or reactive arthritis after *Campylobacter* enteritis, whether or not they possess HLA B27 antigen.[39]

An association with meningism has been described, and exceptionally, a genuine *C. jejuni* meningitis may occur.[40]

Urinary tract infections,[41] Reiter's syndrome,[42] and the Guillain-Barré syndrome[43] due to *C. jejuni* have also been reported. A number of infections have been established in neonates, whose mothers, too, had positive fecal culture in most of the cases. *C. jejuni* bacteremia is increasingly reported and it is no longer a "bacteriological curiosity".

Between March 1977 and December 1980, the Communicable Disease Surveillance Centre in the U.K. (Public Health Laboratory Service) received reports of 37 cases of *C. jejuni* bacteremia. Recently Walder et al.[45] reported four cases of *C. jejuni* bacteremia during a 2-year-period from one Swedish hospital.

C. jejuni bacteremia is probably more frequent than is clinically suspected, since extension to the blood is part of the pathogenesis in the early stages of the disease.

There are two reasons for the scarcity of reports on septicemia: the unlikelihood of growing campylobacter by routine blood culture techniques and the rarity with which blood from patients with enteritis, even when febrile, is sent in for culture. *C. jejuni* will be isolated more frequently from the blood of enteritic patients if blood is taken in the early stages of the disease and if the blood cultures are incubated under the right conditions. Hypogammaglobulinemic persons are unusually susceptible to *C. jejuni*[46] just as they are to infection with *G. lamblia* and salmonella. Hitherto only five deaths have been reported from enteritis due to *C. jejuni* — an 8-day-old infant, a 5-month-old baby, and middle-aged chicken farmer with cirrhosis, and two elderly women.

B. Pathology

Information about the minimal dose of the organism and the incubation period of the disease are fragmentary. Most patients have fallen ill from 1 to 7 days after exposure. The organisms are sensitive to acid, but those surviving the gastric barrier then find themselves in a bile-rich, microaerophilic milieu favorable to their survival and multiplication. Symptomatic patients shed 10^6 to 10^9 organisms per gram of stool. The infection may be localized to the upper gastrointestinal tract and *C. jejuni* has been cultivated from ileal aspirates of children with *Campylobacter* enteritis.[47] Postmortem examinations of patients who died of *C. jejuni* enteritis showed hemorrhagic lesions in the jejunum and ileum,[15] and recently several investigators have showed that the sites of tissue injury may include not only the jejunum and ileum but also the colon.

The histological picture in rectal biopsies from *Campylobacter* colitis is not specific, but shows the same pattern of inflammation as is seen in the infective colitis due to shigella, salmonella, *Entameba histolytica*, or *C. difficile*.[48]

The predominant features of such an infective colitis are acute inflammation, edema, and focal collections of polymorphs in the mucosa. The architecture of the crypts is preserved and mucus depletion in usually minimal (Figure 1). By contrast, in acute ulcerative colitis there is a diffuse increase in chronic inflammatory cells (particularly plasma cells) in the mucosa with a variable number of acute inflammatory cells in the lamina propria, the crypt

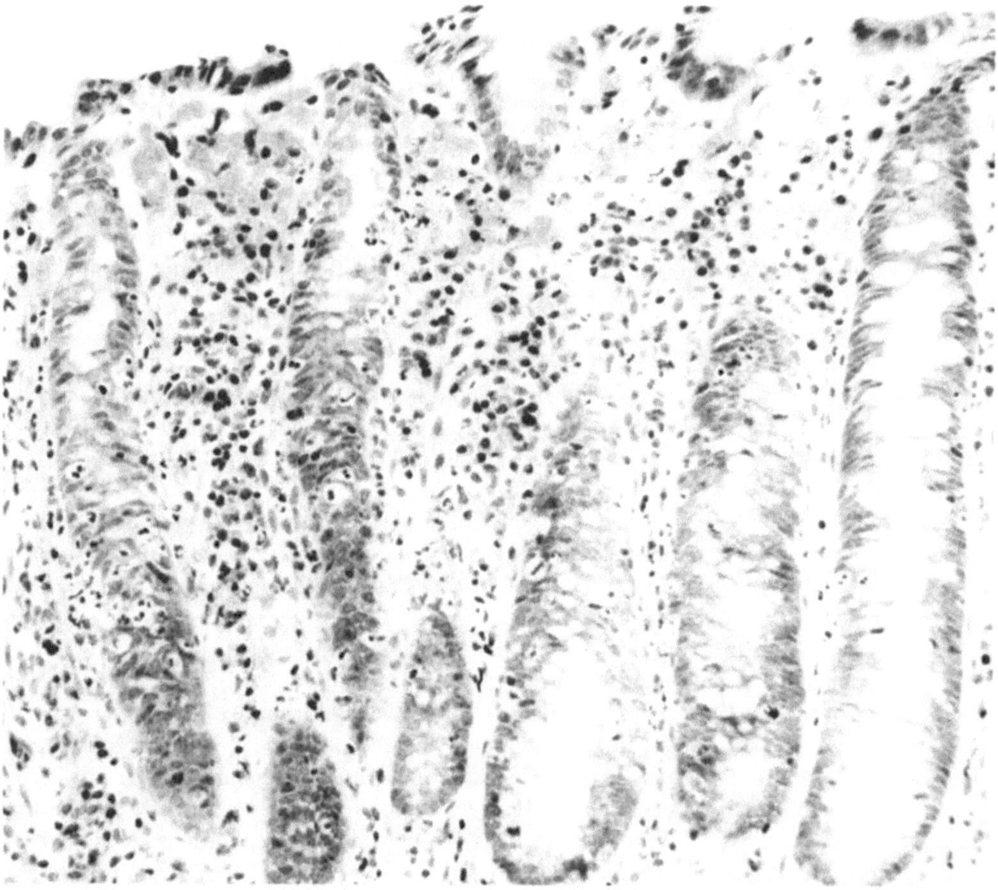

FIGURE 1. Rectal biopsy appearances in campylobacter colitis. There is a severe acute inflammation with polymorphs in the lamina propria, infiltrating the crypt epithelium. Considering the degree of inflammation the goblet-cell population is well preserved. (H.E., magnification × 250.)

Table 1
DIFFERENTIAL DIAGNOSIS OF INFECTIVE COLITIS, ULCERATIVE COLITIS, AND CROHN'S DISEASE IN RECTAL BIOPSIES

Histological features	Infective colitis	Active ulcerative colitis	Crohn's disease
Mucosal edema	+ +	+	+ +
Vascularity	+	+ + +	+
Mucosal inflammation			
Polymorphs	+ + +	+ +	+
Mononuclear cells	±	+ + +	+ + +
Distortion of crypt architecture	−	+ + +	+
Mucus depletion	±	+ + +	±

epithelium, and the lumen of the crypts (in some phases forming crypt abscesses). Mucus depletion is marked and there is a distortion of the crypt architecture. Crohn's colitis with rectal involvement is characterized by a normal goblet-cell population, preservation of crypt architecture, and infiltration of the epithelium by neutrophils. In the lamina propria there

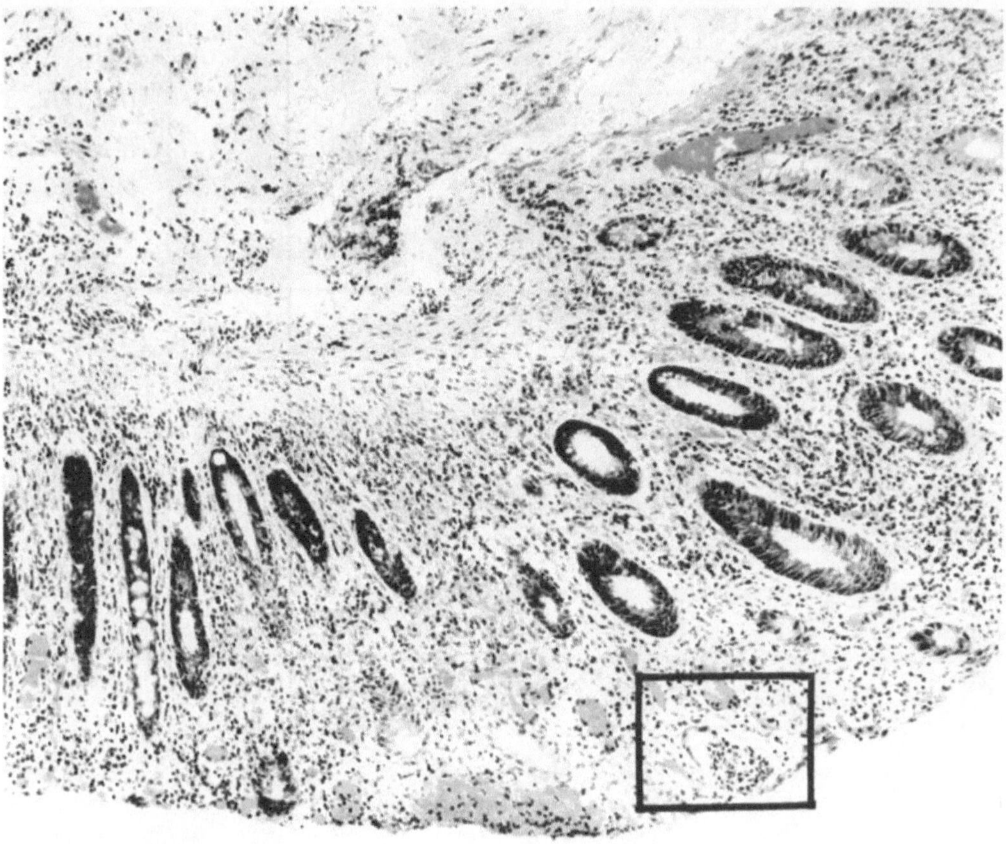

FIGURE 2. Rectal biopsy appearance in campylobacter colitis. Increased cellularity of the lamina propria with polymorphonuclear leukocytes. There is an invasion of crypts and formation of crypt abscesses. (H.E., magnification × 80.)

are increased numbers of both lymphocytes and plasma cells and the inflammation involves the submucosa. The presence of granulomata strongly supports the diagnosis. Table 1 summarizes the features which aid in the differential diagnosis of infective colitis and inflammatory bowel disease.[48]

Though in the majority of cases of *Campylobacter* colitis the diagnosis of an infective form of colitis is usually apparent from the pattern of inflammation, there is an overlap with the appearance seen in the inflammatory bowel disease. In some of the severe cases of campylobacter colitis, rectal biopsy may show marked mucus depletion and crypt abscess formation, and thus may mimic acute ulcerative colitis of short duration where abnormalities of crypt architecture have not developed (Figures 2, 3, and 4). Of the eight patients with *Campylobacter* colitis described by Lambert et al.,[30] three had features suggestive of inflammatory bowel disease. Other workers have described similar findings.[49,50]

C. Treatment

Antibiotics are usually prescribed in *C. jejuni* enterocolitis — often unnecessarily. The patient will usually recover after 1 or 2 days when given an antibiotic. But in most cases a spontaneous recovery would have occurred without antibiotics after onset of the disease. In the absence of chemotherapy, the feces of patients remain positive for about 2 to 7 weeks after the illness. However mild cases excrete the organism for only a few days and there is

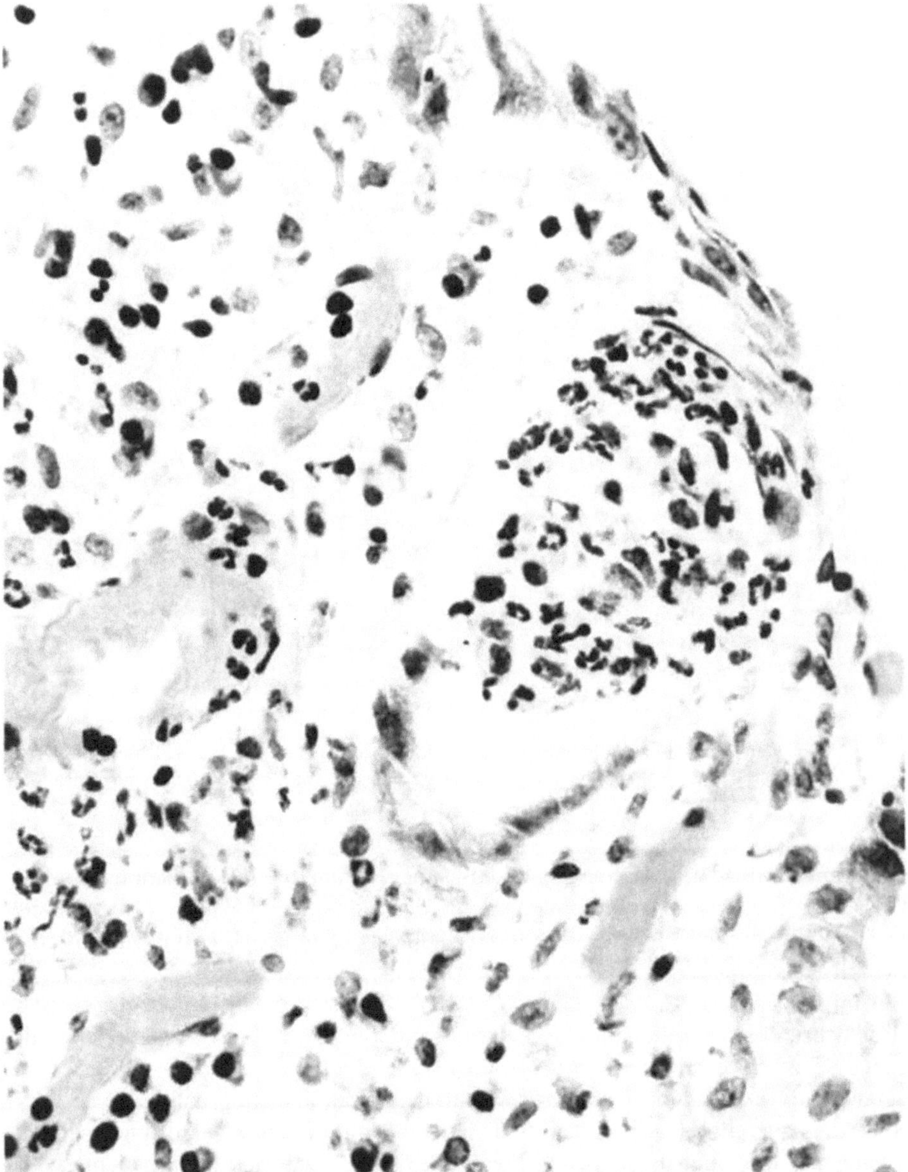

FIGURE 3. A higher magnification of the area marked in Figure 2, showing a crypt abscess. (H.E., magnification × 500.)

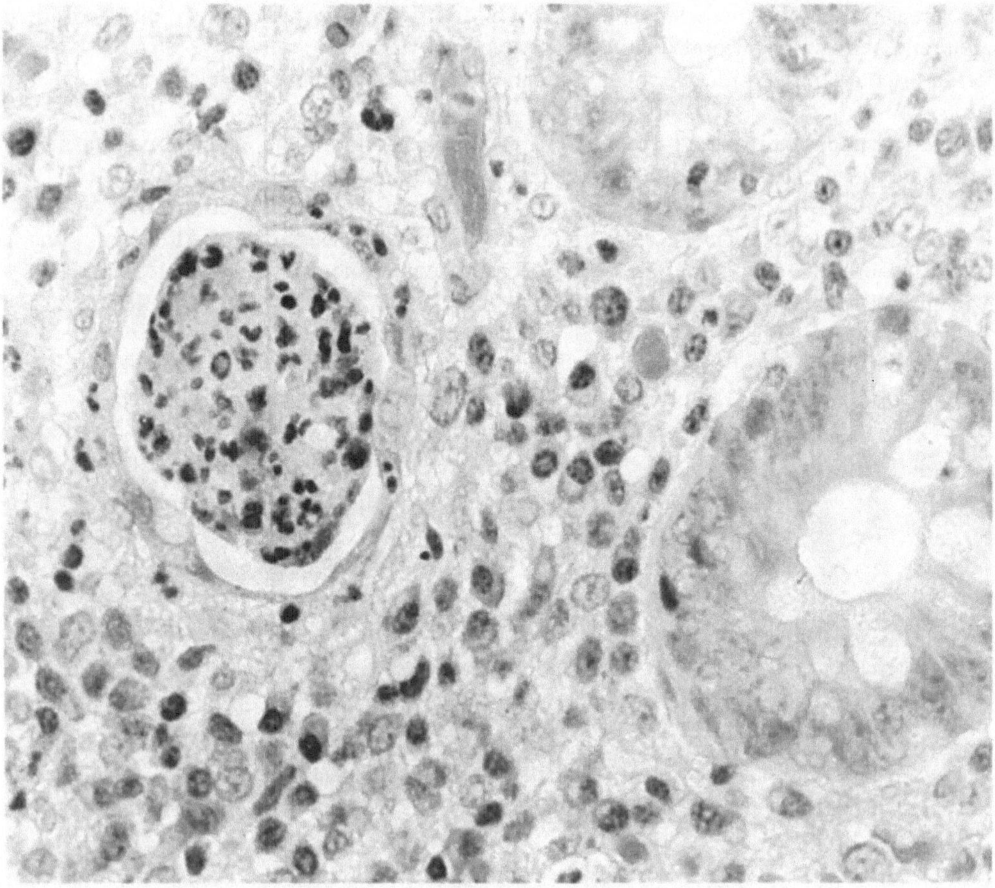

FIGURE 4. Rectal biopsy appearance in campylobacter colitis. There is a mixed inflammatory infiltrate in the lamina propria comprising polymorphs, lymphocytes, and plasma cells. There is some depletion of goblet cells. A crypt abscess is seen in the center. A repeat biopsy 1 month later revealed an entirely normal rectal mucosa. (H.E., magnification × 500.)

always the occasional patient who excretes for a longer period. Erythromycin shortens the fecal excretion of *C. jejuni* in adult patients. It promptly eradicates *C. jejuni* from the feces but without altering the natural course of uncomplicated *Campylobacter* enteritis, if the treatment is not started until four or more days after the onset of symptoms.[51]

Giving patients antibiotics to prevent them passing on the infection to others is usually not indicated, as this form of transmission is infrequent. A nonspecific supportive and symptomatic treatment should be given, as for any other case of diarrhea. If the abdominal pain is severe, or a possibility of a complication exists, it is preferable to administer an antibiotic. In vitro furazolidone, erythromycin, doxycycline, chloramphenicol, and the aminoglycosides are the most active compounds. Almost all the strains of *C. jejuni* are resistant to penicillin G, cephalosporin, colistin, and trimethoprim. Erythromycin stearate 500 mg b.i.d. is a treatment regime that has been used with success. It is preferable to carry out sensitivity tests since cases of erythromycin and of tetracycline resistance have been described.[52]

In the industrialized countries, dehydration due to *C. jejuni* is infrequent, but exceptionally, fluid and electrolyte replacement is necessary in infants. The best treatment for *C. jejuni* infections in developing countries could well prove to be different. Factors such as a low socioeconomic status and malnutrition may well determine the severity of a *C. jejuni* infection and its great prevalence there in very young children. Vomiting and watery diarrhea are frequent and sometimes an oral rehydration is required in the children.

For an effective prevention we must elucidate the local environmental factors that affect transmission.

Antibiotics should be reserved for very severe cases. It is certain that education and better hygiene have far greater roles to play in reducing infections than have antibiotics.

In *Campylobacter* septicemia gentamicin is the drug of choice, but tetracycline, erythromycin, and chloramphenicol are good alternatives. It should be noted that the penicillins and cephalosporins are totally ineffective.[53] Chloramphenicol should be considered in patients with meningitis, since it is difficult to attain an adequate concentration of aminoglycosides in the CSF. The prognosis in septicemic cases is difficult to estimate, but most patients recover after a few days of treatment.

An unfavorable course is likely to be due to the presence of some underlying disease.

REFERENCES

1. **Vinzent, R., Dumas, J., and Picard, N.,** Septicémie grave au cours de la grossesse due à un vibrion. Avortement consécutif, *Bull. Acad. Méd.*, 131, 90, 1947.
2. **Bejot, J.,** La campylobactériose humaine ou vibriose à *Campylobacter fetus, Med. Mal. Infect.*, 3, 281, 1973.
3. **Bokkenheuser, V.,** Vibrio infections in man, *Am. J. Epidemiol.*, 91, 400, 1970.
4. **Kahler, R. L. and Sheldon, H.,** *Vibrio fetus* infection in man, *N. Engl. J. Med.*, 262, 1218, 1960.
5. **Albeaux-Fernet, M., Chabot, J., and Devaux, J. P.,** Uncas de septicémie à *Vibrio foetus*, Evolution mortelle par aggravation d'un cardiopathie pré-existante, *Bull. Mem. Soc. Méd. Hôp. Paris*, 74, 782, 1958.
6. **Auquier, L., Chrétien, J., and Hodara, M.,** Septicémie avec endocardite à *Vibrio foetus, Bull. Mem. Soc. Méd. Hôp. Paris*, 72, 580, 1956.
7. **Loeb, H., Bettag, J. L., Yung, N. K., King, S., and Bronsky, D.,** *Vibrio fetus endocarditis, Am. Heart J.*, 71, 381, 1966.
8. **Eden, A. N.,** *Vibrio foetus* meningitis in a newborn infant, *J. Pediatr.*, 61, 33, 1962.
9. **Fleurette, J., Flandrois, J. P., and Diday, M.,** Méningites et diarrhées à *Vibrio foetus*: à propos d'un cas chez un nourisson, *Presse Méd.*, 79, 480, 1971.
10. **Burgert, W. and Hagstrom, J. W.,** *C. Vibrio foetus* meningoencephalitis, *Arch. Neurol. (Chicago)*, 10, 196, 1964.
11. **Kilo, C., Hagemann, P. O., and Marzi, J.,** Septic arthritis and bacteremia due to *Vibrio fetus*. Report of an unusual case and review of the literature, *Am. J. Med.*, 38, 962, 1965.
12. **Killam, H. A. W., Crowder, J. G., White, A. C., and Edmonds, J. H.,** Pericarditis due to *Vibrio fetus, West. J. Med.*, 120, 200, 1974.
13. **Butzler, J. P.,** Infection with Campylobacters, in *Modern Topics in Infection*, Williams, J. D., Ed., Heinemann, London, 1978, 214.
14. **Caroli, J., Leymarios, J., and Thibault, P.,** Infection à *Vibrio foetus* au cours d'une cirrhose métaictérique, *Rev. Méd. Chirg. Mal. Foie, Rate, Pancréas*, 32(3), 55, 1957.
15. **King, E. O.,** Human infection with *Vibrio foetus* and a closely related *Vibrio, J. Infect. Dis.*, 101, 119, 1957.
16. **King, E. O.,** The laboratory recognition of *Vibrio foetus* and a closely related *Vibrio* isolated from cases of human vibriosis, *Ann. N.Y. Acad. Sci.*, 98, 700, 1962.
17. **Auguste, C., Buttiaux, R., and Tacquet, A.,** Septicémie à *Vibrio foetus* chez un gastrectomisé en état de carence avec hypoprotidémie, *Arch. Mal. App. Dig.*, 43, 861, 1954.
18. **Eden, A. N.,** Perinatal mortality caused by *Vibrio foetus, J. Pediatr.*, 68, 297, 1966.
19. **Butzler, J. P., Dekeyser, P., Detrain, M., and Dehaen, F.,** Related vibrio in stools, *J. Pediatr.*, 82, 493, 1973.
20. **Skirrow, M. B.,** Campylobacter enteritis, a new disease, *Br. Med. J.*, 2, 9, 1977.
21. **Severin, W. P. J.,** Campylobacter enteritis, *Ned. T. Gen.*, 122, 499, 1978.
22. **Lindquist, B., Kjellander, J., and Kosunen, T.,** Campylobacter enteritis in Sweden, *Br. Med. J.*, 1, 300, 1978.

23. **Blaser, M. J., Berkowitz, J., Laforce, F. M., Cravens, J., Reller, L. B., and Wang, W. L.,** Campylobacter enteritis: clinical and epidemiological features, *Ann. Int. Med.*, 91, 179, 1979.
24. **Bokkenheuser, V. D., Richardson, N. J., Bryner, J. H., Roux, D. J., Schutte, A. B., Koornhof, H. J., Treiman, I., and Hartman, E.,** Detection of enteric campylobacteriosis in children, *J. Clin. Microbiol.*, 9, 227, 1979.
25. **Karmali, M. A. and Fleming, P. C.,** Campylobacter enteritis in children, *J. Pediatr.*, 94, 527, 1979.
26. **De Mol, P. and Bosmans, E.,** Campylobacter enteritis in Central Africa, *Lancet*, 1, 604, 1978.
27. **Butzler, J. P. and Skirrow, M. B.,** Campylobacter enteritis, *Clin. Gastroenterol.*, 8, 737, 1979.
28. **Lauwers, S., De Boeck, M., and Butzler, J. P.,** Campylobacter enteritis in Brussels, *Lancet*, 1, 604, 1978.
29. **Butzler, J. P.,** New prospects for treatment and prevention, in *Acute Enteric Infections in Children*, Holme, T., Holmgren, J., Merson, M., and Möllby, R., Eds., Elsevier Sequoia, New York, 1981, 63.
30. **Lambert, M. E., Schofield, P. F., Ironside, A. G., and Mandal, B. K.,** Campylobacter colitis, *Br. Med. J.*, 1, 857, 1979.
31. **Colgan, T., Lambert, J. R., Newman, A., and Luk, S. C.,** *Campylobacter jejuni* enterocolitis. A clinical-pathological study, *Arch. Pathol. Lab. Med.*, 104, 571, 1980.
32. **Blaser, M. J., Parsons, R. B., and Wang, W. L. L.,** Acute colitis caused by *Campylobacter foetus* ss. *jejuni*, *Gastroenterology*, 78, 448, 1980.
33. **Loss, M. J., Mangla, J. C., and Pereira, M.,** Campylobacter colitis presenting as inflammatory bowel disease with segmental colonic ulcerations, *Gastroenterology*, 79, 138, 1980.
34. **Newman, A. and Lambert, J. R.,** *Campylobacter jejuni* causing flare-up in inflammatory bowel diseases, *Lancet*, 2, 919, 1980.
35. **Mandal, B. K.,** Colitis and *Campylobacter* infection. A clinico-pathological study, in *Campylobacter: Epidemiology, Pathogenesis, and Biochemistry*, Newell, D. G., Ed., MTP Press, Lancaster, U.K., 1982, 145.
36. **Mertens, A. and De Smet, M.,** Campylobacter cholecystitis, *Lancet*, 1, 1092, 1979.
37. **Quinn, T., Corey, L., Chaffee, R. G., Schuffer, M. D., and Holmes, K. K.,** Campylobacter proctitis in homosexual men, *Ann. Int. Med.*, 93, 458, 1980.
38. **Lambert, M., Marion, E., Coche, E., and Butzler, J. P.,** Campylobacter enteritis and erythema nodosum, *Lancet*, 1, 1409, 1982.
39. **Berden, J. H., Muytjens, H. I., and Van de Putte, I. B. A.,** Reactive arthritis associated with *Campylobacter jejuni* enteritis, *Br. Med. J.*, 1, 380, 1979.
40. **Norrby, R., McCloskey, G., Zackrisson, G. and Falsen, E.,** Meningitis caused by *Campylobacter fetus* spp. *jejuni*, *Br. Med. J.*, 280, 1164, 1980.
41. **Davies, J. S. and Penfold, J. B.,** *Campylobacter jejuni* infection, *Lancet*, 1, 1091, 1979.
42. **Matti Saari, K. and Kauranen, O.,** Ocular inflammation in Reiter's syndrome associated with *Campylobacter jejuni* enteritis, *Am. J. Ophthalmol.*, 90, 572, 1980.
43. **Rhodes, K. M. and Tattersfield, A. E.,** Guillain-Barré syndrome associated with Campylobacter infection, *Br. Med. J.*, 282, 173, 1982.
44. **Karmali, M. A. and Tan, Y. C.,** Neonatal Campylobacter enteritis, *Can. Med. Assoc. J.*, 122, 192, 1980.
45. **Walder, M., Lindberg, A., Schalen, C., and Öhman, L.,** Five cases of *Campylobacter jejuni/coli* bacteremia, *Scand. J. Infect. Dis.*, 14, 201, 1982.
46. **Glover, S. G., Smith, C. C., Reid, T. M. S., and Khaud, R. R.,** Opportunistic Campylobacter bacteremia in a patient with malignant histocytic medullary reticulosis, *J. Inf.*, 4, 175, 1982.
47. **Cadranel, S., Rodesch, P., Butzler, J. P., and Dekeyser, P.,**Enteritis due to related Vibrio in children, *Am. J. Dis. Child.*, 126, 152, 1973.
48. **Day, D. W., Mandal, B. K., and Morson, B. C.,** The rectal biopsy appearances in salmonella colitis, *Histopathology*, 2, 117, 1978.
49. **Willoughby, C. P., Piris, J., and Truelove, S. C.,** Campylobacter colitis, *J. Clin. Pathol.*, 32, 986, 1979.
50. **Duffy, M. C., Benson, J. B., and Rubin, S. J.,** Mucosal invasion in Campylobacter enteritis, *Am. J. Clin. Pathol.*, 73, 706, 1980.
51. **Anders, B. J., Lauer, B. A., Paisley, J. W., and Reller, L. B.,** Double-blind placebo-controlled trial of erythromycin for treatment of campylobacter enteritis, presented at the 21st Intersci. Conf. Antimicrobial Agents and Chemotherapy, Chicago, November 4 to 6, 1981.
52. **Vanhoof, R., Vanderlinden, M. P., Dierickx, R., Lauwers, S., Yourassowsky, E., and Butzler, J. P.,** Susceptibility of *Campylobacter fetus* subsp. *jejuni* to 29 antimicrobial agents, *Antimicrob. Agents Chemother.*, 14, 553, 1978.
53. **Butzler, J. P., Dekeyser, P., and Lafontaine, T.,** Susceptibility of related vibrios and *Vibrio fetus* to 12 antibiotics, *Antimicrob. Agents Chemother.*, 5, 86, 1974.
54. **Mandal, B., De Mol, P., and Butzler, J. P.,** unpublished observations.
55. **Mandal, B.,** unpublished observation, 1982.

Chapter 3

CAMPYLOBACTER JEJUNI IN TRAVELLERS' DIARRHEA

Peter Speelman and Marc J. Struelens

TABLE OF CONTENTS

I. Introduction 34

II. Epidemiology 34

III. Clinical Features 36

IV. Serology 37

V. Conclusion 37

References 37

I. INTRODUCTION

Newcomers in tropical countries run a special risk of developing diarrhea, especially when they come from countries with high standards of hygiene and sanitation. Many expatriates have frequent episodes of diarrhea during their first year abroad, although these episodes become less frequent within a month of their arrival. The etiology of these diarrheal episodes is diverse. In many studies enterotoxigenic *Escherichia coli* (ETEC) is the most common cause of travellers' diarrhea, although *Shigella, Salmonella*, rotavirus, Norwalk agent, *Giardia lamblia* and *Entamoeba histolytica* are also incriminated. In most studies a large group of patients have no detectable pathogen. Until recently, *Campylobacter jejuni* was not cultured from travellers so it was not considered in the differential diagnosis. However, recent studies suggest that *C. jejuni* is a good candidate for travellers' diarrhea, since it occurs with high prevalence in the tropics and has a short incubation period. In this chapter we will review the evidence for *C. jejuni* as a causative agent of travellers' diarrhea.

II. EPIDEMIOLOGY

In two studies in developed countries, *C. jejuni* was isolated from travellers with diarrhea: 20% of 1336 British patients infected with *C. jejuni* had a history of a recent journey abroad and 73% of 277 Swedish patients with *C. jejuni* were returning from Europe, Africa, Asia, the Middle-East, South America, Canary Islands, and Hawaii.[1,2] Most of these travellers developed symptoms 1 or 2 days after a week-long trip, suggesting short incubation time (1 to 6 days).

Two prospective studies have examined the occurrence of *C. jejuni* in travellers with diarrhea. In a study of 35 U.S. Peace Corps Volunteers in rural Thailand, *C. jejuni* was isolated from one volunteer.[3] In a second study of 64 Panamanian tourists travelling in Mexico for 2 weeks, *C. jejuni* was isolated from 3 travellers with diarrhea.[4]

We have studied 269 travellers and expatriate residents in Dhaka who consulted our clinic at the International Center for Diarrheal Disease Research, Bangladesh (ICDDR,B) for diarrhea during the period July 1981 to June 1982.[5] *C. jejuni* and *Shigella* sp. were each isolated as the sole enteropathogen from 41 patients (15.2%). Other agents identified included *G. lamblia* (5.2%) and *E. histolytica* (4.8%), and *Vibrio cholerae* non 01, *Plesiomonas shigelloides* and *Aeromonas hydrophila* together were isolated from 13 patients (4.8%). Five patients (2%) had mixed infections. In 140 cases (52%) the etiology remained unknown. No attempt was made to detect ETEC or rotavirus infection.

The age and sex distribution of patients infected with *C. jejuni, Shigella* sp., and other agents were not significantly different. Patients with diarrhea associated with *C. jejuni* developed their illness earlier after arrival in Bangladesh than patients with shigellosis or diarrhea of other etiology (Figure 1).

There was little seasonality in the total number of patients seen (Figure 2). Shigellosis was more common in the hot dry season similar to the seasonal pattern seen in the local population. Patients with *C. jejuni* were more commonly seen during the winter. However, the number of patients was too small to allow any conclusion. Moreover, four patients with diarrhea caused by *C. jejuni* were involved in a small outbreak in February 1982.

This outbreak occurred in a group of 36 expatriates who attended a 5-day conference at a resort hotel. Out of 30 participants interviewed, 18 had developed diarrhea, usually with constitutional symptoms, 2 to 10 days after arrival, with a mode incubation period of 5 to 6 days. Fecal cultures were obtained within 2 weeks from 27 participants. Six out of 17 cultures from symptomatic and 2 out of 10 from asymptomatic individuals were positive for *C. jejuni*; 4 different O-serotypes were found, and mixed serotypes infection was documented in one patient. No other pathogens were detected. Convalescent sera were obtained from

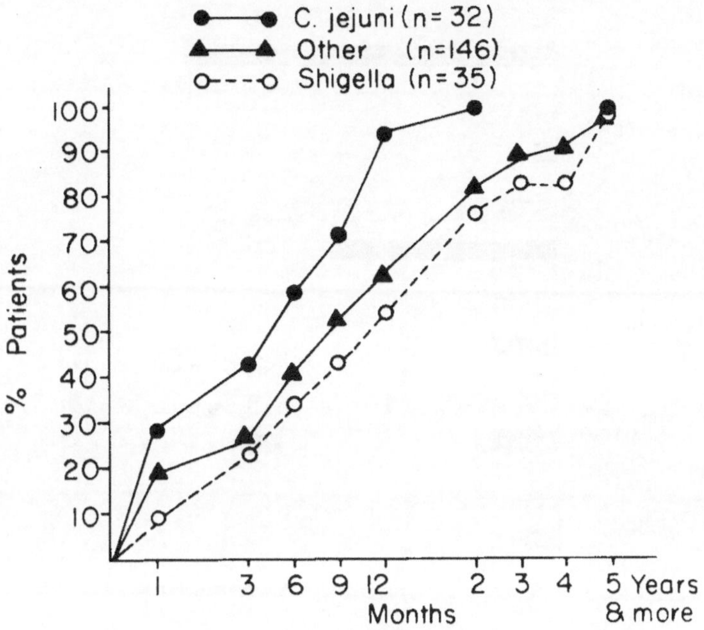

FIGURE 1. Proportion of patients and duration of stay in Bangladesh preceding the diarrheal episode, by etiology. (*Note:* At 12 months the percentage of patients with *C. jejuni* is significantly different from those with *Shigella* and other pathogens. Kolmogorov Smirnov test, $p < 0.01$.)

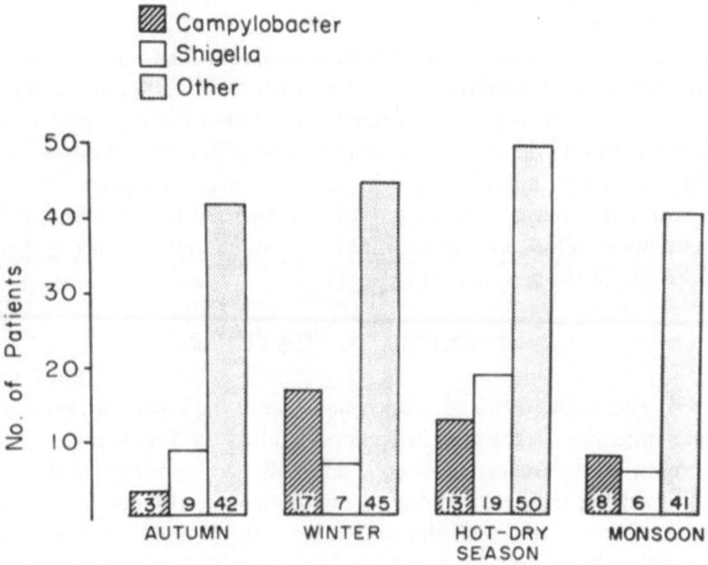

FIGURE 2. Seasonal occurrence of travellers' diarrhea caused by *Campylobacter jejuni*, *Shigella* sp., and other pathogens.

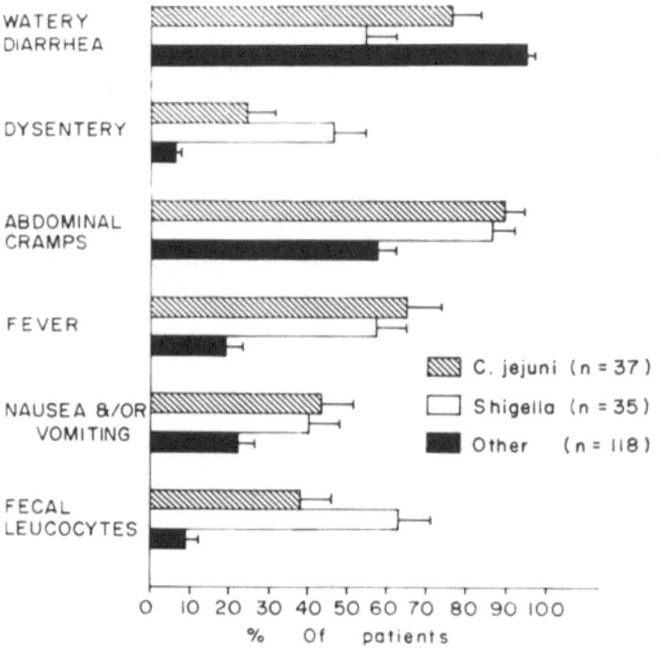

FIGURE 3. Percentage (plus 1 × standard error) of patients with clinical features in travellers' diarrhea, by etiology.

16 patients; 7 patients had complement fixing and/or agglutinin antibodies for *C. jejuni*. All 7 patients had gastrointestinal complaints. Investigations suggested a point source outbreak, although the source was not detected.

To examine the importance of asymptomatic infection with *C. jejuni*, we cultured stool specimens from 196 healthy expatriates; 4 were positive for *C. jejuni*, a prevalence 1/7 of that found in travellers with diarrhea and much lower than the high prevalence of asymptomatic infection reported in the indigenous population by Blaser[6] and Glass.[7] This difference may reflect a higher case-to-infection ratio in travellers which suggests that immunity to *C. jejuni* may be acquired. This hypothesis is further supported by the observation that significantly more patients developed *C. jejuni* diarrhea than shigellosis during their first year in Bangladesh (95% vs. 55%; $p < .01$) (Figure 1).

III. CLINICAL FEATURES

Diarrhea (98%) and fever (69%) of sudden onset were the most common clinical features of campylobacter infection in Swedish travellers, studied by Svedhem and Kaijser.[2] Of the patients, 25% complained of tenesmus or colic and 20% had vomiting. Abdominal tenderness was found in 30% of the patients and only a few patients had blood in the stools.

In our study of 41 travellers with diarrhea caused by *C. jejuni* the onset of illness was found to be sudden and associated with abdominal pain (89%), fever (65%), and nausea or vomiting (43%)[5] (Figure 3). Half of our patients also complained of myalgias and headache. Thirty-eight percent of the patients had fecal leucocytes and erythrocytes on microscopic examination, whereas only 24% of them had noted the presence of blood and/or mucus.

We compared symptoms and signs of patients with diarrhea caused by *C. jejuni*, *Shigella* sp., and other pathogens. Patients with *C. jejuni* and *Shigella* were significantly more likely to complain of fever, abdominal cramps, bloody-mucoid stool and nausea or vomiting, and to have fecal leucocytes and erythrocytes than were patients with other pathogens. Infections with *C. jejuni* were less frequently associated with dysenteric stool than infections with *Shigella* sp. (Figure 3).

Although the spectrum of illness caused by *C. jejuni* does not allow distinction from diarrhea of other causes, abdominal cramps and fever with the presence of fecal leucocytes and erythrocytes in travellers with diarrhea should strongly suggest a diagnosis of *Shigella* or *C. jejuni* enterocolitis.

IV. SEROLOGY

Several studies have shown that campylobacter diarrhea is usually followed by a good serum antibody response. In our study in Dhaka, we were able to obtain sera, prior to infection, from 8 patients. Complement fixing or agglutinin antibodies could not be detected in these sera; convalescent sera collected 1 to 8 weeks after onset of illness had complement-fixing antibodies (titer 1/10 to 1/120 in 21 out of 29 (72%) and agglutinin antibodies (titer 1/160 to 1/5120 in 14 out of 21 sera (66%).[8] The serum antibody titer usually rises by the end of the first week of symptoms, peaks by the third week, and then decreases slowly to nondetectable levels by the third to sixth month.

Although our series is small, convalescent antibody response was significantly more common among diarrheal patients than among asymptomatic excretors, with a clear trend for those with more invasive illness, as indicated by fecal leucocytes and erythrocytes, to show an antibody-response more frequently than patients with watery diarrhea; similar observation in shigellosis has been reported.[9] The infrequent antibody response after asymptomatic infection may indicate colonization with less invasive *C. jejuni* strains; however, several isolates from asymptomatic excretors and cases of diarrhea or dysentery share at least the same O-serotype.[8]

V. CONCLUSION

The information available to date clearly indicates that *C. jejuni* is a causative agent in travellers' diarrhea. Its importance for travellers to different areas of the world remains to be investigated. The clinical presentation and severity of *C. jejuni* diarrhea among travellers does not appear to differ from the campylobacteriosis acquired in developed countries. Most diarrheal episodes caused by *C. jejuni* are uncomplicated, self-limited, and do not require specific therapy. However, prolonged or recurrent episodes of diarrhea, as well as invasive enterocolitis mimicking shigellosis, are not uncommon. In our opinion, based on anecdotal evidence, these patients may benefit from antibiotic therapy. The differential diagnosis of acute enterocolitis, characterized by the presence of fecal leucocytes and erythrocytes, requires specialized microbiological or serological facilities. In practice, where *Shigella* and *C. jejuni* are prevalent, an empirical chemotherapy effective in both infections would be appropriate and should be evaluated.

REFERENCES

1. **Communicable Disease Surveillance Centre (P.H.L.S.) and Communicable Diseases (Scotland) Unit,** Campylobacter infections in Britain, 1977, *Brit. Med. J.*, 1, 1357, 1978.
2. **Svedhem, A. and Kaijser, B.,** *Campylobacter fetus* subspecies *jejuni*: a common cause of diarrhea in Sweden, *J. Infect. Dis.*, 142, 353, 1980.
3. **Echeverria, P., Blacklow, N. R., Sanford, L. B., and Cukor, G. G.,** Travelers' diarrhea among American Peace Corps volunteers in rural Thailand, *J. Infect. Dis.*, 143, 767, 1981.

4. **Ryder, R. W., Oquist, C. A., Greenberg, H., Taylor, D. N., Qrskov, F., Qrskov, I., Kapikian, A. Z., and Sack, R. B.,** Travelers' diarrhea in Panamanian tourists in Mexico, *J. Infect. Dis.*, 144, 442, 1982.
5. **Speelman, P., Struelens, M. J., Sanyal, S. C., and Glass, R. I.,** Detection of *Campylobacter jejuni* and other potential pathogens in travellers with diarrhea in Bangladesh, *Scand. J. Gastroenterol.*, 18(Suppl. 84), 19, 1983.
6. **Blaser, M. J., Glass, R. I., Huq, M. I., Stoll, B. J., Kibriya, G. M., and Alim, A. R. M. A.,** Isolation of *Campylobacter fetus* subsp. *jejuni* from Bangladeshi children, *J. Clin. Microbiol.*, 12, 744, 1980.
7. **Glass, R. I., Stoll, B. J., Huq, M. I., Struelens, M. J., Blaser, M., and Kibriya, A. K. M. G.,** Epidemiological and clinical features of endemic *Campylobacter jejuni* in Bangladesh, *J. Infect. Dis.*, in press.
8. **Struelens, M. J., Speelman, P., and Lauwers, S.,** Serological response in travellers to Bangladesh infected with *Campylobacter jejuni*, Submitted for publication.
9. **Dupont, H. L., Hornick, R. B., Dawkins, A. T., Snyder, M. J., and Formal, S. B.,** The response of man to virulent *Shigella flexneri 2a*, *J. Infect. Dis.*, 119, 296, 1969.

Chapter 4

ISOLATION OF *CAMPYLOBACTER JEJUNI* FROM HUMAN FECES

H. Goossens, M. De Boeck, H. Van Landuyt, and J. P. Butzler

TABLE OF CONTENTS

I. Introduction 40

II. Bacteriological Diagnosis 40
A. Collection, Transport, and Storage of Samples 40
B. Direct Examination 40
C. Primary Isolation 41

III. Inoculation 42
A. Direct Inoculation 42
B. Indirect Inoculation 42
C. Media 42

IV. Incubation 43
A. General Growth Conditions 43
B. Optimal Temperature 44
C. Incubation Methods 44
1. The Poly-Bag System 44
2. The Campy-Pak System 44
3. The Evacuation-Replacement System 45
4. Fortner's Principle 45
5. The Candle Jar System 45
D. Examination of Plates 45
E. Storage of Isolates 46

V. Identification 46

Appendix: Media 47

References 49

I. INTRODUCTION

The idea of a selective culture medium for *Campylobacter* was first of all proposed by the veterinary surgeons. Since 1949[1] the importance of infection due to *C. fetus* in cattle was described by Dutch authors. No way of diagnosing existed for the male contaminator; trials of isolating the bacteria in the preputial sac have been carried out. Florent,[2] in 1956, used blood agar medium to which he added brilliant green. Plastridge and Koths[3] used the selection of certain antibiotics: bacitracin 2 IU/mℓ and novobiocin 2 μg/mℓ. Bisping et al.[4] improved the medium by using a thioglycollate base and a third antibiotic, polymyxin. Plummer et al.,[5] had the idea of slowly centrifuging the stool, then filtering it by means of a Millipore® filter 0.65 μg: only campylobacters go through the filter, whereas most other bacteria are stopped.

A selective medium was not felt to be needed in human pathology because of the ignorance of the pathogenic character of *C. jejuni*. The enteritis that campylobacter provoked was part of the large numbers of enteric infections with negative copra culture.

Meanwhile King,[6] confronted with cases of septicemia due to *C. jejuni*, proposed the hypothesis that the origin was digestive, and that this organism had a frequent implication in diarrhea.

In 1968, one case of septicemia due to *C. jejuni* appeared in St. Pierre Hospital in Brussels.[7] Belgian scientists looked for it in the stools, using a combination of the filtration technique and antibiotics. They succeeded in making the first isolation of *C. jejuni* in human stools.[8]

An approach was made in Australia in 1972. Having remarked upon the notable resistance to cephalothin of a *Campylobacter* strain isolated from a blood culture, Slee[9] looked for it with success in diarrheal stools by arranging discs containing this antibiotic on the blood agar.

A better knowledge of antibiotic resistance patterns[10] allowed more selective media to be made for the isolation of *C. jejuni* and thus omitted the preliminary filtration.

II. BACTERIOLOGICAL DIAGNOSIS

A. Collection, Transport, and Storage of Samples

Either fresh *stool specimens* or *rectal swabs* can be used for culture.

Fecal samples should be kept at +4°C for a maximum of 24 to 48 hr if there is to be a delay before cultures are set up. Tanner and Bullin[11] noted that 2/7 of the stool specimens became negative after 24 hr at ambient temperature, whereas those stored at +4°C remained positive. Patients suffering from acute diarrhea can excrete from 1×10^6 to 1×10^9 CFU (colony forming units) of *C. jejuni* per gram of feces.

From such stool specimens stored at 4°C, Blaser et al.[12] recovered *Campylobacter* up to 3 weeks. Portions from the same specimens kept at 25°C and exposed to air, dried within 96 hr and no viable organism could be recovered after 1 week.

The results of *rectal swabs* for the isolation of *C. jejuni* are not very satisfying, even when a buffered glycerol-saline transport medium is used. Rectal swabs should be conveyed in transport medium, preferably semisolid Cary-Blair medium or a fluid selective "enrichment medium" (see below and Appendixes G, H, I, and J). Cary-Blair medium with reduced agar concentration (see Appendix A) was found to be the best of six transport media for the preservation of *C. jejuni/C. coli* in turkey cecal specimens held at 35°C;[13] a commercial, modified Stuart medium and buffered glycerol-saline performed poorly.

B. Direct Examination

Routine examination of stool specimens has proved to be valuable for rapid presumptive diagnosis, particularly in cases of clinical urgency, e.g., a patient presenting with colitis.[14]

Examination of diarrheal fecal specimens by *dark-field* or *phase-contrast* microscopy can be used to diagnose *Campylobacter* enteritis, provided the stools are not more than a few hours old. In this way, Karmali and Fleming[14] were able to find *C. jejuni* in 75% of patients with *Campylobacter* enteritis found to be positive by copra-culture. Also, when campylobacters were seen in the stools, they always grew upon culture;[14] but observers must be practiced in the art. Phase-contrast microscopy should be performed on organisms, which have been suspended in broth (e.g., brucella, Mueller-Hinton, Todd-Hewitt, T-soy) as opposed to distilled water.[15,16] Salinity also appears to inhibit their motility.[17]

Under phase contrast, *Campylobacter* can be recognized by their characteristic morphology and an extreme rapid darting and oscillating motility. The short forms are more active. Despite their small size, they can be seen by dark-field illumination with a lower objective, and if the light intensity is increased for a few moments, the extra heat produced slows them down. The spiral form and axial spinning motion can then be discerned when scrutinized with high-power objectives.

Stained preparations of stool specimens can also be helpful in an early presumptive diagnosis.[18]

Some authors have used Gram stain, but a crystal violet or Vago stain (technique: after fixation with alcohol on a slide the stool is first colored with eosine 2% for a period of 4 min, and then after rinsing with water it is stained with crystal violet for 2 min) to better show up the morphology of the *Campylobacter*, as they do not take up stain readily. *Campylobacter* appear as slender, Gram-negative comma-, S-, or spiral-shaped organisms. Direct microscopy, such as methylene blue stain, is also of value for detecting red cells and neutrophils, which are present in the feces of 75% or more of patients with *Campylobacter* enteritis.[19,20] They are a sure indication of an acute inflammatory process.

Finally, if the Gram stain reveals a significant leukocytosis in association with small curved Gram-negative rods, and the normal flora appears to be reduced, *Campylobacter* enteritis is strongly suggested. However, it should be remembered that good bacteriological diagnosis is based on a positive stool culture.

C. Primary Isolation

The isolation of *C. jejuni* from feces is a fairly simple procedure which can be done by most hospital laboratories. *C. jejuni* can be isolated from fecal specimens if both selective media that reduce the growth of competing microorganisms and microaerophilic inoculation conditions are used.

Initially, *filtrates of fecal specimens* passed through Millipore® filters (mean pore size of 0.65 μm) in the manner described by Dekeyser et al.[8] and Butzler et al.[21] were cultured on plain blood media. It is a physical method that depends upon differential filtration: most campylobacter are small enough to pass through a 0.65 μm filter, whereas other bacteria usually cannot. Since this procedure is cumbersome and time-consuming it is not recommended for routine isolation.

A significant advance in the isolation of *C. jejuni* from stools occurred in 1977 with Skirrow's report[22] of a blood agar medium, incorporated with multiple antibiotics, which results in the suppression of normal fecal flora and allows *C. jejuni* to grow.

Nowadays, *selective media* are used, since the fecal flora is sufficiently inhibited by the various combinations of antimicrobials in these media (see Appendix, B, C, D, E, and F).

Latterly *fluid "enrichment" media* have been developed, which have increased sensitivity considerably: these are essential for the culture of milk, water, and other samples in which there may be only a few organisms. They embody various formulations, mostly based on the selective agents used in Skirrow's medium. We list herein four media that have given good results (Appendix, items G, H, I, and J).

III. INOCULATION

Routine processing of rectal swabs or stool specimens can include direct inoculation of the specimens onto a selective medium and/or indirect inoculation into Campy-Thio with subsequent subculture to a selective medium after overnight refrigeration.

However, isolation rates of *C. jejuni* from human fecal specimens were equivalent after broth enrichment (thioglycollate medium containing antibiotics) and direct inoculation on two brucella blood agar media containing ferrous sulfate, sodium metabisulfite, and sodium pyruvate, identical concentrations of vancomycin and trimethoprim, and different concentrations of polymyxin B and cephalothin.[23]

For a routine laboratory, it should be sufficient if only a direct inoculation of the feces onto a selective medium is carried out.

A. Direct Inoculation

Rectal swabs — They can be applied directly to 1/4 of a selective agar plate and then streaked for separation with a loop; or they can be squeezed in 1 or 2 mℓ of broth (e.g., Mueller-Hinton), which is then inoculated and isolated on a selective medium, with a loop.

Stool — A loopful of diarrheal stool can be inoculated and streaked for isolation immediately on the agar, but if solid stool is being cultured saline should first be added to the feces sample to obtain a smooth, heavy suspension (a swab can be used to mix the water with the feces); a loopful of this suspension is then inoculated onto a selective medium.

B. Indirect Inoculation

Rectal swab — Place the swab about 1 cm into the Campy-Thio (see Appendix J) and twirl the swab (this allows the inoculation of the organism into a zone of reduced oxygen).

Stool — A few drops (3 to 5) of liquid stool can be placed in the Campy-Thio, but when it is a solid stool, again a suspension should first be made before a few drops can be added to the thioglycollate medium.

All Campy-Thios are refrigerated overnight and subcultured the next day. To subculture, the tip of a Pasteur pipette should be placed about 2 cm below the surface of the Campy-Thio, and a sample should be continually withdrawn as the tip is slowly brought to the surface. A few drops are then placed on a selective medium and streaked for isolation.

C. Media

Currently different types of selective media are in common use. The characteristics of each medium will be discussed.

Skirrow's medium (Appendix, B) — Contains vancomycin, polymyxin B, and trimethoprim (TMP) as the selective antimicrobials. It has been said that it is essential to use lysed horse blood in order to ensure neutralization of TMP antagonists that are present in most media. However, Bopp et al.[24] were able to prove that blood from other animal species could be substituted for lysed horse blood in Skirrow's medium without compromising the activity of TMP. It is suggested that there exists a synergism between TMP and one or more of the other antimicrobial agents used in Skirrow's medium. If a different basal medium from that specified is used, e.g., Columbia agar, it may be necessary to increase the concentration of polymyxin B.

Butzler's medium Oxoid (Appendix, C) — Butzler's medium is in some ways different from Skirrow's medium. Since TMP has been replaced by cefazolin and novobiocin, it can be prepared with sheep or other animal blood. Also, novobiocin suppresses more *Proteus* spp. and cycloheximide is added to the medium to suppress yeast.

It is more selective for the isolation of *C. jejuni* from human feces. Therefore Butzler's medium is preferable to Skirrow's medium for use in developing countries, because extra selectivity makes plate reading easier. In our experience, it was noted that the amount of agar added to the medium should be reduced to 10.5 g instead of 20 g/ℓ; thus making the colonies spread out better.

Campy-BAP (Appendix, D) — Vancomycin, polymyxin B, and TMP of Skirrow's medium are combined with cephalothin and amphotericin B of Butzler's medium (cycloheximide has been replaced by amphotericin B because of its better activity on yeast).

These three selective media are the most widely used today. They are commercially available from BBL, Remel, Gibco, and Scott. In addition, lyophilized vials of each of the three antimicrobial formulations are also commercially available from Oxoid Ltd. (SR 69: Skirrow, SR 85: Butzler, SR 98: Blaser-Wang) if laboratories elect to make their own media.

Two other media have recently been described:

Preston medium (Appendix, E) — It is similar to Skirrow's medium, but a nutrient base that contains fewer TMP antagonists has been used. Also rifampicin proved to be more successful than vancomycin and less expensive than bacitracin for the inhibition of *Bacillus* spp. that swarm on culture media. This medium, when compared with Skirrow's medium, proved to be more selective and is more successful for the isolation of campylobacters from environmental, animal, poultry, and human sources. The FBP supplement (see Appendix), was also found to improve qualitative growth of *Campylobacter* ssp. on primary isolation.

Butzler's medium Virion (Appendix, F) — This medium has been developed because Skirrow's medium, Butzler's medium Oxoid, and Campy-BAP lack sufficient inhibition of pseudomonas and some other Enterobacteriaceae. A comparative study between Butzler's medium Oxoid and Butzler's medium Virion has recently been done. Although both media were comparable in their isolation of *C. jejuni*, Butzler's medium Virion was superior because the competing fecal flora were best suppressed by this new selective medium: on Butzler's medium Oxoid, no growth of fecal flora was seen in 47.2% of the examined stools, whereas on the new medium no growth was seen in 72.2%; also, normal stool grew into the fourth quadrant of the plates with Butzler's medium Oxoid in 3.6%, but in only 0.8% of the plates with Butzler's medium Virion. Therefore, this new medium allows easier reading of plates, which might be very important to investigators who have little or no experience with the isolation of campylobacter from stools.

IV. INCUBATION

A. General Growth Conditions

The single most important feature of campylobacters in relation to cultivation is their sensitivity to oxygen.[25]

They are strictly *microaerophilic:* oxygen is essential for growth, yet as oxygen concentration increases, the viable count is reduced. Thus, campylobacters are inhibited at the normal atmospheric oxygen level. The optimum level of oxygen required for growth has been reported to be 6%.[26]

It is postulated that microaerophilic bacteria are more sensitive than other oxygen-dependent bacteria to toxic forms of oxygen (superoxide anions, peroxide)[25] that occur in aerobic culture media, and that compounds which enhance the aerotolerance of microaerophilic bacteria do so by quenching these toxic forms of oxygen. It has been suggested that substances that neutralize these oxygen derivates should be added to the basal medium. George et al.[25] described a supplement consisting of 0.05% each of ferrous sulfate, sodium metabisulfite, and sodium pyruvate (Appendix: *FBP supplement).* These compounds are presented together as Oxoid SR 84 supplement. The authors concluded that, even in the absence of precise atmospheric conditions, the addition of this supplement increased the probability of isolating strains of *Campylobacter* species. Those laboratories that can only use candle jars (where the oxygen level is approximately 17%, see below), or use very approximate gas mixtures, would particularly be able to benefit from the addition of the growth supplement.

In a recent study, we compared Butzler's selective medium (15 g/ℓ agar), with and without FBP supplement, incubated in a candle jar at 42°C. As a standard technique, we also incubated plates in a special incubator, where 2/3 of the air is extracted by means of a vacuum pump and refilled with a mixture of 95% N_2 and 5% CO_2. All cultures were incubated at 42°C and 3331 stools were tested. We found no difference in isolation rate of *C. jejuni* between the candle jar system and the standard incubation technique. We concluded that the isolation of *Campylobacter* on a selective medium with (but also without) growth supplement in a candle jar is a reliable and easy method.[36]

B. Optimal Temperature

Plates streaked for the isolation of *C. fetus* subsp. *jejuni* should be incubated at 42°C for 48 hr. The organism will also grow at 37°C. The higher temperature, however, allows for better growth and at the same time appears to inhibit some of the fecal flora. The disadvantage of incubation at 42°C is that *C. fetus* subsp. *intestinalis* usually will not grow. However, *C. fetus* subsp. *intestinalis* is rarely isolated in stools:[27] it is associated with disease in immune-compromised hosts and is basically an opportunistic organism.

The advantage of incubation at 42°C, a temperature at which *C. jejuni* will usually begin to grow in 24 instead of 48 hr, outweighs the disadvantage of missing most *C. fetus* subsp. *intestinalis*.

C. Incubation Methods

The atmosphere for incubation should always contain 5% oxygen and 10% carbon dioxide to satisfy the microaerophilic and capnophilic nature of *C. jejuni*.

Various methods for creation of this atmosphere are available to clinical microbiology laboratories. Their value as well as possible restrictions with respect to costs, laboratory size, number of specimens received, and work-flow will be discussed.

1. The Poly-Bag System

This system is comprised of a small polyethylene bag (approximately 20 × 38 cm), a tank of gas containing 5% oxygen, 10% carbon dioxide, and 85% nitrogen, and a rubber band. Campylo plates are placed inside the bag. The bag is inflated by using the tank with a regulator and short hose, then collapsed once or twice to flush out ambient atmosphere, reinflated, tied off with a rubber band, and placed in the incubator at the desired temperature.

It is important that the bag, when fully inflated, contain only 6 to 8 plates; organisms growing on the selective medium will further deplete the already-reduced oxygen concentration to a level that is not suitable for the growth of *Campylobacter* sp. As long as the number of plates does not exceed half the volume of the fully inflated bag, the oxygen concentration remains adequate.

The poly-bag system is extremely cost effective and is readily adaptable to both high and low volume labs. In the latter case, where a small number of specimens are processed, as few as one or two plates can be incubated per bag. The bag may be opened and re-gassed to add more plates. The re-gassing does not disrupt the growth of *C. jejuni* on plates already incubated, since the desired atmosphere is always achieved immediately.

2. The Campy-Pak System (BBL)

This consists of a Gas Pak® jar and a gas-generating envelope. Up to eight plates are placed inside the jar along with a gas-generating envelope which is activated by the addition of water. The lid is subsequently fastened. An anaerobic jar, but without catalyst, can also be used. The system is well suited to high volume labs and for culturing Campy-Thios in the indirect inoculation method.

Even though a single jar can be opened to add new specimens, this approach necessitates the use of a new envelope each time the jar is opened. More important is the 1-hour delay following the activation of a new envelope, before suitable atmospheric conditions are again achieved.

3. The Evacuation-Replacement System

A torbal jar or a special incubator can be used. Two thirds of the air is extracted by means of a vacuum pump. This reduces the pressure to about 260 mmHg. The atmospheric pressure is then restored by introducing a gas mixture of 95% nitrogen and 5% carbon dioxide. The N_2/CO_2 mixture is to be preferred rather than a hydrogen/carbon dioxide mixture, because the final gas mixture containing hydrogen is potentially explosive. The final atmosphere contains approximately 6% oxygen, 6% carbon dioxide, and 88% nitrogen.

4. Fortner's Principle

This torbal jar, which utilizes the ability of a rapidly growing facultative anaerobe to reduce oxygen via a closed system was successfully used by Karmali and Fleming for the primary isolation of campylobacters from feces.[28]

A strain of *Providencia rettgeri* that was resistant to the selective agents used in campylobacter selective medium was placed on one half of the selective agar and a fecal sample was streaked on the other half. The two halves of the plate were then sealed with autoclave tape and incubated.

An alternative method was also used, in which the *P. rettgeri* (or *Escherichia coli*) and fecal samples were streaked on separate plates, which were then incubated via a polyethylene bag (Ziploc® type) that was sealed after expulsion of as much air as possible.

The advantage of this method is that it requires no jar, vacuum pumps, or gas resources. A disadvantage of this system is that additional incubation time may be required for the growth of campylobacters because the facultative organism must first grow and use up the oxygen before its concentration is low enough.

5. The Candle Jar System

The candle serves two functions: it initially reduces the oxygen level to approximately 17% and also provides CO_2. This system also relies on the presence of facultative breakthrough organisms growing on the selective Campy plate to further reduce the oxygen (a modified Fortner principle).

Although inexpensive, a disadvantage is that the total incubation time must be extended to 72 hr.

In addition, a 42°C incubator is mandatory for growth. Pure cultures usually do not do well in this system.

D. Examination of Plates

Plates should be examined at *24 and 48 hr* and then returned quickly to the appropriate atmosphere to ensure viability of the more oxygen-sensitive strains.

Two distinct colony types have been described.[29] One type is nonhemolytic, flat, glossy to dark grey and effuse, with a tendency to spread along the direction of the tracks of the inoculating wire. It appears as though the colony was touched with a bacteriological needle, drawn across the medium, and reincubated. When well spaced, they resemble droplets of fluid or "honey drops" that have been splattered on the agar. The diameter can reach to 10 mm. Some are very irregular with a raised point in the center of the colony.

The other type is round, convex, and smooth with an entire edge and are 1 to 2 mm in diameter. A small percentage of isolates may appear tan and slightly pink. However, the type of colony obtained seems to be related to the freshness of the medium. Reduction in the moisture content of the medium produces a profound effect on the colony morphology of *C. jejuni*. Fresh medium produces flat, grayish, spreading colonies with an irregular shape and watery appearance. Colonies grown on plates dried at 30°C for 48 hr are round and convex with an entire edge. Plates dried for only 24 hr produce colonies of an intermediate nature.[30]

Of course, it is also possible that other factors such as concentration of agar, the state of nutrients, or buildup of toxic products could have a similar effect.[31] For this reason, we have decreased the agar concentration to 10.5 g/1000 mℓ in Butzler's medium (see Appendix, C and F). It is recommended that plates be refrigerated at 4°C and held in the dark to avoid any effect of peroxides that might be produced by exposure to light.[31]

On highly selective media, the organism tends to grow into the third and fourth areas of streaking, whereas most of the normal stool flora which might break through is confined to the first quadrant of the plate. However, plates should be examined with caution since small colonies may only appear into the first quadrant surrounded by fecal flora. Primary cultures from diarrheal stools generally yield a much more profuse growth than those from normal stools.

Once familiarity with the morphology of *C. jejuni* is gained, it is easy to recognize them on plates.

E. Storage of Isolates

Problems have been noted with long-term *storage* of *C. jejuni*, including *lyophilization*. The organism has been very well maintained by culture in *thioglycollate semisolid medium* stored in liquid nitrogen or at −70°C for at least a year. Campylobacters can be stored satisfactorily for up to 6 months at room temperature in *Tarrozi's liver broth*.[32] Cultures in semisolid yeast extract-aspartate nutrient broth (YNAAB, Appendix L) kept at 37°C are viable after 1 year.

Freeze dried cultures have been reported to survive up to 8 years.[32] Good recovery can also be obtained after several months from cultures stored in *FBP broth* (Appendix K): for preservation by freezing, glycerol should be added to this medium (Appendix, K).

Although *C. fetus* subsp. *intestinalis* withstands shipping well, subsp. *jejuni* does not survive long on solid or in liquid or in semisolid media. *Campylobacter* cells in older cultures become coccoid and nonviable.[31] There is a need for reliable *transport media,* for sending cultures of *C. jejuni* to reference laboratories for confirmation and to other laboratories for teaching quality assurance and scientific exchange of information.

Wang et al.[33] reported an *enriched brucella medium* for storage and transport of cultures of *C. fetus* subsp. *intestinalis* and subsp. *jejuni* (Appendix, M). This medium was successfully used to send cultures of *C. jejuni* in the mail with a time in transit under 3 weeks. *FBP broth* (Appendix K) can also be used with success for storage and transport of cultures of *C. jejuni*.

We reported very good results with the *following procedure*:[36] a pure *Campylobacter* culture from a selective medium is suspended with a swab in a Campy-Thio liquid medium which is then incubated for 24 hr at 42°C. This thioglycollate medium is then dispensed in 3-mℓ amounts into upright tubes (15 by 50 mm); finally, the liquid medium in the tube is covered to the top with solid paraffin (Merck Art. 7153). The tube is closed with a lid which is screwed down. We still found positive cultures after 9 weeks when stored at 4°C and after 3 weeks when kept at room temperature.

V. IDENTIFICATION

Any colony resembling *Campylobacter* on a selective plate, grown at 42°C, should be smeared and *stained* with a strong stain such as crystal violet, carbol fuchsin, or Vago, as

Campylobacter do not take up stain readily. *Campylobacter* can be seen as slender, Gram-negative, spiral or S-shaped organisms with tapering ends. Occasionally, spiral morphology is not obvious and in some cases spindle-shaped bacilli dominate. A highly characteristic feature of campylobacters is that they degenerate into coccoid forms after a few days of culture,[31] especially when grown on solid media. Generally, these coccoid forms have lost their motility and fail to subculture.

In addition to the characteristic morphology, phase-contrast microscopy, which will show a *typical darting motility*, could be worth performing. However, the characteristic S-shaped or spiral morphology of *C. jejuni/C. coli* is the best guide to identification. It has also been noted that in wet films prepared from agar cultures for dark-ground or phase-contrast microscopy, many bacteria will be tethered to one another and the characteristic rapid darting and jerking motion will not be seen to best advantage.

Positive *oxidase and catalase test,* performed on strains from plates that have been incubated at 42°C, can also complete the presumptive identification of *C. jejuni*.

In the experience of investigators familiar with the organism, its morphology is usually sufficiently characteristic to allow its isolation and identification. In hospital laboratories that routinely culture stools for salmonella and shigella, confirmatory identification tests are unnecessary. However, when attempting to isolate the organism from blood or feces using a 37°C- rather than a 42°C-incubator, tests should be done to differentiate *C. jejuni* from *C. fetus*. Temperature tolerance tests are very reliable for this purpose; but the triphenyltetrazolium chloride (TTC) and nalidixic acid tolerance tests are also very useful. Tests that allow identification of *Campylobacter* species and subspecies found in the intestine and that allow us to differentiate between *C. jejuni* and *C. coli* are clearly discussed in Chapter 1.

APPENDIX: MEDIA

Blood Agar

Blood Agar Base No. 2 (Oxoid, CM 271), 1 ℓ; defibrinated horse blood, 50 to 70 mℓ. Add blood to molten agar cooled to 50°C.

Transport Medium

A. Cary-Blair with Reduced Agar[13]

Sodium thioglycollate, 1.5 g; disodium hydrogen phosphate, 1.1 g; sodium chloride, 5 g; agar, 1.6 g; distilled water, 991 mℓ; calcium, chloride (freshly prepared 1% solution), 9 mℓ. Adjust to pH 8.4 and steam for 15 min.

Selective Media

B. Skirrow's Medium[22]

Blood Agar Base No. 2 (Oxoid, CM 271), 1 ℓ; lysed defibrinated horse blood 50 mℓ; vancomycin, 10 mg; polymyxin B, 2500 IU; trimethoprim, 5 mg. Add blood and other ingredients to molten agar after cooling to 50°C.

C. Butzler's Medium Oxoid[20]

Thioglycollate medium USP, 1 ℓ; agar, 10.5 g; sheep blood, 50 to 70 mℓ; bacitracin, 25,000 IU; novobiocin, 5 mg; actidione (cycloheximide), 50 mg; colistin, 10,000 units; and cefazolin, 15 mg. Add blood and other ingredients to the sterile molten agar cooled to 50°C.

D. Campy-BAP[19]

Brucella Agar Base (Oxoid, CM 169), 1 ℓ; sheep blood, 100 mℓ; vancomycin, 10 mg; trimethoprim, 5 mg; polymyxin B, 2500 IU; amphotericin B, 2 mg; and cephalothin, 15 mg. Add blood and other ingredients to molten agar after cooling to 50°C.

E. Preston Medium[34]

Nutrient Broth No. 2 (Oxoid, CM 67), 1 ℓ; New Zealand agar, 10.2 g; lysed defibrinated horse blood, 50 mℓ; rifampicin, 10 mg; polymyxin B, 5000 IU; trimethoprim, 10 mg; cycloheximide, 100 mg. Add blood and other ingredients to molten agar after autoclaving and cooling to 50°C.

F. Butzler's Medium Virion

Columbia Agar Base (Oxoid, CM 331), 1 ℓ; sheep blood, 50 to 70 mℓ; cefoperazone, 15 mg; rifampicin, 10 mg; colistin, 10,000 units; and amphotericin B, 2 mg. Add blood and other ingredients to the sterile molten agar cooled to 50°C.

Enrichment Media

G. Liquid Enrichment Medium (LEM)[35]

Nutrient Broth No. 2 (Oxoid), 1 ℓ; lysed defibrinated horse blood, 70 mℓ; vancomycin, 10 mg; polymyxin B, 2500 IU; trimethoprim, 25 mg. Add blood and other ingredients to broth after autoclaving and cooling.

H. Preston Medium

Prepared as a broth and supplemented with FBP supplement (see Appendix, E).

I. Waterman's Enrichment Broth

Diagnostic Thioglycollate Broth (Lab. M.), 1 ℓ; lysed defibrinated horse blood, 50 mℓ; vancomycin, 40 mg; trimethoprim, 20 mg; polymyxin B, 10,000 IU; and actidione, 100 mg. Add blood and other ingredients after autoclaving and cooling.

J. Campy-Thio

Thioglycollate broth with 0.16% agar and the antimicrobials as used in Campy-BAP (see Appendix, D).

Storage Media

K. FBP Broth

Nutrient Broth No. 2 (Oxoid CM 67), 1 ℓ; agar, 1.2 g; ferrous sulfate ($FeSO_4 \cdot 7H_2O$), 0.5 g; sodium metabisulfite, 0.5 g; sodium pyruvate, 0.5 g. This medium may be used as a suspending medium for storage of *Campylobacters* at temperatures below freezing if 150 mℓ glycerol is substituted for 150 mℓ water.

L. Yeast Extract-Aspartate Nutrient Broth (YNAAB)

Bacteriological Peptone (Oxoid), 5 g; Lab. Lemco (Oxoid), 4 g; yeast extract (Difco), 1 g; sodium chloride, 5 g; potassium L-aspartate, 2 g; agar (Oxoid Technical Agar No. 3), 2 g; demineralized water, 1 ℓ. Adjust to pH 7.4.

M. Enriched Brucella Medium[33]

Brucella broth (BBL), with 0.5% agar (Difco); 10% defibrinated sheep blood is added aseptically after autoclaving the base medium. This medium can then be dispensed in 4-mℓ amounts into upright tubes (16 × 125 mm).

FBP Supplement[25]

Ferrous sulfate ($FeSO_4 \cdot 7H_2O$), 0.125 g; sodium metabisulfite, 0.125 g; and sodium pyruvate, 0.125 g; sufficient to supplement 500 mℓ of a nutrient medium.

REFERENCES

1. **Stegenga, Th. and Terpstra, J. I.,** Over *Vibrio foetus* infecties bij het rund en enzootische steriliteit, *Tijdschrift Diergeneesk.*, 74, 293, 1949.
2. **Florent, A.,** Les deux Vibrioses génitales de la bête bovine: la Vibriose vénérienne due à *V. foetus venerialis* et la Vibriose d'origine intestinale due à *V. foetus intestinalis, Proc. 16th Int. Vet. Congr.*, 2, 489, 1959.
3. **Plastridge, W. N. and Koths, M. E.,** Antibiotic sensitivity of bovine fibrios, *Am. J. Vet. Res.*, 22, 864, 1961.
4. **Bisping, W., Freytag, U., and Krauss, H.,** Feststellung der Vibrionenhepatitis der Hünner in Nordwestdeutschland, *Berl. Münch. Tierarztl. Wschr.*, 76, 456, 1963.
5. **Plummer, G. J., Duvall, W. C., and Shepler, V. M.,** A preliminary report on a new technic for isolation of *Vibrio fetus* from carrier bulls, *Cornell Vet.*, 52, 110, 1962.
6. **King, E. O.,** Human infections with *Vibrio fetus* and a closely related vibrio, *J. Infect. Dis.*, 101, 119, 1957.
7. **Butzler, J. P.,** Campylobacter, Een Miskende Enteropathogene Kiem, thesis, Free University of Brussels, Brussels, 1974.
8. **Dekeyser, P., Goussuin-Detrain, M., Butzler, J. P., and Sternon, J.,** Acute enteritis due to related Vibrio: first positive stool culture, *J. Infect. Dis.*, 125, 390, 1972.
9. **Slee, K. J.,** Human Vibriosis, an endogenous infection, *Aust. J. Med. Technol.*, 3, 7, 1972.
10. **Vanhoof, R., Vanderlinden, M. P., Dierckx, R., Lauwers, S., Yourassowsky, E., and Butzler, J. P.,** Susceptibility of *Campylobacter fetus* subsp. *jejuni* to 29 antimicrobial agents, *Antimicrob. Agents Chemother.*, 14(4), 553, 1978.
11. **Tanner, E. I. and Bullin, C. H.,** Campylobacter enteritis, *Br. Med. J.*, 2 (6086), 579, 1977.
12. **Blaser, M. J., Hardesty, H. L., Powers, B., and Wang, W. L.,** Survival of *Campylobacter fetus* subsp. *jejuni* in biological milieus, *J. Clin. Microbiol.*, 11(4), 309, 1980.
13. **Luechtefeld, N. W., Wang, W. L., Blaser, M. J., and Reller, L. B.,** Evaluation of transport and storage techniques for isolation of *Campylobacter fetus* subsp. *jejuni* from turkey cecal specimens, *J. Clin. Microbiol.*, 13, 438, 1981.
14. **Karmali, M. A. and Fleming, P. C.,** Campylobacter enteritis in children, *J. Paediatr.*, 94, 527, 1979.
15. **Chester, B. and Poulos, E. G.,** Rapid presumptive identification of vibrios by immobilization in distilled water, *J. Clin. Microbiol.*, 11, 537, 1980.
16. **Kaplan, R. L., Lenette, E. H., Balows, A., Hausler, W. J., Jr., and Truant, J. P., Eds.,** *Manual of Clinical Microbiology*, 3rd ed., American Society of Microbiology, Washington, D.C., 1980, 235.
17. **Steele, T. W. and McDermot, S.,** Campylobacter enteritis in South Australia, *Med. J. Aust.*, 2, 204, 1978.
18. **Kwintowski, J. E., Kaplan, R. L., and Landar, W.,** Campylobacter Enteritis: Clinical Features and Laboratory Identification, Tech. Bull. 3, Illinois Department of Public Health, Chicago, 1979.
19. **Blaser, M. J., Berkowitz, I. D., Laforce, F. M., Cravens, J., Reller, L. B., and Wang, W. L.,** Campylobacter enteritis: clinical and epidemiological features, *Ann. Intern. Med.*, 91, 179, 1979.
20. **Butzler, J. P. and Skirrow, M. B.,** Campylobacter enteritis, *Clin. Gastroenterol.*, 8(3), 737, 1979.
21. **Butzler, J. P., Dekeyser, P., Detrain, M., and Dehaene, F.,** Related Vibrio in stools, *J. Pediatr.*, 82, 318, 1973.
22. **Skirrow, M. B.,** Campylobacter enteritis: a "new" disease, *Br. Med. J.*, 2, 9, 1977.
23. **Gilchrist, M. J. R., Grewell, M., and Washington, J. A.,** Evaluation of media for isolation of *Campylobacter fetus* subsp. *jejuni* from fecal specimens, *J. Clin. Microbiol.*, 14(4), 393, 1981.
24. **Bopp, C. A., Wells, J. G., and Barrett, T. J.,** Trimethoprim activity in media selective for *Campylobacter jejuni, J. Clin. Microbiol.*, 16(5), 808, 1982.
25. **George, H. A., Hoffman, P. S., Smibert, R. M., and Krieg, N. R.,** Improved media for growth and aerotolerance of *Campylobacter fetus, J. Clin. Microbiol.*, 8(1), 36, 1978.
26. **Kiggins, E. M. and Plastridge, W. N.,** Effect of gaseous environment on growth and catalase content of *Vibrio fetus* cultures of bovine origin, *J. Bacteriol.*, 72, 397, 1956.
27. **Lauwers, S., De Boeck, M., and Butzler, J. P.,** Campylobacter enteritis in Brussels, *Lancet*, 1, 604, 1978.
28. **Karmali, M. A. and Fleming, P. C.,** Application of the Fortner principle to isolation of campylobacter from stools, *J. Clin. Microbiol.*, 10, 245, 1979.
29. **Smibert, R. M.,** Vibrio fetus var. intestinalis isolated from fecal and intestinal contents of clinically normal sheep: biochemical and cultural characteristics of microaerophilic vibrios isolated from the intestinal contents of sheep, *Am. J. Vet. Res.*, 26, 320, 1965.
30. **Buck, G. E. and Kelly, T.,** Effect of moisture content of the medium on colony morphology of *Campylobacter fetus* subsp. *jejuni, J. Clin. Microbiol.*, 14(5), 585, 1981.
31. **Smibert, R. M.,** The genus *Campylobacter, Ann. Rev. Microbiol.*, 32, 673, 1978.

32. **Ullmann, U.,** Methods in *Campylobacter,* in *Methods in Microbiology,* Bergan, T. and Norris, J. R., Eds., Academic Press, New York, 1979, 435.
33. **Wang, W. L., Leuchtefeld, N. W., Reller, L. B., and Blaser, M. J.,** Enriched Brucella medium for storage and transport of cultures of *Campylobacter fetus* subsp. *jejuni, J. Clin. Microbiol.,* 12(3), 479, 1980.
34. **Bolton, F. J. and Robertson, L.,** A selective medium for isolating *Campylobacter jejuni,* in *Campylobacter: Epidemiology, Pathogenesis, and Biochemistry,* Newell, G. D., Ed., MTP Press, Lancaster, U.K., 1982.
35. **Bruce, D., Zochowski, W., and Fleming, G. A.,** Campylobacter infection in cats and dogs, *Vet. Rec.,* 107, 200, 1980.
36. **Goossens, H., De Boeck, M., Van Landuyt, H., and Butzler, J. P.,** unpublished data.

Chapter 5

SEROTYPING *CAMPYLOBACTER JEJUNI* AND *CAMPYLOBACTER COLI* ON THE BASIS OF THERMOSTABLE ANTIGENS

S. Lauwers and J. L. Penner

TABLE OF CONTENTS

I. Introduction 52

II. Serotyping 52
  A. General Considerations 52
  B. Historical Development 53
  C. Serotyping on the Basis of the Thermostable Antigens 54
  D. Nomenclature of Serotypes 55
  E. Individual Serotyping Schemes for *C. jejuni* and *C. coli* 55

III. Application of Serotyping 56
  A. Prevalence of Serotypes in Human Isolates 56
  B. Investigation of Outbreaks 57
  C. Serotypes of Isolates From Animals 57
  D. Clinical Studies 57

IV. Discussion 58

References 58

I. INTRODUCTION

Less than 10 years after Butzler et al.[1] identified a pathogenic role for the "related vibrios",[2] the group of bacteria has become recognized as a major cause of human enteritis.[3,4] They surpass in frequency of isolation the well-known *Salmonella*, *Shigella* and enteropathogenic *Escherichia coli*.[5] Not surprisingly, a keen interest in the epidemiology of *Campylobacter* infections has emerged and considerable attention is currently focused on methods for subspecific classification. Methods that have proven effective for serotyping other Gram-negative species are now under evaluation for their effectiveness in discriminating among the strains of *Campylobacter jejuni* and *Campylobacter coli*, the two species now recognized to occur in the "related vibrios".

II. SEROTYPING

A. General Considerations

Antigens for subspecific classification of Gram-negative bacteria have generally included those associated with the lipopolysaccharide (LPS) of the outer membrane, the flagella, and the material found in the capsule. The Kauffmann-White scheme for *Salmonella* is based on antigens from each of the three structural components and are known, respectively, as the O, H, and K antigens.[6] The *Salmonella* serotyping scheme is a triumph of classical serology and is widely held to be the model to follow in developing serotyping schemes for other gram-negative species. Indeed, most of the serotyping schemes for other species of the Enterobacteriaceae have been modeled closely along the lines of the *Salmonella* scheme. In these schemes the O, H, and K antigens are distinguished from each other largely by differences in their tolerance to heat treatment. Their immunological specificities are determined using bacterial cell suspensions with the antigens *in situ* and the preferred test is the agglutination reaction performed on glass slides, in tubes, or in wells of microtitration plates.

The O antigens are characterized as thermostable. Their antigenicity is essentially unaltered after heating for 2 hr at 100 or 120°C. Flagella are protein in composition and thermolabile. Such simple separation of O and H antigens may be complicated by the presence of K antigens. These may be polysaccharide or protein and heat treatment may cause their release from the bacterial cell or the destruction of their specificity. There are, however, several capsular types that differ from one to another in their thermolability. Difficulties in serotyping arising from the involvement of the K antigens are generally recognized and for some species, at least, the thermolability of the K antigens have been determined and K antisera have been prepared. Strains of such species may be typed on the basis of O, H, and K antigens. However, the dependence on heat stability as a major criterion for characterizing the three antigen classes has led to some interpretative problems. Thermolabile O antigens were not recognized under the rigid definition and the thermolabile *Salmonella* 0:5 antigen was classified, along with the Vi antigen, as a K antigen.[6] The discovery of pili (fimbriae) was made after the H antigens of the *Salmonella* and other Gram-negative species had been defined and thus earlier workers were unaware of their participation in cross reactions among different H serotypes.[7] In other cases the pili mimicked capsules in their ability to inhibit O agglutination reactions.[8]

The O antigens, largely because of their endotoxic properties, have been the subject of intensive investigation at the molecular level.[9] The molecular structure of several *Salmonella* O antigens have been elucidated.[10] In the course of these investigations procedures for extracting and purifying LPS have been developed.[11,12] With purified extracted antigens, techniques more precise than the agglutination reaction may be exploited in defining the immunological specificities of the O antigens. Improvements in the staining procedures allow clearer visualization of LPS after polyacrylamide gel electrophoresis.[13] These and other recently developed technologies are expected to lead to new insights on the molecular biology of the O antigen and could considerably influence the approach taken by a microbiologist faced with the task of developing a new scheme for serotyping other species of Gram-negative bacteria. Because of the availability of new techniques and procedures, the traditional approaches to the development of schemes for O serotyping could become somewhat less appealing, and it should not be surprising that researchers would avail themselves of the prevailing knowledge and methodology and design fresh approaches to the question of serotyping. The option of working with extracted antigens rather than with antigens located *in situ* could be particularly attractive because of the wider array of immunological and biochemical procedures with which the extracted antigens can be investigated. This is particularly relevant to serology of *Campylobacter* for which the classical approach to developing O serotyping schemes has not been spectacularly successful.

B. Historical Development

In 1971 ten strains that would now be classified as *C. jejuni* or *C. coli* were shown by Berg et al.[14] to have a thermostable antigen different in specificity from those found in *C. fetus* subsp. *fetus* and *C. fetus* subsp. *venerealis*. The antigen was designated serotype "C" thereby extending the scheme of Morgan[15] which consisted of serotypes "A" and "B" for strains of the other two species. A thermolabile antigen was also identified in six of the ten strains when they were tested in seven antisera directed against thermolabile antigens. Because the "C" antigen was demonstrable by agglutination of boiled or autoclaved cell suspensions, it was taken, without further evidence, to be an O antigen. An account of the nature of the thermolabile antigens was not given. This study, to the authors' knowledge, is the first attempt at defining antigens useful in serotyping the *C. jejuni*-*C. coli* group. The recent realization of the importance of these bacteria in human disease prompted a renewed interest in these particular antigens and investigators began to assess the feasibility of producing serotyping schemes.

Some groups used as their model the scheme of Berg et al.[14] and prepared antisera against live or formalized bacterial suspensions and tested them for specificity using agglutination reactions and suspensions of live, formalized, or heated cells. Berg et al.[14] reported that boiled suspensions showed a tendency towards autoagglutination and Itoh et al.[16] reported that agglutination tests with autoclaved cells showed no immunological specificity. Lack of specificity with heated cell suspensions was also reported by Penner and Hennessy.[17] Moreover, Kosunen reported that antisera against autoclaved suspensions gave extensive cross reactions when tested by tube agglutination or by coagglutination and that no detectable reactions occurred when the slide agglutination was used.[18] Abbott et al.[19] reported that it was necessary to cross absorb each of their 'O' antisera (prepared against organisms heated for 15 min at 100°C) because of the many cross reactions. From these observations the view evolved that the agglutination test was unsatisfactory for identifying the thermostable antigens of heat-treated cell suspensions. On the other hand, it became clear that live or formalin-treated suspensions were more satisfactory immunogens than were heat-altered cell suspensions for producing specific antisera. With the more specific antisera, serotyping schemes could be designed to type on the basis of thermolabile antigens in the manner described by Berg et al.[14] Such antisera were absorbed with heated homologous cell suspensions to remove antibody directed against the thermostable antigens and, in some cases, also with heterologous live or formolized cells to remove antibodies against cross-reacting components. Schemes such as those described by Itoh et al.[16] and Lior et al.[20] are based on the thermolabile antigen. Details of these schemes are found in another chapter.

Another scheme that has been proposed is based on the use of antisera that are apparently not absorbed to remove antibody against thermostable components and thus agglutinations of cells occur on the basis of both thermolabile and thermostable antigens.[21] It should be noted that there is currently insufficient evidence to conclude that the material referred to as the thermolabile antigen is a single entity. Research at the molecular level is expected to ascertain its composition and determine the extent to which the capsule, flagella, and perhaps proteins from the outer membrane contribute to its antigenicity.

Because of difficulties experienced with the agglutination reaction when applied to heated cells, novel approaches had to be considered for defining the thermostable antigens. Earlier groups of workers had demonstrated that thermostable antigens could be extracted from *C. fetus* subsp. *venerealis* using methods routinely employed to extract O antigens from Enterobacteriaceae. Among these were Ristic and co-workers[22-24] who extracted their thermostable antigen by ethanol precipitation from supernatant obtained from cell suspensions. Antibodies in rabbit antisera prepared against their antigen were detectable by agglutination reactions, the double-diffusion technique of Ouchterlony, passive hemagglutination, and by complement fixation. Newsam and St. George[25] obtained a thermostable fraction from autoclaved cells and detected antibodies against the material by passive hemagglutination. In other studies by Winter[26] and McCoy et al.[27] trichloroacetic acid and phenol water were used to extract endotoxic LPS from *C. fetus* and antibodies in rabbit antisera were demonstrated by passive hemagglutination and double-diffusion techniques. Bokkenheuser[28] also used the passive hemagglutination test and reported that it was more sensitive than the agglutination test in his study on *C. fetus* subsp. *fetus*.

Because of the high endorsement given by these investigators to the use of extracted antigens and the passive hemagglutination technique and because of the limitations of the agglutination reaction when used with heated cell suspensions, the feasibility of applying their methods to the study of the thermostable antigens of *C. jejuni* and *C. coli* was investigated by two groups working independently. Lauwers et al.[29] and Penner and Hennessy[17] demonstrated that thermostable antigens could be readily extracted and that their specificities could be defined with rabbit antisera using passive hemagglutination. Furthermore, serotyping schemes were developed that differentiated effectively among isolates of both species.

Unequivocal evidence is not available that the thermostable antigens used as differentiating markers in these typing schemes are O antigens. It needs to be emphasized, however, that the known properties of the thermostable antigens are common to the O antigens of Enterobacteriaceae. It should also be pointed out that the presence of O antigens in *C. jejuni* and *C. coli* has been confirmed by the work of Naess and Hofstad[30] who determined the chemical composition of lipopolysaccharide extracted from strains of the two species. It can be confidently predicted that future studies linking the results of the biochemical studies with the results of the serological studies will demonstrate convincingly whether the thermostable antigens are or are not O antigens.

C. Serotyping on the Basis of the Thermostable Antigens

Although the thermostable antigens extractable from *C. jejuni* and *C. coli* have not been completely characterized, considerable progress may be reported on the development of schemes that utilize them as the basis for serotyping. Two independently developed schemes have been described and passive hemagglutination was used in both studies to define the immunological specificities of the thermostable antigens. The extraction of antigens for sensitizing the erythrocytes was performed according to the well-known procedures established for Enterobacteriaceae by Neter et al.[31] in 1952 and has been described in detail in one study.[17] Briefly, saline suspensions of the bacteria are heated for 1 hr at 100°C, cooled, and centrifuged. The supernatant containing the thermostable antigen in solution is used for sensitizing erythrocytes for passive hemagglutination. Sheep erythrocytes were used in one study[17] and human Rh-negative erythrocytes in the other.[29] The antisera used for serotyping were produced in rabbits by immunizing with live cells intravenously only[17] or with five subcutaneous inoculations with Freund's complete adjuvant followed by intravenous injections.[29] At least 50 serotypes of thermostable antigens have been identified.

The serotyping procedure involves the titration of each antiserum against erythrocytes sensitized with antigen extracted from the isolate for which the serotype is to be determined. The isolate is assigned to the serotype corresponding to the antiserum or antisera in which agglutination of the erythrocytes occurs to homologous titer or to within one, two, or three dilutions of homologous titer. Most of the isolates encountered at this point have reacted in only one tiserum, but some showed reactions in two or more antisera indicating that the specificity of the thermostable antigen is different from that of the serotype reference strains. Some multiply-reacting isolates show strong reactions (up to, or within one or two dilutions of homologous titer) in each of the antisera in which a reaction occurs, but others react strongly in one antiserum and weakly in the others. The weak reactions may be considered as minor but they are not unimportant because epidemiologically linked isolates have been found to be identical in such minor reactions. Multiply-reacting isolates are new serotypes and antisera against them may be absorbed with cell suspensions of the cross-reacting strains to produce monospecific antisera. At this point, use of cross-absorbed antisera has been limited to typing multiply-reacting isolates from one outbreak of *C. jejuni* enteritis in order that the outbreak strain could be more precisely defined.

The task of titrating all the antisera each time an isolate is to be typed is somewhat labor-intensive and automated procedures for performing some of the procedures, such as serial dilutions, are necessary when large numbers of isolates are routinely serotyped. One of the authors (S. Lauwers) prepares five pools of antisera and screens each isolate in the pools before titration of individual antisera. This procedure is both practical and time saving.

D. Nomenclature of Serotypes

Investigations of the specificities of the *Campylobacter* antigens are underway in several laboratories and in order to avoid confusion in the literature it was resolved at a round table discussion on antigenic typing[32] that the investigators would provisionally identify their serotypes by using as a prescript a series of letters from their last name to precede the number assigned to the serotype. In order to expedite progress towards a single unified antigenic scheme, agreement was reached among investigators to exchange strains to make comparisons of the serotypes of the individual schemes. This work has not been completed, but some of the more common serotypes described by Lauwers et al.[29] and Penner and Hennessy[17] have been compared. Pen serotypes 1,2,4,5,8,11,14,18,21,28,30,31, and 33 have been found by cross-absorption and cross-titration tests to correspond respectively to Lau serotypes 2,1,25,32,46,4,24,7,38,42,11,6, and 51.[40] It is anticipated that further work of this nature will achieve comparisons of all serotype strains, not only from these two schemes but also from other independently developed schemes.

E. Individual Serotyping Schemes for *C. jejuni* and *C. coli*

The development of serotyping schemes was initiated before the two-species concept for thermophilic campylobacters was introduced and serotype reference strains of both species were included in the schemes. Results of serotyping large numbers of isolates led to the observation that isolates belonging to *C. jejuni* generally reacted in antisera prepared against strains of the same species. Similarly, *C. coli* isolates were found, in most cases, to react in antisera against *C. coli* serotypes. Data in connection with this aspect from one laboratory[40] are presented in Table 1. Four of the 222 (2%) that reacted in *C. jejuni* antisera, were hippurate negative and therefore classified as *C. coli* (see Table 1). Such isolates need to be examined further to establish that they are not *C. jejuni* which have lost their ability to hydrolyze hippurate. Except for one case, serotype Lau 20, a similar pattern of distribution was seen for serotyping with *C. coli* antisera. The possibility that serotype reference strain Lau 20 could be a hippurate-negative *C. jejuni* also begs further investigation. Alternatively, the isolates that react in antisera of the other species could be new serotypes that have, in

Table 1
RELATION BETWEEN BIOTYPE AND SEROTYPE[40]

Serotype (Lau)	No. strains (patients)	Hippurate test +	Hippurate test −
Reference Strain Hippurate Positive			
1	65 (43)	65	0
2	41 (26)	41	0
3	12 (10)	11	1
4	14 (9)	14	0
6	17 (11)	16	1
7	3 (2)	3	0
13	3 (3)	3	0
15	12 (9)	12	0
17	3 (2)	3	0
18	2 (2)	2	0
25	40 (27)	38	2
26	4 (3)	4	0
32	4 (2)	4	0
38	2 (1)	2	0
Totals	222 (151)	218	4
Reference Strain Hippurate Negative			
5/8	12 (10)	6	6
11	3 (3)	0	3
14	12 (10)	0	12
20	16 (16)	16	0
42	2 (2)	0	2
Totals	45 (41)	22	23

addition to the observed reaction, a major specificity not identified by the available antisera. Antisera need to be prepared against such isolates to resolve this question. However, these observations provided strong indications that each species possessed its own particular set of specificities for the thermostable antigen. The observation that other species have distinctive sets of antigenic specificities has been previously noted.[33] The evidence for the separate sets of thermostable antigenic specificities shown in Table 1 may also be cited as support for the two species classification scheme.

Serotyping of each species may be performed separately. Only those isolates that do not type in antisera of one species need to be tested in antisera of the other species. This reduces substantially the number of titrations necessary to type an isolate once its hippurate reaction is known.

III. APPLICATION OF SEROTYPING

A. Prevalence of Serotypes in Human Isolates

During the development of the serotyping schemes isolates from hospitals were routinely serotyped. Although the major objectives of this work were to evaluate the effectiveness of the schemes and find new serotypes, an equally important objective was to determine the distribution of serotypes among human isolates. A total of 870 isolates from 627 patients at St. Pierre Hospital, Brussels were typed. Clustering of isolates due to 15 small family outbreaks was noted. Taking into account only one isolate from each cluster the serotype distribution was determined: 40 serotypes were recognized among the isolates. Of the isolates, 50% fell into the most common types (Lauwers' serotypes 1,2,25,3,6, and 5/8) and 25% of these belonged to the first two serotypes. Significant variation in serotype distribution from year to year was not seen in this hospital. The serotype distribution of isolates collected in other areas of Belgium were markedly similar to that found for the Brussels hospital. Moreover, 50% of the 81 isolates from Bruges, Belgium, 37% of the 57 isolates from Sweden, 35% of the 135 isolates from Norway, and 30% of the 22 from West Germany fell into the six common serotypes. Isolates from Bangladesh, Thailand, Nigeria, Zaire, and Peru were more frequently untypable. Percentages untypable ranged from 25 to 45% in contrast to 18% for isolates from Brussels. The two serotypes most common in Brussels (Lau 1 and Lau 2) were encountered in each of these widely separated countries. A distinctive feature in collections from these places appears to be a much higher incidence of two particular serotypes (Lau 13, Lau 14) that are only infrequently encountered in Europe. A high frequency of isolation of the two serotypes most common in Brussels has also been found in studies conducted in Canada. McMyne et al.[34] found this to be the case for isolates collected at a public health laboratory and Karmali et al.[35] found a high frequency of these types in a pediatric hospital.

B. Investigation of Outbreaks

Considerable interest in the epidemiology of outbreaks of campylobacter enteritis prompted the application of serotyping in the investigation of several outbreaks. There were 9 isolates from 7 patients in an outbreak in a nursery in Italy and 15 isolates from 14 patients in a milk-borne outbreak in Switzerland that all belonged to serotype Lau 2, but in Sweden 17 isolates from the same number of patients associated with a food-borne outbreak were serotype Lau 1.[40] In British Columbia, Canada, seven isolates from six patients in a community outbreak of gastroenteritis were identified by McMyne et al.[34] as serotype Pen 1 (Lau 2) and six family outbreaks investigated by the same group of workers were found to be caused by different serotypes of *C. jejuni*.

C. Serotypes of Isolates From Animals

In most animal species that have been examined, either *C. jejuni* or *C. coli* have been isolated from some individuals of each species. From some animals, both *C. jejuni* and *C. coli* have been isolated. In one study it was shown that 50% of the 61 isolates from cows belonged to serotypes (Lau 1, Lau 2, Lau 5/8) that are also common among human isolates. *C. jejuni* serotypes common among human isolates were also found in 33 dog and 33 poultry isolates, but 18 of 20 pig isolates were *C. coli* and 10 of this species serotyped in *C. coli* antisera — 5 were serotype Lau 35 and 5 were Lau 43. One *C. jejuni* isolate from a pig belonged to serotype Lau 4. In a study of zoo animals by Leuchtefeld et al.,[36] fecal specimens from 12 primates, 2 felids, 13 hoofed animals, 6 birds, 1 panda, and 1 reptile were cultured for *Campylobacter*. Only 5.6% of the specimens from healthy animals were positive, but 31.8% of the specimens from diarrheic animals were positive. At the time of study the hippurate hydrolysis test was not performed, thus precluding further classification. The isolates were serotyped, however, and some, but not all, belonged to serotypes common among human isolates. In a later study, Luechtefeld and Wang[37] did perform the hippurate test on isolates from both wild and domestic animals and found that the *C. coli* (hippurate negative) isolates belonged to serotypes that were, in most cases, uncommon among human isolates.

D. Clinical Studies

The occurrence of multiple bouts of campylobacter enteritis in two immune-deficient patients was investigated by Ahnen and Brown.[38] From one patient two isolates each from separate episodes of diarrhea were found to be different in serotype suggesting separate infections, but four isolates recovered during separate episodes over a 17-month period from a second patient were all of one serotype. Whether these episodes were due to reinfection or to relapses because of a chronic carrier could not be resolved but it was clear to the authors that hypogammaglobulinemic patients were unusually susceptible to infection with *C. jejuni*. However, in the study of patients by Karmali et al.[39] it was evident that separate episodes of enteritis were due to reinfections. Evidence for this was established by demonstrating that the isolates recovered during the different episodes were different in serotype.

IV. DISCUSSION

It is clearly evident that the many immunological specificities of the thermostable antigen fulfill the role of a marker for differentiating isolates of *C. jejuni* and *C. coli*. This has been established in two independently developed schemes in each of which more than 50 serotypes have been defined. The applications of the schemes in investigating outbreaks have illustrated their effectiveness in separating outbreak isolates from unrelated isolates. In prevalence studies conducted in widely separated places the same serotype (Lau 1-Pen 2) has been found to be the most frequently occurring. Additional work to compare the two schemes need to be completed to produce a single scheme for serotyping on the basis of the thermostable antigen. Such a scheme promises to have a major role in future studies on the epidemiology of infections caused by these bacteria. Studies at the molecular level will, no doubt, lead to insights on the biological and biochemical properties of the thermostable antigen and lead to insights on the function of the antigen in the cell.

REFERENCES

1. **Butzler, J. P., Dekeyser, P., Detrain, M., and Dehaen, F.**, Related vibrio in stools, *J. Pediatr.*, 82, 318, 1973.
2. **King, E. O.**, Human infections with *Vibrio fetus* and a closely related vibrio, *J. Infect. Dis.*, 101, 119, 1959.
3. **Butzler, J. P. and Skirrow, M. B.**, Campylobacter enteritis, *Clin. Gastroenterol.*, 8, 737, 1979.
4. **Svedhem, A. and Kaijser, B.**, *Campylobacter fetus* subspecies *jejuni:* A common cause of diarrhea in Sweden, *J. Infect. Dis.*, 142, 353, 1980.
5. **Karmali, M. A. and Fleming, P. C.**, Campylobacter enteritis in children, *J. Pediatr.*, 94, 527, 1979.
6. **Kauffmann, F.**, *The Bacteriology of the Enterobacteriaceae*, Williams & Wilkins, Baltimore, 1966.
7. **Duguid, J. P. and Campbell, I.**, Antigens of the type-1 fimbriae of *Salmonella* and other Enterobacteria, *J. Med. Microbiol.*, 2, 353, 1969.
8. **Aleksic, S., Rhode, R., Aleksic, V., and Muller, G.**, A new fimbrial antigen as a cause for a complete O-Inagglutinability of various *Arizona* strains, *Zbl. Bakt. Hyg.*, 241(Abstr.), 427, 1978.
9. **Luderitz, O.**, Endotoxins and other cell wall components of Gram-negative bacteria and their biological activities, in *Microbiology — 1977*, Schlesinger, D., Ed., American Society of Microbiology, Washington, D.C., 1977, 239.
10. **Luderitz, O., Staub, A. M., and Westphal, O.**, Immunochemistry of O and R antigens of *Salmonella* and related *Enterobacteriaceae*, *Bacteriol. Rev.*, 30, 192, 1966.
11. **Westphal, O., Luderitz, O., and Bister, F.**, Uber die extraktion von Bacterien mit Phenol-Wasser, *Z. Naturforsch. Teil B*, 7B, 148, 1952.
12. **Leive, L. and Shovlin, K.**, Physical, chemical, and immunological properties of lipopolysaccharide released from *Escherichia coli* by ethylenediaminetetraacetate, *J. Biol. Chem.*, 243, 6384, 1968.
13. **Tsai, C.-M. and Frasch, C. E.**, A sensitive silver stain for detecting lipopolysaccharides in polyacrylamide gels, *Anal. Biochem.*, 119, 115, 1982.
14. **Berg, R. L., Jutila, J. W., and Firehammer, B. D.**, A revised classification of *Vibrio fetus*, *Am. J. Vet. Res.*, 32, 11, 1971.

15. **Morgan, W. J. B.,** Studies on the antigenic structure of *Vibrio fetus, J. Comp. Pathol. Therap.*, 69, 125, 1959.
16. **Itoh, T., Saito, K., Yanagawa, Y., Sakai, S., and Ohashi, M.,** Serologic typing of thermophilic campylobacters isolated in Tokyo, in *Campylobacter: Epidemiology, Pathogenesis, and Biochemistry*, Newell, D. G., Ed., MTP Press, Lancaster, U.K., 1982, 106.
17. **Penner, J. L. and Hennessy, J. N.,** Passive hemagglutination technique for serotyping *Campylobacter fetus* subsp. *jejuni* on the basis of soluble, heat-stable antigens, *J. Clin. Microbiol.*, 12, 732, 1980.
18. **Firehammer, B. D. and Davies, J.,** General discussion and editorial comment, Section IV: Serology and serotyping, in *Campylobacter: Epidemiology, Pathogenesis, and Biochemistry*, Newell, D. G., Ed., MTP Press, Lancaster, U.K., 1982, 122.
19. **Abbott, J. D., Dale, B., Eldridge, J., Jones, D. M., and Sutcliffe, E. M.,** Serotyping of *Campylobacter jejuni/coli, J. Clin. Pathol.*, 33, 762, 1980.
20. **Lior, H., Woodward, D. L., Edgar, J. A., Laroche, L. J., and Gill, P.,** Serotyping of *Campylobacter jejuni* by slide agglutination based on heat-labile antigenic factors, *J. Clin. Microbiol.*, 15, 761, 1982.
21. **Rogol, M., Sechter, I., Braunstein, I., and Gerichter, Ch. B.,** Provisional antigenic scheme for *Campylobacter jejuni*, in *Campylobacter: Epidemiology, Pathogenesis, and Biochemistry*, Newell, D. G., Ed., MTP Press, Lancaster, U.K., 1982, 98.
22. **Ristic, M. and Brandly, C. A.,** Characterization of *Vibrio fetus* antigens. I. Chemical properties and serological activities of a soluble antigen, *Am. J. Vet. Res.*, 20, 148, 1959.
23. **Ristic, M. and Brandly, C. A.,** Characterization of *Vibrio fetus* antigens. II. Agglutination of polysaccharide-sensitized sheep erythrocytes by specific antiserums, *Am. J. Vet. Res.*, 20, 154, 1959.
24. **Ristic, M. and Walker, J.,** Characterization of *Vibrio fetus* antigens. III. Study of serologic and biophysical properties of a polysaccharide fraction in a hemolytic system, *Am. J. Vet. Res.*, 21, 884, 1960.
25. **Newsam, I. D. B. and St. George, T. D.,** Diagnosis of bovine vibriosis. III. Indirect haemagglutination using untanned sheep erythrocytes, *Aust. Vet. J.*, 43, 283, 1967.
26. **Winter, A. J.,** An antigenic analysis of *Vibrio fetus*. III. Chemical, biologic, and antigenic properties of endotoxin, *Am. J. Vet. Res.*, 27, 653, 1966.
27. **McCoy, E. C., Doyle, D., Burda, K., Corbiel, L. B., and Winter, A. J.,** Superficial antigens of *Campylobacter (Vibrio) fetus:* Characterization of an antiphagocytic component, *Infect. Immunol.*, 11, 519, 1975.
28. **Bokkenheuser, V.,** *Vibrio fetus* infections in man: a serological test, *Infect Immunol.*, 5, 222, 1972.
29. **Lauwers, S., Vlaes, L., and Butzler, J. P.,** Campylobacter serotyping and epidemiology, *Lancet*, 1, 158, 1981.
30. **Naess, V. and Hofstad, T.,** Isolation and chemical composition of lipopolysaccharide from *Campylobacter jejuni, Acta Pathol. Microbiol. Immunol. Scand. Sect. B*, 90, 135, 1982.
31. **Neter, E., Bertram, L. F., Zah, D. A., Merdoch, M. R., and Argerman, C. E.,** Studies on hemagglutination and hemolysis by *Escherichia coli* antisera, *J. Exp. Med.*, 96, 1, 1952.
32. **Feldman, R. A.,** Round table discussion on antigenic typing, in *Campylobacter, Epidemiology, Pathogenesis and Biochemistry*, Newell, D. G., Ed., MTP Press, Lancaster, 1982, 124.
33. **Penner, J. L. and Hennessy, J. N.,** Reassignment of the intermediate strains of *Proteus rettgeri* biovar 5 to *Providencia stuartii* on the basis of the somatic (0) antigens, *Int. J. Syst. Bacteriol.*, 27, 71, 1977.
34. **McMyne, P. M. S., Penner, J. L., Mathias, R. G., Black, W. A., and Hennessy, J. N.,** Serotyping of *Campylobacter jejuni* isolated from sporadic cases and outbreaks in British Columbia, *J. Clin. Microbiol.*, 16, 281, 1982.
35. **Karmali, M. A., Penner, J. L., Fleming, P. C., Williams, A., and Hennessy, J. N.,** The serotype and biotype distribution of clinical isolates of *Campylobacter jejuni-coli* over a three-year period, *J. Infect. Dis.*, in press.
36. **Luechtefeld, N. W., Cambre, R. C., and Wang, W.-L. L.,** Isolation of *Campylobacter fetus* subsp. *jejuni* from 300 animals, *J. Am. Vet. Med. Assoc.*, 179, 1119, 1981.
37. **Luechtefeld, N. W. and Wang, W.-L. L.,** Hippurate hydrolysis by and triphenyltetrazolium tolerance of *Campylobacter fetus, J. Clin. Microbiol.*, 15, 137, 1982.
38. **Ahnen, D. J. and Brown, W. R.,** Campylobacter enteritis in immune-deficient patients, *Ann. Intern. Med.*, 96, 187, 1982.
39. **Karmali, M. A., Kosoy, M., Newman, A., Tischler, M., and Penner, J. L.,** Reinfection with *Campylobacter jejuni*, (Letter), *Lancet*, 2, 158, 1981.
40. **Lauwers, S.,** unpublished data.

Chapter 6

SEROTYPING OF *CAMPYLOBACTER JEJUNI* AND *C. COLI* BY SLIDE AGGLUTINATION BASED ON HEAT-LABILE ANTIGENIC FACTORS

Hermy Lior

TABLE OF CONTENTS

I. Introduction 62

II. Methodology 62

III. Serotyping of *C. jejuni* and *C. coli* Isolates 68

IV. Biotyping 74

References 74

I. INTRODUCTION

The recognition of *Campylobacters (Vibrio fetus)* as important animal pathogens at the beginning of the century stimulated serological studies designed to elucidate the epidemiology of these organisms.

Following the report of Smith and Reagh in 1903[1] on the different behavior of the somatic and flagellar antigens of salmonella, Smith and Taylor[2] studied the serological characteristics of 22 strains of *C. fetus (V. fetus)* isolated from calves. Their finding that all the strains were serologically alike was confirmed later by Smith and Orcutt[3] who also reported that the bovine strains also shared a common "H" flagellar antigen. However the significance of these results was greatly diminished, as the isolates were obtained from a single large herd and were not, therefore, completely representative.

The first report showing the antigenic differentiation among *C. fetus* strains isolated from sheep was published in 1946 by Blakemore and Gledhill.[4]

Subsequent reports on the serology of *C. fetus* have, in main, included only direct agglutination in crude, unabsorbed antisera without recourse to specific absorbed materials.[5-9] Price et al. in 1955[8] reported that the serological differences of the "H" flagellar antigens of 30 *C. fetus* strains were not sufficiently clear-cut to allow a separation into serogroups.

The study of the heat-stable (O) and heat-labile (H) antigens was continued by Wiidik and Hlidar[9] who were also the first to suggest the presence on the surface of *C. fetus* of a heat-labile capsular (K) antigen. By 1957, King[10] recognized that the *C. fetus* isolated from man was different biochemically and in growth temperature characteristics from another group of vibrios isolated from man which she designated "related vibrios", and was able to show that they possessed different serological specificities from *C. fetus*.

The serological studies initiated by von Mitscherlich and Liess[11] were continued by Morgan[12] and White and Walsh[13] who used absorbed sera against heat-stable (O) and heat-labile (H) antigens. These studies were further advanced by Berg et al.[14] who classified campylobacters into three groups, A, B, and C on the basis of heat-stable antigens. Using antisera against whole cells absorbed with homologous heated and live cross-reactive antigens, Berg et al.[14] were able to show the presence of seven thermolabile antigenic factors. Ten *C. jejuni* strains ("related vibrios") were found to belong to heat-stable serogroup C and five strains studied further were found to possess a heat-labile factor 1. The occurrence of multiple thermolabile antigens on a single strain in all serogroups except for serogroups C prevented a clear differentiation of the strains by this serotyping system. With the development of isolation techniques from stool specimens,[15] and the recognition of *C. jejuni* as an important agent in human diarrheal disease, the serological differentiation of these organisms became an important event in the epidemiology of the disease. The serotyping of *C. jejuni* using a slide agglutination technique with live bacteria, and 6 antisera prepared from human (3) and nonhuman isolates (3), was reported by Butzler and Skirrow.[16] Of the 90 human isolates investigated, 84 were found typable — 16 strains reacted only with the antisera prepared with the human isolates, 9 strains agglutinated in the sera prepared with the nonhuman isolates, and 59 isolates cross reacted with both groups of antisera.

These and previous studies[14] have shown that the differentiation of these organisms can be achieved by a rapid, easy-to-perform slide agglutination technique. Lior et al.[17] using a methodology similar to the one used for the serotyping of other enteric pathogens, developed a serotyping scheme for *C. jejuni* and *C. coli* using live cultures and absorbed, specific antisera for the detection of heat-labile antigenic factors.

II. METHODOLOGY

Vaccine suspensions containing 10^{10} bacteria/mℓ used to innoculate rabbits were prepared from smooth colonies of reference strains inoculated on Mueller-Hinton broth (Oxoid Ltd., London, England) containing 1.0 to 1.25% agar (Difco Laboratories, Detroit, Mich.) and incubated for 48 hr at 37°C in the gas mixture of 5% O_2, 10% CO_2, and 85% N_2 as recommended by Skirrow.[18]

New Zealand white rabbits weighing 2.7 to 3.0 kg each were injected intravenously at 4- to 5-day intervals for 4 weeks with increasing doses (0.5 to 2.5 mℓ) of bacterial suspension in phosphate buffered saline (PBS) pH 7.2 containing 0.5% formalin.

The rabbits were exsanguinated 7 to 10 days after the last injection and the sera preserved with 1:10,000 Merthiolate® at 4°C.

Homologous and heterologous titers were determined by slide agglutination and by tube titration. Tube titrations were performed using equal amounts of twofold serum dilutions in PBS and formalized bacterial suspensions, washed three times, diluted to an optical density (OD) of 0.45 at 530 nm, and incubated for 4 hr in a 50°C water bath. A positive reaction occurred within 4 hr as loose, floccular aggregates of typical flagellar agglutination. Homologous and heterologous titers are shown in Table 1.

Slide agglutinations were performed on glass slides by emulsifying a small loopful of bacteria (grown on fresh, moist Mueller-Hinton agar plates containing 5% sheep blood) into a small drop (10 to 20 μℓ) of PBS to which a drop (50 μℓ) of diluted antisera was added. Autoagglutinability was determined by emulsifying a small loopful of bacteria into a drop of PBS. Presence or lack of agglutination was determined 30 to 45 sec after the addition of the antisera.

The use of slide agglutination techniques with living *C. jejuni* strains have been hindered by the tendency of these organisms to autoagglutinate in saline or aggregate as sticky and rubbery masses. Butzler,[19] using bacterial suspensions in formalin-saline, reported a 50% reduction in the number of autoagglutinable strains, while Lior et al.[17] reported only 4% rough cultures following repeated subculturing of the strains on fresh moist media incubated at 37°C. The sticky, rubbery characteristics reported by Butzler and Skirrow[16] with *C. jejuni* has been previously described with *Neisseria meningitidis*[20] and also with some halophilic organisms.[21] The extracellular DNA-associated slime was found responsible for this phenomenon and the autoagglutinability described with some *N. gonorrhoeae* strains has been eliminated by the use of nucleases.[22] In our laboratory we were successful in eliminating the stringy characteristic displayed by some *C. jejuni* by the use of a DNase solution, suggesting that the extracellular DNA-slime contributes to the stringy-rubbery autoagglutinability observed with some *C. jejuni*. A small loopful of the sticky growth was emulsified into a small drop (10 to 20 μℓ) of a 0.1% DNase solution in PBS prior to the addition of antisera. A smooth, homogeneous suspension was obtained within 5 to 10 sec of mixing.

Homologous and heterologous reactions of the antisera were determined on slide agglutination with live reference strains using crude unabsorbed antisera diluted 1:5 in PBS (Table 2). The cross reactivity among campylobacters was reported by Price et al.,[8] Juhler,[23] and Ristic et al.[24] as a result of the sharing by *C. fetus* of heat-stable somatic antigens; Abbott et al.,[25] Kosunen et al.,[26] and Lior et al.[17] showed that both heat-stable and heat-labile antigens were responsible for the cross reactions observed with *C. jejuni*.

A marked decrease in the number of heterologous cross reactions was reported by Lior et al.[17] after the removal of antibodies to homologous heated antigens (Table 3). The few remaining cross reactions seen with live reference strains (Table 3) were removed by absorption with live strains until all heterologous reactions were eliminated. Briefly, the absorptions were performed by mixing 3 to 5 mℓ of crude antisera, diluted 1:2 in PBS, with washed, homologous heated or heterologous unheated bacterial suspensions obtained from 10 to 15 large 150 × 15 Petri plates. The mixture was incubated at 37°C for 2 hr followed by overnight refrigeration at 4°C and then centrifuged for 30 min at 5000 × gravity. The supernatant was saved and rechecked in order to determine that the removal of the cross-

Table 1
HOMOLOGOUS AND HETEROLOGOUS TITERS[a] OF *C. JEJUNI* AND *C. COLI* ANTISERA

Reference[b] strains	Unabsorbed antisera																	
	1	2	4	5	6	7	8	9	10	11	12	13	14	15	16	17	18	19
1	6,400			100											200			
2	200	3,200																
4			1,600															
5				6,400	200	100									200			
6				200	1,600													
7						3,200												
8							12,800	200										
9								12,800									100	
10		400	200						3,200									
11										3,200		100				800	100	
12											3,200							
13												3,200						
14													12,800					
15														6,400		400		
16															12,800			
17														100		25,600		
18	200																1,600	
19																		1,600
20						100					100							
21					100						200		200	100				
22																		
23		200																
24					200						400						800	
25							200											
26						200												
27																		
28																		
29							400											
30							200											
31								200										
32		200																
33		200		200				200			200			200				
34																		
35																		
36	200													400		3,200		
37						200												

Reference[b] strains	Unabsorbed antisera																	
	20	21	22	23	24	25	26	27	28	29	30	31	32	33	34	35	36	37
1		200																
2							200						200					
4																		
5	100						800											
6									400									200
7								800										
8																		
9							200											
10													800					
11			100		200								1,600				200	
12	200	100			1,600	200	200		200									
13																		
14																		
15																	200	
16																		
17																		
18									400									
19																		
20	6,400						200	1,600	1,600								800	
21		1,600					400							400				
22			1,600															
23				6,400														
24		200			6,400				3,200									
25	400					3,200	200		200					200		200		
26							12,800											
27								12,800										
28									3,200									
29										3,200								
30			200				200				3,200							
31												12,800			200			
32													1,600	200				
33													200	6,400				
34															6,400	6,400		
35															200	25,600		
36							200						200				12,800	
37																		3,200

[a] Tube titrations incubated at 50° for 4 hr.
[b] Formalinized antigens.

Table 2
HOMOLOGOUS AND HETEROLOGOUS SLIDE AGGLUTINATION REACTIONS OF *C. JEJUNI* AND *C. COLI* UNABSORBED ANTISERA[a]

	Antisera																																			
Reference live strains	1	2	4	5	6	7	8	9	10	11	12	13	14	15	16	17	18	19	20	21	22	23	24	25	26	27	28	29	30	31	32	33	34	35	36	37
1	4+	3+	—	—	—	4+	—	—	—	2+	—	—	—	—	—	—	3+	—	—	2+	—	2+	2+	—	2+	—	—	—	—	—	—	—	—	—	4+	—
2	3+	4+	—	—	—	—	—	—	—	3+	—	—	—	—	—	2+	—	—	—	—	—	—	—	—	2+	—	—	—	—	—	2+	—	—	—	2+	—
4	—	—	4+	—	4+	—	—	—	—	—	—	—	—	—	—	2+	—	—	—	—	—	—	—	—	—	—	—	—	—	—	—	—	—	—	—	—
5	—	—	—	4+	4+	—	—	—	—	—	2+	—	—	—	—	—	3+	—	3+	—	—	—	—	—	3+	—	—	—	—	—	—	—	—	—	—	—
6	—	—	—	4+	4+	—	—	—	—	—	—	—	—	—	—	—	—	—	—	—	—	—	—	—	2+	—	—	—	—	—	—	—	—	—	—	—
7	—	—	—	—	—	4+	—	—	—	—	—	—	—	—	—	4+	2+	—	—	—	—	—	—	—	3+	—	—	—	—	—	—	—	—	—	—	3+
8	—	—	—	—	—	—	4+	—	—	—	—	—	—	—	—	—	—	—	—	—	—	—	—	—	—	—	—	—	—	—	—	—	—	—	—	—
9	—	—	—	—	—	—	—	4+	—	—	—	—	—	—	3+	3+	—	—	—	—	—	—	—	—	—	—	—	—	—	—	—	—	—	—	—	—
10	—	2+	—	—	—	—	—	—	4+	—	—	—	—	—	—	—	—	—	—	—	—	—	—	—	—	—	—	—	—	—	—	—	—	—	2+	3+
11	2+	—	—	—	—	—	—	—	—	4+	—	2+	—	—	—	3+	—	—	—	—	—	—	—	—	—	—	—	—	—	—	3+	—	—	—	—	—
12	—	—	—	—	—	—	—	—	—	—	4+	—	—	—	—	—	—	—	3+	—	—	—	—	—	—	—	—	—	—	—	—	—	—	—	—	—
13	—	—	—	—	—	—	—	4+	—	3+	—	4+	—	2+	2+	3+	—	—	—	—	—	—	—	—	—	—	—	—	—	—	—	—	—	—	—	—
14	—	—	—	—	—	—	—	—	—	—	—	—	4+	—	—	—	—	—	—	—	—	—	—	—	—	—	—	—	—	—	—	—	—	—	—	—
15	—	—	—	—	—	—	—	—	—	—	—	—	—	4+	—	—	—	—	—	—	—	—	—	—	—	—	—	—	—	—	—	—	—	—	—	—
16	—	—	—	—	—	—	—	—	—	2+	—	—	—	—	4+	—	—	—	—	—	—	—	—	—	—	—	—	—	—	—	—	—	—	4+	—	
17	3+	3+	—	—	—	—	—	—	—	2+	—	—	—	—	—	4+	—	—	—	—	—	3+	—	—	—	—	—	—	—	—	—	—	—	—	—	4+
18	—	—	—	—	—	—	—	—	—	—	—	—	—	—	—	—	4+	3+	—	—	—	—	—	—	—	4+	—	—	—	—	—	—	—	—	—	—
19	—	—	—	—	—	—	—	—	—	—	—	—	—	—	—	—	—	4+	—	—	—	—	—	—	—	—	—	—	—	—	—	—	—	—	—	
20	—	—	—	—	—	—	—	—	—	—	3+	—	—	—	—	—	—	—	4+	—	—	—	—	—	—	—	—	—	—	—	—	—	—	—	—	—
21	3+	—	—	—	—	—	—	—	—	—	—	—	—	—	2+	2+	—	—	—	4+	—	—	—	—	2+	—	—	—	—	—	—	—	—	—	—	2+
22	—	—	—	—	—	—	—	—	—	—	—	—	—	—	—	—	—	—	—	—	4+	—	—	—	—	—	—	—	2+	—	—	—	—	—	—	—
23	—	—	—	—	—	—	—	—	—	—	—	—	—	—	—	—	—	—	—	—	—	4+	—	—	—	3+	—	—	—	—	—	—	—	—	—	—
24	—	—	—	—	—	—	—	—	—	—	—	—	—	—	—	—	2+	—	—	—	—	—	4+	—	—	—	—	—	—	—	—	—	—	—	—	—
25	—	—	—	—	—	—	—	—	—	—	—	—	—	—	—	—	—	—	—	—	—	—	—	4+	—	—	—	—	—	—	—	—	—	—	4+	2+
26	—	—	—	—	—	—	—	—	—	—	—	—	—	—	—	—	—	—	—	—	—	—	—	—	4+	4+	—	—	—	—	—	—	—	—	—	—
27	—	—	—	—	—	—	—	—	—	—	—	—	—	—	—	—	—	—	2+	—	—	—	—	—	4+	4+	—	—	—	—	—	—	—	—	—	—
28	—	—	—	—	—	—	—	—	—	—	—	—	—	—	—	—	—	—	—	—	—	—	—	—	—	—	4+	—	2+	—	—	—	—	—	—	—
29	—	—	—	—	—	—	—	—	—	—	—	—	—	—	—	—	—	—	—	—	—	—	—	—	—	—	—	4+	—	—	—	—	—	—	2+	—
30	—	—	—	—	—	—	—	—	—	—	—	—	—	—	—	—	—	—	—	—	—	—	—	—	2+	—	—	—	4+	—	—	—	—	—	—	—
31	—	—	—	—	—	—	—	—	—	—	—	—	—	—	—	—	—	—	—	—	—	—	—	—	—	—	—	—	—	4+	3+	—	2+	2+	—	2+
32	—	—	—	—	—	—	—	—	—	3+	—	—	—	—	—	—	—	—	—	—	—	—	—	—	—	—	—	—	—	—	4+	2+	—	—	2+	—
33	—	—	—	—	—	—	—	—	—	—	—	—	—	—	—	—	—	—	—	—	—	—	—	—	—	—	—	—	—	—	—	4+	—	—	—	—
34	—	—	—	—	—	—	—	—	—	—	—	—	—	—	—	—	—	—	—	—	—	—	—	—	—	—	—	—	—	—	—	—	4+	4+	—	—
35	—	—	—	—	—	—	—	—	—	—	—	—	—	—	—	—	—	—	—	—	—	—	—	—	—	—	—	—	—	—	—	—	4+	4+	—	—
36	—	—	—	—	—	—	—	—	—	—	—	—	—	—	—	—	—	—	—	—	—	—	—	—	—	—	—	—	—	—	2+	—	—	—	4+	—
37	—	—	—	—	—	2+	—	—	—	—	—	—	—	—	—	2+	—	—	—	—	—	—	—	—	—	—	—	—	—	—	2+	—	—	—	—	4+

[a] Antisera diluted 1:5 in PBS.

Table 3
HOMOLOGOUS AND HETEROLOGOUS SLIDE AGGLUTINATION REACTIONS OF *C. JEJUNI* AND *C. COLI* ANTISERA[a] ABSORBED WITH HOMOLOGOUS HEATED ANTIGENS

Reference live strains	Antisera																																		
	1	2	4	5	6	7	8	9	10	11	12	13	14	15	16	17	18	19	20	21	22	23	24	25	26	27	28	29	30	31	32	33	34	35	36	37
1	4+	—	—	—	—	4+	—	—	—	—	—	—	—	—	—	—	—	—	—	2+	—	—	—	—	2+	—	—	—	—	—	—	—	—	—	—	—
2	2+	4+	—	—	—	—	—	—	—	3+	—	—	—	—	—	—	—	—	—	—	—	—	—	—	—	—	—	—	—	—	—	—	—	—	—	—
4	—	—	4+	—	4+	—	—	—	—	—	—	—	—	—	—	2+	—	—	—	—	—	—	—	—	—	—	—	—	—	—	—	—	—	—	—	—
5	—	—	—	4+	4+	—	—	—	—	—	2+	—	—	—	—	—	—	—	2+	—	—	—	—	—	3+	—	—	—	—	—	—	—	—	—	—	—
6	.	—	—	3+	4+	—	—	—	—	—	—	—	—	—	—	—	—	—	—	—	—	—	—	—	—	—	—	—	—	—	—	—	—	—	—	—
7	—	—	—	—	—	4+	—	—	—	—	—	—	—	—	—	—	—	—	—	—	—	—	—	—	—	—	—	—	—	—	—	—	—	—	—	2+
8	—	—	—	—	—	—	4+	—	—	—	—	—	—	—	—	—	—	—	—	—	—	—	—	—	—	—	—	—	—	—	—	—	—	—	—	—
9	—	—	—	—	—	—	—	4+	—	—	—	—	—	—	—	2+	—	—	—	—	—	—	—	—	—	—	—	—	—	—	—	—	—	—	—	—
10	—	—	—	—	—	—	—	—	4+	—	—	—	—	—	—	—	—	—	—	—	—	—	—	—	—	—	—	—	—	—	—	—	—	—	—	2+
11	—	—	—	—	—	—	—	—	—	4+	—	—	—	—	—	—	—	—	—	—	—	—	—	—	—	—	—	—	—	—	2+	—	—	—	—	—
12	—	—	—	—	—	—	—	—	—	—	4+	—	—	—	—	—	—	—	—	—	—	—	—	—	—	—	—	—	—	—	—	—	—	—	—	—
13	—	—	—	—	—	—	—	—	—	—	—	4+	—	—	—	—	—	—	—	—	—	—	—	—	—	—	—	—	—	—	—	—	—	—	—	—
14	—	—	—	—	—	—	—	—	—	—	—	—	4+	—	—	—	—	—	—	—	—	—	—	—	—	—	—	—	—	—	—	—	—	—	—	—
15	—	—	—	—	—	—	—	—	—	—	—	—	—	4+	—	—	—	—	—	—	—	—	—	—	—	—	—	—	—	—	—	—	—	—	—	—
16	—	—	—	—	—	—	—	—	—	—	—	—	—	—	4+	—	—	—	—	—	—	—	—	—	—	—	—	—	—	—	—	—	—	—	—	—
17	—	—	—	—	—	—	—	—	—	—	—	—	—	—	—	4+	—	—	—	—	—	3+	—	—	—	—	—	—	—	—	—	—	—	—	—	3+
18	—	—	—	—	—	—	—	—	—	—	—	—	—	—	—	—	4+	3+	—	—	—	—	—	—	—	—	—	—	—	—	—	—	—	—	—	—
19	—	—	—	—	—	—	—	—	—	—	—	—	—	—	—	—	—	4+	—	—	—	—	—	—	—	—	—	—	—	—	—	—	—	—	—	—
20	—	—	—	—	—	—	—	—	—	—	—	—	—	—	—	—	—	—	4+	—	—	—	—	—	—	—	—	—	—	—	—	—	—	—	—	—
21	—	—	—	—	—	—	—	—	—	—	—	—	—	—	—	—	—	—	—	4+	—	—	—	—	—	—	—	—	—	—	—	—	—	—	—	—
22	—	—	—	—	—	—	—	—	—	—	—	—	—	—	—	—	—	—	—	—	4+	—	—	—	—	—	—	—	—	—	—	—	—	—	—	—
23	—	—	—	—	—	—	—	—	—	—	—	—	—	—	—	—	—	—	—	—	—	4+	—	—	—	—	—	—	—	—	—	—	—	—	—	—
24	—	—	—	—	—	—	—	—	—	—	—	—	—	—	—	—	—	—	—	—	—	—	4+	—	—	—	—	—	—	—	—	—	—	—	—	—
25	—	—	—	—	—	—	—	—	—	—	—	—	—	—	—	—	—	—	—	—	—	—	—	4+	—	—	—	—	—	—	—	—	—	—	—	—
26	—	—	—	—	—	—	—	—	—	—	—	—	—	—	—	—	—	—	—	—	—	—	—	—	4+	—	—	—	—	—	—	—	—	—	—	—
27	—	—	—	—	—	—	—	—	—	—	—	—	—	—	—	—	—	—	—	—	—	—	—	—	4+	4+	—	—	—	—	—	—	—	—	—	—
28	—	—	—	—	—	—	—	—	—	—	—	—	—	—	—	—	—	—	—	—	—	—	—	—	—	—	4+	—	—	—	—	—	—	—	—	—
29	—	—	—	—	—	—	—	—	—	—	—	—	—	—	—	—	—	—	—	—	—	—	—	—	—	—	—	4+	—	—	—	—	—	—	—	—
30	—	—	—	—	—	—	—	—	—	—	—	—	—	—	—	—	—	—	—	—	—	—	—	—	2+	—	—	—	4+	—	—	—	—	—	—	—
31	—	—	—	—	—	—	—	—	—	—	—	—	—	—	—	—	—	—	—	—	—	—	—	—	—	—	—	—	—	4+	—	—	—	—	—	—
32	—	—	—	—	—	—	—	—	—	—	—	—	—	—	—	—	—	—	—	—	—	—	—	—	—	—	—	—	—	—	4+	—	—	—	—	—
33	—	—	—	—	—	—	—	—	—	—	—	—	—	—	—	—	—	—	—	—	—	—	—	—	—	—	—	—	—	—	—	4+	—	—	—	—
34	—	—	—	—	—	—	—	—	—	—	—	—	—	—	—	—	—	—	—	—	—	—	—	—	—	—	—	—	—	—	—	—	4+	4+	—	—
35	—	—	—	—	—	—	—	—	—	—	—	—	—	—	—	—	—	—	—	—	—	—	—	—	—	—	—	—	—	—	—	—	3+	4+	—	—
36	—	—	—	—	—	—	—	—	—	—	—	—	—	—	—	—	—	—	—	—	—	—	—	—	—	—	—	—	—	—	—	—	—	—	4+	—
37	—	—	—	—	—	—	—	—	—	—	—	—	—	—	—	—	—	—	—	—	—	—	—	—	—	—	—	—	—	—	—	—	—	—	—	3+

[a] Antisera diluted 1:5 in PBS.

reacting antibodies was complete. The antisera so prepared contained antibodies only to heat-labile antigenic factors — possible capsular and flagellar antigens.

The presence of capsular (K) antigens in *C. fetus* was suggested by Wiidik and Hlidar[9] as a result of the inagglutinability of *C. fetus* live cells in antisera prepared with heated bacterial suspensions. McCoy et al.[27] studied the surface antigens of *C. fetus* subsp. *fetus* (*C. fetus* subsp. *intestinalis*) and found a discrete structure on the outer surface of the cell. This microcapsule, glycoprotein in nature, was associated with inagglutinability of live cells in antisera prepared with heated antigens and is therefore believed to represent the capsular "K" antigens.

The study directed at flagellar antigens has shown that antisera prepared with whole or solubilized flagella (flagellins) cause motility inhibition and clumping of bacterial cells.[28,29] Ullman[30] reported on the different amino acid compositions of the flagella belonging to various serotypes of *C. fetus*, and with the help of specific absorbed antisera was able to show common and specific flagellar antigenic factors.

The determination of the precise nature and role of the heat-labile antigenic factors — the capsular (K) and flagellar (H) antigens — is presently being investigated in our laboratory. Preliminary results indicate that the immunogenic response in rabbits to whole-cell antigens appears to be mainly directed at flagellar factors, although antibodies to heat-stable and other heat-labile somatic antigens are also present. Aflagellate mutants and partially purified flagella are being used as antigens in the preparation of absorbed antisera which will permit the study of the heat-labile factors of *C. jejuni*.

III. SEROTYPING OF *C. JEJUNI* AND *C. COLI* ISOLATES

The serotyping scheme developed by Lior et al.[17] was extended from 21 serogroups to presently comprise 36 serogroups (Table 4). A list of the reference strains is presented in Table 5. Serological screening of the *C. jejuni* and *C. coli* isolates is performed with 6 polyvalent antisera pools followed by agglutination with monovalent antisera and confirmation of the serotype with specific absorbed antisera. All antisera were standardized for slide work with the homologous reference strain, the working dilution being the dilution preceding the last dilution at which a 3+ to 4+ agglutination is obtained within 30 to 45 sec.

A major factor in the reproducibility of serotyping results may depend on the purity of cultures investigated. Experience with salmonella has shown the occurrence of multiple serotypes in contaminated material, and therefore serological investigations are performed on single colonies. This may present a difficulty with *C. jejuni* because of the characteristic confluent "lawn" type of growth exhibited by these organisms, especially on fresh, moist media. As moist media promote the confluent growth of *C. jejuni*, dried blood agar plates incubated at 30°C for 24 hr[31] or media containing 2% agar[32] should be used. For serotyping purposes, single colonies should then be transferred to a fresh, moist Mueller-Hinton blood agar plate and incubated for 48 hr at 37°C.

Of the 1504 *C. jejuni* and *C. coli* cultures investigated — 1163 from human and 341 from nonhuman sources — 85% (1273 cultures) were typable in single sera (Table 4) or in various pairs of antisera (Table 6).

Among *C. jejuni* and *C. coli* strains from all sources, 12 serogroups (1, 2, 4, 5, 6, 7, 8, 9, 11, 17, 20, and 21), each comprising 20 or more isolates, were encountered frequently and represented 75% (960) of typable strains; 63% (806 isolates) belong to the 7 most common serogroups, each comprised of 50 or more isolates. There were 174 strains (11%) found to be untypable and are being continuously investigated with antisera to new serotypes as they become available; 4% (58 strains) were rough, autoagglutinable in saline.

Of the human isolates, 85% (994) were typable: 850 cultures belonged to 35 serogroups

Table 4
SEROGROUPS[a] OF *C. JEJUNI* AND *C. COLI* ISOLATED FROM HUMAN AND NONHUMAN SOURCES SUBMITTED TO LCDC, OTTAWA[b]

Serogroup	Human sources (%)	Nonhuman sources						Total (%)
		Chicken	Turkey	Cattle	Swine	Others	Subtotal	
1	137(15)	5	3	6	4	—	18	155(14)
2	83(10)	12	12	—	6	4[c]	34	117(11)
4	167(20)	25	1	7	2	4[d]	39	206(19)
5	33(4)	1	—	1	2	—	4	37(3)
6	20(2)	3	—	1	—	—	4	24(2)
7	98(12)	5	—	25	1	3[e]	34	132(12)
8	40(5)	5	15	2	1	4[f]	27	67(6)
9	37(4)	5	1	3	—	1[g]	10	47(4.5)
10	10	—	—	—	1	—	1	11
11	55(7)	9	—	—	2	4[h]	15	70(6.5)
12	5	—	1	—	5	—	6	11
13	9	—	—	—	4	—	4	13
14	1	—	—	—	—	—	—	1
15	12	—	—	—	—	—	—	12
16	6	—	—	—	—	—	—	6
17	50(6)	1	3	—	2	3[i]	9	59(5.5)
18	11	3	—	—	—	2[j]	5	16
19	2	1	—	—	—	—	1	3
20	3	3	6	—	12	—	21	24(2)
21	13	3	6	—	—	—	9	22(2)
22	3	—	—	—	—	—	—	3
23	3	—	—	—	—	2[k]	2	5
24	1	—	—	—	—	—	—	1
25	—	—	—	—	—	1[l]	1	1
26	3	—	—	1	—	1[m]	2	5
27	4	—	—	—	—	1[n]	1	5
28	4	—	—	—	—	—	—	4
29	5	—	—	—	—	—	—	5
30	1	—	—	—	—	—	—	1
31	2	—	—	—	—	—	—	2
32	2	—	—	—	—	—	—	2
33	2	—	—	—	—	—	—	2
34	1	—	—	—	—	1[o]	1	2
35	8	—	—	—	—	1[p]	1	9
36	16(2)	—	—	—	—	—	—	16
37	3	—	—	—	—	—	—	3
Total	850	81	48	46	42	32	249	1099

[a] Agglutination in single sera.
[b] Includes isolates from U.S., South Africa, Yugoslavia, and Israel.
[c] 1 Lemur
2 Dogs
1 Nonhuman
[d] 2 Dogs
2 Avian
[e] 2 Dogs
1 Sheep
[f] 3 Dogs
1 Cat
[g] 1 Sheep
[h] 1 Goat
3 Rabbits
[i] 1 Pigeon
2 Dogs
[j] 2 Dogs
[k] 2 Ducks
[l] 1 Duck
[m] 1 Monkey
[n] 1 Gazelle
[o] 1 NARTC
[p] 1 NARTC

Table 5
LIST OF *C. JEJUNI*, *C. COLI* AND NARTC REFERENCE STRAINS

Serotype	Strain	Biotype	Source
LIO 1	134	1	Human
LIO 2	195/1780	1	Human
LIO 4	1/ NTCC11168	1	Human
LIO 5	170	1	Human
LIO 6	6	2	Human
LIO 7	35	1	Human
LIO 8	52	3	Human
LIO 9	88	1	Human
LIO 10	142	1	Human
LIO 11	244	1	Human
LIO 12	264	3	Human
LIO 13	343/T015	1	Human
IO 14	348/T020	3	Human
LIO 15	388	2	Human
LIO 16	728	1	Human
LIO 17	556	1	Chicken
LIO 18	563	2	Chicken
LIO 19	544	2	Chicken
LIO 20	602	3	Swine
LIO 21	699	3	Chicken
LIO 22	918/T040	1	Human
LIO 23	720	1	Human
LIO 24	1213/T0349	3	Human
LIO 25	1228	3	Duck
LIO 26	913/T035	1	Nonhuman
LIO 27	919/T041	1	Nonhuman
LIO 28	1180	1	Human
LIO 29	1982/BR11	3	?
LIO 30	1215/T052	2	Human
LIO 31	729	3	Human
LIO 32	910/T031	1	Human
LIO 33	1545	1	Human
LIO 34	1556	4	Human
LIO 35	1728	4	Nonhuman
LIO 36	2074	1	Human
LIO 37	2024/J200	1	Human

and 144 isolates showing agglutination in various pairs of antisera belonged to 33 subserogroups; 72% (720 strains) belonged to 10 common serogroups, each represented by 20 or more isolates (1, 2, 4, 5, 6, 7, 8, 9, 11, and 17) (Table 7). Of the human strains, 590 (60%) belonged to the 6 most common serogroups, each represented by 50 or more isolates. Serogroup 4, with 167 isolates (17%), was most common among *C. jejuni* isolated from human cases of gastroenteritis, followed by serogroup 1 (137 isolates — 14%), serogroup 7 (98 isolates — 10%), serogroup 2 (83 isolates — 8%), serogroup 11 (55 isolates — 6%), and serogroup 17 (50 isolates — 5%).

In the nonhuman group, 84% of the chicken isolates found typable belonged to 14 serogroups and 11 subserogroups: 31% of the chicken isolates belonged to serogroup 4, followed by serogroup 2 (15%), and serogroup 11 (11%). These three serogroups are commonly found among human isolates. Of the isolates from turkey, 73% were also found to belong to 6 of the 10 common serogroups; 31% belonged to serogroup 8 followed by serogroup 2 (25%), serogroup 1 (6%), and serogroup 17 (6%). Of *C. jejuni* isolated from cattle, 98% belonged to 7 common human serogroups; 54% belonged to serogroup 7, 15% to serogroup

Table 6
SUBSEROGROUPS[a] OF *C. JEJUNI* ISOLATED FROM HUMAN AND NONHUMAN SOURCES

Subserogroups	Human sources	Nonhuman sources: Chicken	Turkey	Swine	Other	Subtotal	Total
1,2	47	1	1	2	—	4	51
1,4	18	1	—	—	—	1	19
1,7	1	—	—	—	—	—	1
1,9	5	—	—	—	—	—	5
1,10	8	—	—	1	—	1	9
1,11	—	—	—	2	—	2	2
1,12	—	—	—	1	—	1	1
1,17	3	—	1	—	1[b]	2	5
1,18	2	—	—	—	—	—	2
1,21	1	—	—	—	—	—	1
2,4	4	—	—	—	—	—	4
2,8	2	—	—	—	—	—	2
2,9	—	—	4	—	—	4	4
2,11	3	1	—	1	—	2	5
2,17	1	—	—	—	—	—	1
2,18	1	1	—	—	—	1	2
2,21	2	—	1	—	—	1	3
4,7	3	—	—	—	—	—	3
4,8	1	1	—	—	—	1	2
4,9	1	—	—	—	—	—	1
4,10	2	—	—	—	—	—	2
4,11	—	1	—	—	—	1	1
4,17	3	—	—	—	—	—	3
4,21	5	1	—	—	—	1	6
5,6	15	1	—	1	—	2	17
5,7	—	—	—	—	1[c]	1	1
5,10	1	—	—	—	—	—	1
6,11	1	—	—	—	—	—	
6,21	1	—	—	—	—	—	1
7,10	—	—	—	1	—	1	1
7,18	1	—	—	—	—	—	1
7,21	1	—	—	—	—	—	1
8,9	—	1	—	1	—	2	2
8,10	1	—	—	—	—	—	1
8,11	—	1	—	—	—	1	1
8,12	1	—	—	—	—	—	1
9,11	5	1	—	—	—	1	6
9,28	1	—	—	—	—	—	1
16,21	1	—	—	—	—	—	1
18,19	1	—	—	—	—	—	1
21,28	1	—	—	—	—	—	1
Total	144	11	7	10	2	30	174

[a] Agglutination in pairs of sera.
[b] Pigeon.
[c] Sheep.

4, and 13% belonged to serogroup 1. Of the 70 swine isolates studied, 74% (52 strains) were typable and belonged to 12 serogroups (42 strains) and 8 subserogroups (10 strains); 48% of isolates belonged to 8 common serogroups (1, 2, 4, 5, 7, 8, 11, and 17).

A total of 15% of the human isolates and 11% of the nonhuman isolates found typable were agglutinated by various pairs of absorbed antisera, indicating the presence of two

Table 7
DISTRIBUTION OF TEN MOST COMMON *C. JEJUNI* SEROGROUPS

Serogroup	Human sources	Nonhuman sources				
		Chicken	Turkey	Cattle	Swine	Others
1	137	5	3	6	4	—
2	83	12	12	—	6	2 Dogs
4	167	25	1	7	2	2 Dogs
5	33	1	—	1	2	—
6	20	3	—	1	—	—
7	98	5	—	25	1	2 Dogs 1 Sheep
8	40	5	15	2	1	3 Dogs 1 Cat
9	37	5	1	3	—	1 Sheep
11	55	9	—	—	2	1 Sheep 1 Goat 3 Rabbits
17	50	1	3	—	2	2 Dogs 1 Pigeon
Total 10 serogroups	720(85%)	71(88%)	35(73%)	45(98%)	20(48%)	20(63%)
Total all serogroups	850	81	48	46	42	32

antigenic factors (Table 6) and were designated as subserogroups. Strains of *C. fetus* possessing multiple antigenic factors were described by Berg et al.[14]

All ten common serogroups identified among human isolates were also found among the nonhuman isolates, bringing additional evidence that chickens, turkey, cattle, and swine are important reservoirs for human infection. Of the 35 serogroups identified among human isolates, 18 were also identified among nonhuman isolates. Some of the serogroups such as 15, 16, 22, 28, and 36 have been identified so far only among human isolates and are consistent with the person-to-person spread reported by some workers.[33-36] The serotyping results of 32 family outbreaks and 7 community outbreaks are presented in Table 8.

IV. BIOTYPING

While the differentiation within the genus *Campylobacter* has been reported by Veron and Chatelain[37] by tolerance to various chemicals, very few markers are available for the differentiation within the *C. jejuni* group.

In our studies we have integrated the biotyping scheme described by Skirrow and Benjamin[38] with our serotyping scheme (Table 9). This has allowed a further differentiation of the serogroups into four biotypes: biotype 1 and 2, typical of *C. jejuni*; biotype 3, typical of *C. coli*; and biotype 4 representing the nalidixic acid-resistant-thermophilic campylobacters (NARTC).

A total of 78% of the strains studied so far, including most of the human, chicken, cattle, and 50% of the turkey isolates, belong to biotype 1, while 6% of isolates, mostly from human sources, belong to biotype 2 and 16% of the strains, including all the swine isolates, 50% of the turkey strains, some chicken, and about 10% of the human isolates did not hydrolyze hippuric acid and belong to biotype 3. Most of these isolates, probably *C. coli*, were found agglutinable by *C. jejuni* antisera, indicating a strong serological relationship between the two species. The finding within this biotype that seven of the serogroups

Table 8
SEROTYPING OF *C. JEJUNI* ISOLATED FROM RELATED CASES AND OUTBREAKS

No. of foci	No. of cases	Serogroup
4	2 Each	1
4	2 Each	1,2
2	2 Each	2
1	3	4
4	2 Each	4
1	3	7
4	2 Each	7
3	2 Each	8
1	2	9
1	2	11
2	2 Each	17
5	2 Each	U.I.
1 Milk-borne	13/27	1
1 Water-borne	6	1,2
1	3/?	1
1	3/?	1

Note: U.I. = under investigation.

identified among human isolates were also identified among the swine isolates brings additional evidence of the role played by swine in human infection.

Two biotype 4 strains, the reference strain for serogroups 31 and 34, were isolated for the first time from human cases of gastroenteritis.

The integration of the biotyping and serotyping schemes resulted in a further differentiation of *C. jejuni* isolates, especially of the common serogroups 1,2,5,7,8, and 11. The subdivision of these serogroups provides additional epidemiological information. It is interesting to note that all isolates identified to date as serogroup 6 belong to biotype 2, as did the NTCC #11392 type strain for biotype 2. Serogroups 12, 14, and 20 have been identified so far only among hippurate-negative *C. coli* isolated from both human and nonhuman sources.

Table 9
DISTRIBUTION OF SEROTYPES AND BIOTYPES OF *C. JEJUNI*, *C. COLI*, AND NARTC ISOLATED FROM HUMAN AND NONHUMAN SOURCES SUBMITTED TO LCDC, OTTAWA[a]

	C. jejuni										*C. coli*					NARTC
	Biotype 1 (Hip+,H_2S−)					Biotype 2 (Hip+,H_2S+)					Biotype 3 (Hip−,H_2S−)					Biotype 4 (Hip−,H_2S+)
Serogroup	Human	Chicken	Turkey	Swine	Other	Human	Chicken	Turkey	Swine	Other	Human	Chicken	Turkey	Swine	Other	
1	76	1	1	—	5 Cattle	2	—	—	—	—	6	—	2	1	—	—
2	54	8	5		2 Dogs											
4	121	8	1	—	7 Cattle 1 Nonhuman	—	—	—	—	—	—	1	—	—	—	—
5	23	1	—	—	1 Cattle	—	—	—	—	—	1	—	—	1	—	—
6	—	—	—	—	—	16	1	—	—	—	—	—	—	—	—	—
7	62	1	—	—	25 Cattle 1 Dog 1 Sheep	3	—	—	—	—	1	1	—	—	—	—
8	19	2	11	—	2 Cattle	2	—	—	—	2 Dogs	14	2	4	1	—	—
9	24	2	1	—	1 Sheep 3 Cattle	—	—	—	—	—	1	—	—	—	—	—
10	4	—	—	—	—	—	—	—	—	—	1	—	—	—	—	—
11	31	5	—	—	3 Rabbits	1	3	—	—	1 Goat	1	—	—	1	—	—
12	—	—	—	—	—	—	—	—	—	—	5	—	1	5	—	—
13	4	—	—	—	—	—	—	—	—	—	—	—	—	—	—	—
14	—	—	—	—	—	—	—	—	—	—	1	—	—	—	—	—
15	1	—	—	—	—	2	—	—	—	—	—	—	—	—	—	—
16	4	—	—	—	—	—	—	—	—	—	—	—	—	—	—	—
17	37	2	—	—	1 Pigeon 2 Dogs	—	—	—	—	—	—	—	3	2	—	—
18	10	2	—	—	—	1	1	—	—	2 Dogs	—	—	—	—	—	—
19	2	—	—	—	—	—	1	—	—	—	—	—	—	—	—	—
20	—	—	—	—	—	—	—	—	—	—	3	3	6	12	—	—
21	9	1	5	—	—	—	—	—	—	—	6	2	1	—	—	—
22	1	—	—	—	—	—	—	—	—	—	2	—	—	—	—	—
23	1	—	—	—	—	2	—	—	—	2 Dogs	—	—	—	—	—	—
24	—	—	—	—	—	—	—	—	—	—	1	—	—	—	—	—
25	—	—	—	—	—	—	—	—	—	—	—	—	—	—	1 Duck	—
26	2	—	—	—	2 Cattle 1 Monkey 1 Gazelle	—	—	—	—	—	1	—	—	—	—	—
27	—	—	—	—		3	—	—	—	—	—	—	—	—	—	—
28	4	—	—	—	—	—	—	—	—	—	—	—	—	—	—	—
29	2	—	—	—	—	—	—	—	—	—	3	—	—	—	—	—
30	—	—	—	—	—	1	—	—	—	—	—	—	—	—	—	—
31	1	—	—	—	—	—	—	—	—	—	1	—	—	—	—	—
32	1	—	—	—	—	—	—	—	—	—	—	—	—	—	—	—
33	2	—	—	—	—	—	—	—	—	—	—	—	—	—	—	—
34	—	—	—	—	—	—	—	—	—	—	—	—	—	—	—	1 Human 1 N/S
35	3	—	—	—	—	—	—	—	—	—	2	—	—	—	—	
36	14	—	—	—	—	—	—	—	—	—	1	—	—	—	—	—
37	2	—	—	—	—	—	—	—	—	—	1	—	—	—	—	—
Total	514	33	24	—	59	34	6	—	—	7	58	9	24	29	2	2
All Sources	(%)		78%				6%					16%				

[a] Includes isolates from U.S., South Africa, Yugoslavia, and Israel.

REFERENCES

1. **Smith, T. and Reagh, A. L.**, The agglutination affinities of related bacteria parasitic in different hosts, *J. Med. Res.*, 9, 270, 1903.
2. **Smith, T. and Taylor, M. S.**, Some morphological and biological characters of the spirilla (*V. fetus* n. sp.) associated with disease of the fetal membranes in cattle, *J. Exp. Med.*, 30, 299, 1919.
3. **Smith, T. and Orcutt, M. L.**, Vibrios from calves and their serological relation to *Vibrio fetus*, *J. Exp. Med.*, 45, 391, 1927.
4. **Blakemore, F. and Gledhill, A. W.**, Studies on Vibrionic abortion of sheep. II. The antigenic relationship between strains of *V. fetus*, *J. Comp. Pathol.*, 56, 74, 1946.
5. **Levy, M. L.**, The agglutination test in vibrionic abortion in sheep, *J. Comp. Pathol.*, 60, 65, 1950.
6. **Gallut, J.**, Recherches immunochimiques sur *Vibrio fetus*. I. Etude serologique preliminaire de 10 sources d'origine diverses, *Ann. Inst. Pasteur*, 83, 449, 1952.
7. **Marsh, H. and Firehammer, B. D.**, Serological relationships of 23 ovine and 3 bovine strains of *Vibrio fetus*, *Am. J. Vet. Res.*, 14, 396, 1953.
8. **Price, K. E., Poelma, L. T., and Faber, J. E.**, Serological and physiological relationship between strains of *Vibrio fetus*, *Am. J. Vet. Res.*, 16, 164, 1955.
9. **Wiidik, R. W. and Hlidar, G. E.**, Untersuchungen über die Antigene Strukture von *Vibrio fetus* vom Rind. Das Kapsel-oder K-antigen vom *Vibrio fetus*, *Zbl. Vet. Med.*, 2, 238, 1955.
10. **King, E. O.**, Human infections with *Vibrio fetus* and a closely related Vibrio, *J. Infect. Dis.*, 101, 119, 1957.
11. **von Mitscherlich, E. and Liess, B.**, Die Serologische Differenzierung von *Vibrio fetus* Stämmen, *Dtsch. Tieraerztl. Wochenschr.*, 65, 36, 1958.
12. **Morgan, W. J. B.**, Studies on the antigenic structure of *V. fetus*, *J. Comp. Pathol. Ther.*, 69, 125, 1959.
13. **White, F. H. and Walsh, A. F.**, Biochemical and serologic relationship of isolants of *Vibrio fetus* from man, *J. Infect. Dis.*, 121, 471, 1970.
14. **Berg, R. L., Jutila, J. W., and Firehammer, B. D.**, A revised classification of *Vibrio fetus*, *Am. J. Vet. Res.*, 32, 11, 1971.
15. **Butzler, J. P., Dekeyser, P., Detrain, M., and Dehaen, F.**, Related "vibrio" in stools, *J. Pediatr.*, 82, 493, 1973.
16. **Butzler, J. P. and Skirrow, M. B.**, Campylobacter enteritis, *Clin. Gastroenterol.*, 8, 737, 1979.
17. **Lior, H., Woodward, D. L., Edgar, J. A., LaRoche, L. J., and Gill, P.**, Serotyping of *Campylobacter jejuni* by slide agglutination based on heat-labile antigenic factors, *J. Clin .Microbiol.*, 15, 761, 1982.
18. **Skirrow, M. B.**, Campylobacter enteritis: a "new" disease, *Br. Med. J.*, 2, 9, 1977.
19. **Butzler, J. P.**, Infection with Campylobacters, in *Modern Topics in Infection*, Williams, J. D., Ed., Heinemann, London, 1978, 214.
20. **Catlin, B. W.**, Transformation of *Neisseria meningitidis* by deoxyribonucleates from cells and from culture slime, *J. Bacteriol.*, 79, 579, 1960.
21. **Smithies, W. R. and Gibbons, N. E.**, The deoxyribose nucleic acid slime layer of some halophilic bacteria, *Can. J. Microbiol.*, 1, 614, 1955.
22. **Arko, R. J., Wong, K. H., and Peacock, W. L.**, Nuclease enhancement of specific cell agglutination in a serodiagnostic test for *Neisseria gonorrhoeae*, *J. Clin. Microbiol.*, 9, 517, 1979.
23. **Juhler, H.**, Comparative serological examinations of 21 bovine strains of *Vibrio fetus*. *Nord. Vet. Med.*, 7, 52, 1955.
24. **Ristic, M., White, F. H., Doty, R. B., Herzberg, M., and Sanders, D. A.**, The characteristics of agglutinating antigens of *Vibrio fetus* variants. I. Effect of heat and formalin on serological activity, *Am. J. Vet. Res.*, 18, 764, 1957.
25. **Abbott, J. D., Dale, B., Eldridge, J., Jones, D. M., and Suttcliffe, E. M.**, Serotyping of *Campylobacter jejuni/coli*, *J. Clin. Pathol.*, 33, 762, 1980.
26. **Kosunen, T. U., Danielsson, D., and Kjellander, J.**, Serology of *Campylobacter fetus* ss. *jejuni*, *Acta Pathol. Microbiol. Immunol. Scand. Sect. B*, 90, 191, 1982.
27. **McCoy, E. C., Doyle, D., Burda, K., Corbeil, L. B., and Winter, A. J.**, Superficial antigens of *Campylobacter (Vibrio) fetus*: characterization of an antiphagocytic component, *Infect. Immunol.*, 11, 517, 1975.
28. **McCoy, E. C., Doyle, D., Wiltberger, H., Burda, K., and Winter, A. J.**, Flagella ultrastructure and flagella associated antigens of *Campylobacter fetus*, *J. Bacteriol.*, 122, 307, 1975.
29. **Corbeil, L. B., Schurig, G. D., Duncan, J. R., Corbeil, R. R., and Winter, A. J.**, Immunoglobulin classes and biological function of *Campylobacter (Vibrio) fetus* antibodies in serum and cervicovaginal mucus, *Infect. Immunol.*, 10, 422, 1974.
30. **Ullman, U.**, Biochemical and serological investigations of flagellae of *Campylobacter fetus*, *Zbl. Bakt. Hyg.*, 234(Abstr.), 346, 1976.

31. **Buck, G. E. and Kelly, M. T.,** Effect of moisture content of the medium on colony morphology of *Campylobacter fetus* subsp. *jejuni*, *J. Clin. Microbiol.*, 14, 585, 1981.
32. **Skirrow, M. B. and Benjamin, J.,** '1001' Campylobacters: cultural characteristics of intestinal campylobacters from man and animals, *J. Hyg. (Cambridge)*, 85, 427, 1980.
33. **Cadranel, S., Rodesch, P., Butzler, J. P., and Dekeyser, P.,** Enteritis due to "related Vibrio" in children, *Am. J. Dis. Child.*, 126, 152, 1973.
34. **Butzler, J. P., Dekegel, D., and Hubrechts, J. M.,** Mode of transmission of human campylobacteriosis, *Curr. Chemother.*, 10, 174, 1978.
35. **Severin, W. P. J.,** Campylobacter enteritis, *Ned. Tijdschgeneusk*, 122, 499, 1978.
36. **Blaser, M. J., Waldman, R. J., Barrett, T., and Erlandson, A. L.,** Outbreaks of Campylobacter enteritis in two extended families — evidence for person-to-person transmission, *J. Pediatr.*, 98, 254, 1981.
37. **Veron, M. and Chatelain, R.,** Taxonomic study of the genus *Campylobacter* Sebald and Vernon and designation of the neotype strain for the type species *Campylobacter fetus* (Smith and Taylor) Sebald and Veron, *Int. J. Syst. Bacteriol.*, 23, 122, 1973.
38. **Skirrow, M. B. and Benjamin, J.,** Differentiation of enteropathogenic campylobacter, *J. Clin. Pathol.*, 33, 1122, 1980.

Chapter 7

SUSCEPTIBILITY OF CAMPYLOBACTERS TO ANTIMICROBIAL AGENTS

R. Vanhoof

TABLE OF CONTENTS

I. Introduction 78

II. In Vitro Susceptibility of *C. jejuni* 78
- A. The Penicillins 78
- B. The Cephalosporins 78
- C. The Aminoglycosides 80
- D. The Tetracyclines 80
- E. Macrolides and Lincosamide Antibiotics 81
- F. The Polypeptide Antibiotics 81
- G. Nitrofuran Compounds 81
- H. Sulfonamides, Trimethoprim, and Cotrimoxazole 82
- I. Miscellaneous Antimicrobial Agents 82

III. In Vitro Susceptibility of *C. fetus* subsp. *fetus* 82

IV. Disk Sensitivity Testing with *Campylobacter* 83

V. Conclusions 83

References 85

I. INTRODUCTION

Nowadays campylobacters have become important pathogenic microorganisms, especially since investigators in Brussels[7] facilitated the isolation of the organism by using a selective technique. Later their work was confirmed by Skirrow.[13] Due to the development and application of these improved techniques *C. jejuni* is now regarded as one of the most important causes of bacterial diarrhea in man.

The determination of the susceptibility of campylobacters to different antimicrobial agents may be important for three main reasons:

1. The identification of drugs to which campylobacters are susceptible. These drugs may be useful in the treatment of the disease. In general, *C. jejuni* enteritis is a self-limiting disease, but sometimes the microorganism provokes a more severe and prolonged illness. Furthermore, *C. jejuni* has also been isolated from blood and this indicates that it can be invasive.[5] So, seriously ill and septicemic patients should be treated and the sensitivity tests may guide the therapy.
2. The recognition of antibiotics to which campylobacters are always resistant and which might be used to improve the selectiveness of the isolation medium.
3. The detection of strains with a "special" resistance pattern that possibly might be due to the presence of R-plasmids.

The data available on antibiotic susceptibility of *Campylobacter* will be discussed in this article. However, the interpretation and comparison of results provided by different authors should be made with caution because there are great differences in the techniques used for determining in vitro sensitivity patterns. Some of the data are summarized in Table 1.

II. IN VITRO SUSCEPTIBILITY OF *C. JEJUNI*

A. The Penicillins

In general the penicillins are only moderately active against campylobacters. Most of the strains are reported to be fairly resistant to penicillin G[3,4,10,12,22-24] although Svedhem et al.[14] regarded their population as sensitive. Penicillin G and penicillin V seemed to have comparable activity on *C. jejuni*.[14] Cloxacillin is not active against *C. jejuni*,[10,23] but flucloxacillin does seem to show some activity.[14] Ampicillin is more active than the other penicillins and in most of the studies three quarters of the isolates are inhibited by concentrations easily achievable in the blood.[1,3,4,10,12,14,19,21-24] Vanhoof et al.[21] found a relatively high MIC 90 value for ampicillin in both ape and calf isolates. Ampicillin appeared to be slightly more active than amoxycillin, but both antibiotics showed good bactericidal activity against *C. jejuni*.[22] Fleming and co-workers noticed that ampicillin resistance in *C. jejuni* was associated with the production of a beta-lactamase.[8]

Other penicillins such as carbenicillin,[4,10,14,22-24] mezlocillin,[24] and mecillinam[12,24] have also been investigated, but none of these compounds was very active on campylobacters. Ahonkhai et al.[1] demonstrated, however, that *N*-formimidoyl thienamycin exhibited a high level of activity and that this compound was, in fact, the most active antimicrobial agent in their whole study. All strains were susceptible to 0.03 μg/mℓ.

B. The Cephalosporins

The cephalosporins show only a moderate to poor activity against *C. jejuni*. Karmali et al.[10] found that cephalosporin C was an active compound, with 90% of the strains inhibited by 16 μg/mℓ. However most of the strains are resistant to the different cephalosporins tested, i.e., cephaloridine,[10] cephalothin,[3,4,10,12,23,24] cefazolin,[10,22] and cephalexin.[10,23,24] Cefotax-

Table 1
BACTERIOSTATIC AND BACTERICIDAL ACTIVITY OF 24 ANTIMICROBIAL AGENTS AGAINST 86 STRAINS OF *C. JEJUNI*

	MIC (μg/mℓ of medium)				MBC (μg/mℓ of medium)			
Drug	Range	MIC_{50}[a]	MIC_{90}[b]	Geometric mean	Range	MBC_{50}[a]	MBC_{90}[b]	Geometric mean
Penicillin G	<0.048—50	6.25	12.5	4.60	0.097—>50	12.5	50	8.56
Ampicillin	≤0.048—50	3.12	12.5	2.15	0.097—>50	3.12	25	3.79
Amoxycillin	≤0.048—>50	3.12	12.5	2.53	0.097—>50	3.12	25	3.58
Carbenicillin	<0.048—>50	6.25	25	6.35	≤0.048—>50	12.5	50	10.81
Cefazolin	0.097—>50	25	50	12.90	0.097—>50	50	50	27.44
Cefotaxime	<0.048—50	3.12	6.25	2.00	0.097—>50	3.12	25	3.79
Moxalactam	0.195—>50	12.5	50	10.55	0.78—>50	25	>50	18.40
Erythromycin	≤0.048—>50	0.195	0.78	0.24	≤0.048—>50	0.39	1.56	0.48
Thiamphenicol	≤0.048—>50	1.56	6.25	1.21	≤0.048—>50	1.56	25	2.41
Chloramphenicol	≤0.048—12.5	1.56	3.12	1.26	≤0.048—>50	1.56	25	2.61
Clindamycin	≤0.048—12.5	0.097	0.39	0.13	≤0.048—>50	0.195	12.5	0.32
Gentamicin	≤0.048—0.78	0.097	0.195	0.14	≤0.048—>50	0.195	3.12	0.29
Tobramycin	≤0.048—25	0.39	1.56	0.48	≤0.048—>50	0.78	6.25	0.92
Amikacin	≤0.048—3.12	0.39	0.78	0.39	≤0.048—>50	0.78	3.12	0.87
Tetracycline	0.097—>50	0.195	1.56	0.36	0.097—>50	0.39	6.25	0.61
Doxycycline	≤0.048—>50	0.097	1.56	0.22	≤0.048—>50	0.195	6.25	0.49
Minocycline	≤0.048—>50	≤0.048	0.39	0.12	≤0.048—>50	0.195	1.56	0.28
Colistin	≤0.048—50	0.78	3.12	0.96	0.097—>50	1.56	12.5	2.12
Nalidixic acid	0.195—>50	6.25	50	6.61	0.39—>50	6.25	>50	9.42
Furazolidone	≤0.048—0.78	0.097	0.195	0.09	≤0.048—12.5	0.097	1.56	0.16
Metronidazole	0.097—>50	12.5	50	7.58	0.39—>50	25	50	13.22
Sulfamethoxazole	≤0.97—>1,000	15.6	125	18.94	1.95—>1,000	62.5	>1,000	73.97
Trimethoprim	12.5—>200	200	>200	198.36	25—>200	>200	>200	269.47
Co-trimoxazole	≤0.97—>1,000	31.2	125	31.47	1.95—>1,000	125	>1,000	110.74

[a] MIC_{50} and MBC_{50}: MIC and MBC for 50% of the isolates.
[b] MIC_{90} and MBC_{90}: MIC and MBC for 90% of the isolates.

Adapted from Vanhoof, R., Gordts, B., Dierickx, R., Coignau, H., and Butzler, J. B., *Antimicrob. Agents Chemother.*, 18, 118, 1980. With permission.

Table 2
PREVALENCE OF STRAINS OF *C. JEJUNI* RESISTANT TO ERYTHROMYCIN AND TETRACYCLINE

					Erythromycin resistance		Tetracycline resistance	
Authors	Ref.	Country	Date	No. strains tested	Prevalence %	MIC level[a]	Prevalence %	MIC level[a]
Butzler	4	Belgium	1974	114	0.9	>100	2.0	≥25
Vanhoof	23	Belgium	1978	95	8.4	≥12.5	5.3	≥6.25
Brunton	2	U.K.	1978	407	0.5	100	N.D.[b]	
Walder	24,25	Sweden	1979	98	8.0	—	0	
Vanhoof	22	Belgium	1980	86	2.3	≥25	8.1	≥6.25
Svedhem	14	Sweden	1981	72	5	≥64		
				84			20	≥16
Karmali	10	Canada	1981	209	1	≥256		
				172			12	≥64
Buck	3	U.S.	1982	24	0	—	12.5	≥16
Vanhoof	21	Belgium	1982	426	4.2	≥25	4	≥25

[a] Expressed in μg/mℓ.
[b] N.D.: not done.

ime, a new semisynthetic parental cephalosporin, was found to be fairly active against *C. jejuni* strains.[1,10,17,22,24] Moxalactam, also a third-generation cephalosporin, was significantly less active than cefotaxime.[1,10,22] Furthermore, campylobacters are not susceptible to the newer cephalosporins such as cefoxitin,[10,23] cefamandole,[10] cefuroxime,[14,23,24] and cefoperazone.[1]

C. The Aminoglycosides

The aminoglycosides are considered to be among the most active compounds. In general, gentamicin is regarded as the most potent aminoglycoside antibiotic[1,3,4,10,12,14,21-24] and until now only one isolate showed an unusual resistance level to gentamicin. It had an MIC of 12.5 μg/mℓ.[23] Gentamicin was significantly more active than tobramycin and amikacin although these latter two compounds were also very active in vitro against the *C. jejuni* strains.[22] Vanhoof et al.[22] found one strain completely resistant to tobramycin, (MIC 25 μg/mℓ). Kanamycin was significantly less active than gentamicin.[3,4,10,23] Most of the strains were also sensitive to streptomycin[4] and neomycin.[4,23]

D. The Tetracyclines

Tetracyclines are generally very effective drugs and most of the strains are inhibited by less than 1 μg/mℓ of tetracycline.[3,4,10,12,14,19,21-24] In spite of this apparently good activity many authors have reported the presence of resistant strains (Table 2). In Belgium, Vanhoof et al.[21-23] found in recent surveys that only about 5 to 8% of their isolates showed resistance to tetracycline. In a Swedish study published by Svedhem et al.,[14] 20% of the strains were resistant to tetracycline (MIC > 16 μg/mℓ). In 1979 however, Walder[24] found all his isolates fully susceptible to tetracycline. In Canada about 12% of the isolates showed high-level resistance to tetracycline with MIC values varying from 64 to 256 μg/mℓ.[10] In the U.S. Buck and Kelly[3] found 12.5% (3 strains out of 24) highly resistant to tetracycline. It seems, therefore, that tetracycline resistance in *C. jejuni* is significantly higher in North America than in Europe, although Svedhem et al.[14] reported also a high prevalence of resistant strains. Furthermore it should be noticed that a high proportion of tetracycline-resistant strains can be found in chickens.[14,21] Taylor et al.[15] have shown that tetracycline resistance in strains of *C. jejuni* is plasmid-mediated and transmissible within the *C. jejuni* species and from *C. jejuni* to *C. fetus* subsp. *fetus*. Attempts to transfer these plasmids to *Escherichia coli* were not successful.

Many authors found that doxycycline is more active than tetracycline,[10,22-25] although in the Vanhoof et al. series[22,23] minocycline was the most active compound among the tetracyclines. Although doxycycline and minocycline are more active than tetracycline against *C. jejuni* and may be used as alternatives when conventional tetracycline is considered as inappropriate, it should be kept in mind that there is cross resistance among the different tetracycline compounds.[10,22,23]

E. Macrolides and Lincosamide Antibiotics

In general, erythromycin is regarded as a very active and useful drug and is currently recommended as the drug of choice in the treatment of enteritis due to *C. jejuni*.[5] However, the prevalence of erythromycin-resistant strains (see Table 2) may limit its clinical use. In Canada[10] and the U.K.[2] the frequency of erythromycin resistance is about 1%. This contrasts markedly with the situation in Sweden[14,24,25] where approximately 10% of the strains are resistant to erythromycin. As early as 1974, Butzler et al.[4] reported one strain with a high level of resistance to erythromycin (MIC > 100 μg/mℓ). More recently Vanhoof et al.[21,22] reported a 2 to 4% incidence of erythromycin-resistant strains in Belgium.

Other investigators[3,12] did not find resistant strains at all. The reasons for the geographical differences in frequency of erythromycin resistance are not known. Erythromycin-resistant *C. jejuni* strains may also be found in animals. The frequency of resistant strains in chickens was found to be about 8%.[14,21] Erythromycin-resistant strains were also found among sheep and calf isolates.[21] There is no substantial evidence to date that erythromycin resistance in *C. jejuni* is plasmid-mediated.[16]

Normally, clindamycin also is very active against *C. jejuni*,[3,10,14,21-24] but resistant strains have been found.[10,23,24] Karmali et al.[10] showed that there is cross resistance between macrolides and lincosamide antibiotics. Lincomycin, on the other hand, is mostly inactive.[14,23,24]

F. The Polypeptide Antibiotics

Polymyxin B and colistin (polymyxin E) are two basic polypeptides. They are only poorly absorbed from the gastrointestinal tract. These polymyxins are generally inactive against *C. jejuni*[4,10,23] and there is no significant difference between the activity of polymyxin B and colistin on *C. jejuni*.[23] Karmali et al.[10] and Vanhoof et al.[22] noticed a curious and apparently contradictory effect of polymyxin B and colistin on *C. jejuni* strains. For many of their isolates, the MICs were less than the concentration present in the medium on which they had been isolated originally. This discrepancy is probably due to the difference in techniques used to determine MIC and those used for isolation of the strains from stools. Furthermore, Karmali et al.[10] found that the susceptibility of the strains to polymyxin B was influenced by the size of the inoculum and the temperature of incubation.

Vancomycin, a complex soluble glycopolypeptide, and bacitracin, both show only weak in vitro activity against *C. jejuni*.[1,10,23]

G. Nitrofuran Compounds

In the Karmali et al. study,[10] nitrofurantoin was one of the most active compounds. All their strains were inhibited at a concentration of 2 μg/mℓ or less. Svedhem et al.,[14] however, found their population somewhat less sensitive to this compound. Furazolidone is also a very active drug[21-23] and was found to have a remarkable bactericidal activity on *C. jejuni*.[22] Nifuroxazide, another nonabsorbable nitrofuran derivative, also showed very good bacteriostatic and bactericidal activity against *C. jejuni*.[18]

H. Sulfonamides, Trimethoprim, and Cotrimoxazole

The activity of sulfamethoxazole against *C. jejuni* is rather weak,[10,12,22-24] and the same is true of sulfadiazine.[14] Furthermore, the inhibitory activity of sulfamethoxazole seems to be markedly influenced by the inoculum size.[10]

Trimethoprim is inactive.[10,12,14,22-24] This compound is one of the three antimicrobials incorporated in the selective medium used by Skirrow for the primary isolation of *C. jejuni* from stools.[13] Co-trimoxazole, a combination of sulfamethoxazole and trimethoprim, does not seem to be substantially more active than sulfamethoxazole alone. Approximately 50% of strains are resistant to co-trimoxazole.[22,23]

I. Miscellaneous Antimicrobial Agents

Chloramphenicol is one of the most active compounds against *C. jejuni* and most of the strains are inhibited by 8 μg/mℓ.[1,3,4,10,12,14,19,21-24] Thiamphenicol shows less activity than chloramphenicol.[22,23]

Most *C. jejuni* are highly resistant to novobiocin[10] and rifampin.[1,10,12] In general, all the *C. jejuni* strains are inhibited by nalidixic acid at 40 μg/mℓ,[10,14] although it should be noted that Vanhoof et al.[22,23] found that about 5% of their human strains are resistant to this concentration of nalidixic acid. Furthermore, Butzler and Skirrow[5] stated that these nalidixic acid-resistant strains can commonly be isolated from seagulls.

The distribution of the bacteriostatic activity of metronidazole is very broad[10,22-24] and the activity of tinidazole, another imidazole derivative is marginal too.[14,24]

Bicozamycin, an original compound, the absorption of which is rather limited after oral administration, has a good in vitro bacteriostatic activity on *C. jejuni*.[19]

III. IN VITRO SUSCEPTIBILITY OF *C. FETUS* SUBSP. *FETUS*

Relatively little data are available on the antibiotic susceptibility of *C. fetus* subsp. *fetus*. Most of the antimicrobial susceptibility testing reported has been on limited numbers of strains.

C. fetus subsp. *fetus* is relatively resistant to penicillin.[4,6] Most of the strains are inhibited by achievable serum concentrations of ampicillin[4,6] and ticarcillin.[6] Chow et al.[6] found one strain highly resistant to ampicillin (MIC $\geq$ 160 μg/mℓ). Chow et al.[6] also reported their population of *C. fetus* subsp. *fetus* strains as sensitive to carbenicillin, while the isolates tested by Butzler et al.[4] were markedly less sensitive to this compound. In general *C. fetus* subsp. *fetus* is relatively resistant to the cephalosporins. The following cephalosporins have been tested by different authors: cephalothin,[4,6,9] cefazolin, cefamandole, cephalexin, cefoxitin,[6,9] cefaclor,[6] cephalosporin C, cephaloridine, and cefotaxamine.[9]

Karmali and colleagues[9] demonstrated that *C. fetus* subsp. *fetus* was significantly more susceptible to cephaloridine, cephalothin, and cefazolin than was *C. jejuni*. The difference was less marked with cephalosporin C and cefamandole. With cephalexin, cefotaxime, and cefoxitin there was no difference. They were also able to demonstrate that a rapid species differentiation between *C. jejuni* and *C. fetus* subsp. *fetus* could be achieved by using a disk sensitivity test[9] (see Section IV). *C. fetus* subsp. *fetus* is highly susceptible to tetracycline,[4,6] doxycycline and minocycline,[6] gentamicin,[4,6] amikacin,[6] streptomycin, and neomycin.[4] Minocycline seems to be the most active tetracycline derivative.[6] Furthermore, all isolates are inhibited at readily achievable serum concentrations of clindamycin,[6] chloramphenicol and kanamycin,[4,6] and metronidazole.[6]

Erythromycin does not seem to have a uniform activity on *C. fetus* subsp. *fetus*. Chow et al.[6] found that their strains were relatively rather resistant to erythromycin, with MICs ranging from 0.2 to $\geq$25 μg/mℓ and with a median MIC of 2.5 μg/mℓ, while in the Butzler et al. series,[4] all the strains were fully susceptible (MICs ranging from 0.097 to 1.56 μg/

mℓ). Lincomycin is also relatively inactive against *C. fetus* subsp. *jejuni*. All the strains are resistant to colistin[4] and vancomycin.[6]

IV. DISK SENSITIVITY TESTING WITH *CAMPYLOBACTER*

For various reasons antibiotic susceptibility testing is necessary to guide the treatment of an infectious disease. However, the determination of the MIC values of antimicrobial agents is very often time-consuming and expensive and is not readily practicable in every routine laboratory. So, the disk sensitivity test may be an alternative technique to evaluate the susceptibility of a microorganism to antimicrobial agents. For campylobacters, however, practically no data on antibiotic disk sensitivities have been published. Furthermore, it must be mentioned that the sensitivity results obtained with disk methods may be seriously questioned because of the relatively slow growth and the microaerophilic character of campylobacters.

Vanhoof et al.,[20] in Brussels, investigated the reliability of a standardized disk diffusion method using Bauer and Kirby's technique and Rosco neosensitabs to test the susceptibility of *C. jejuni* strains to different chemotherapeutic agents. A least-square line of regression, assuming that there was a straight-line relation between MIC and the inhibition zone and a correlation coefficient (r) were calculated. In general there was a poor correlation between the MIC and the size of the inhibition zone. The following regression line equations have been calculated (correlation coefficient r between brackets):

Erythromycin	$y = 36.3 - 9.8 \times (r = -0.55)$
Ampicillin	$y = 33.8 - 7.9 \times (r = -0.45)$
Cotrimoxazole	$y = 40.5 - 5.7 \times (r = -0.19)$
Tetracycline	$y = 37.6 - 9.3 \times (r = -0.55)$
Furazolidone	$y = 48.6 - 5.8 \times (r = -0.06)$
Colistin	$y = 36.7 - 2.7 \times (r = -0.11)$

By using the zone-diameter interpretative standards for neosensitabs and the approximate MIC correlates described by the National Committee for Clinical Laboratory Standards, an error rate analysis[11] has been done by Vanhoof et al.[20] (see Figure 1). The error rate is a measure of the percentage of strains answered as false resistant or as false susceptible. In general the error rate was acceptable for all antibiotics except co-trimoxazole. For this compound 15.3% of the strains were reported as false resistant, and this may indicate that the disk diffusion test is not a reliable test to predict the susceptibility of *C. jejuni* to co-trimoxazole. Furthermore, it has been demonstrated that the disk diffusion test using nitrofurantoin to predict the susceptibility to nifuroxazide is also unreliable.[18] Disk sensitivity testing may also be used for species differentiation as has been described by Karmali et al.[9] In their study, the *C. jejuni* strains failed to produce an inhibition zone around a cephalothin disk of 30 μg, while the *C. fetus* subsp. *fetus* strains produced a clear zone of inhibition. The *C. fetus* subsp. *fetus* isolates had zone diameters around the cephalothin disk varying from 15 to 25 mm with a mean of 21 mm. *C. fetus* subsp. *fetus* can therefore be differentiated from *C. jejuni* by its susceptibility to a 30 μg cephalothin disk and its resistance to the 30 μg nalidixic acid disk.

V. CONCLUSIONS

The role of antimicrobial agents in the therapy of diarrhea is somewhat controversial, although many of these drugs are generally accepted as standard treatments. Nowadays, however, chemotherapy for most intestinal infections is very often considered as ineffective and even disadvantageous.

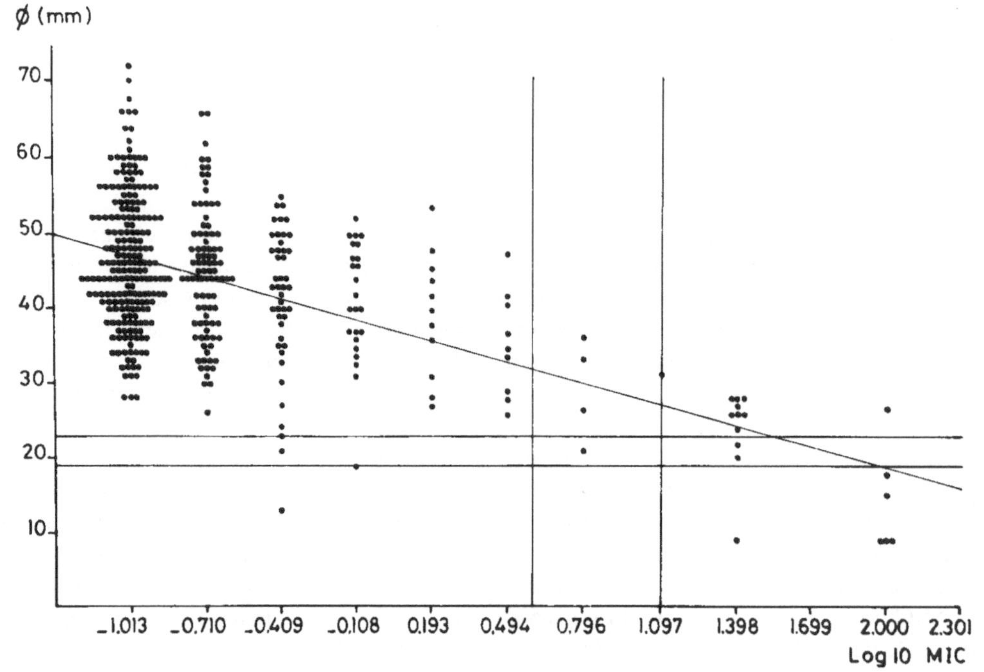

FIGURE 1. Error rate analyses for tetracycline by using a 80-μg Rosco Neosensitab disk.

Enteritis due to campylobacters is usually a mild disease and is self-limiting in most cases. However, antibiotic therapy may be indicated in the more severe cases, in patients with a prolonged illness, and in septicemic patients.

The choice of drug has hitherto been based on in-vitro data only. The current drug of choice is probably erythromycin.[5] Erythromycin is very active on *C. jejuni*, is furthermore nontoxic, and has a narrow spectrum, but the prevalence of resistant strains may be a serious disadvantage. Possible alternative drugs are the aminoglycosides, in particular gentamicin, tobramycin and amikacin, the tetracyclines, clindamycin, furazolidone, and chloramphenicol. All these compounds have a high level of in vitro activity against *C. jejuni*. Aminoglycosides may be the drugs of choice in the septicemic patient. The tetracyclines are contraindicated in children and resistance to tetracycline is plasmid mediated. Furazolidone is a very active compound but exerts only a contact effect in the gut. Clindamycin may be useful but there is cross resistance with erythromycin. The use of chloramphenicol is probably limited because of the risk of serious side effects. Some other possible alternative drugs are rosaramicin,[1] thienamycin,[1] nifuroxazide,[18] bicozamycin,[19] and cefotaxime.[1,10,17,22,24] However, all these newer compounds, as well as the others, need to undergo carefully controlled clinical trials in order to evaluate their usefulness in daily medical practice. Ampicillin, the other broad spectrum penicillins, the sulfonamides, and co-trimoxazole are only moderately active on *C. jejuni*.

Antibiotics with poor in-vitro activity against *C. jejuni* include penicillin G, cloxacillin, the cephalosporins in general, lincomycin, the polymyxins, vancomycin, bacitracin, novobiocin, rifampin, and trimethoprim. Some of these compounds may be incorporated into media used for primary isolation in order to improve the selectiveness.

REFERENCES

1. **Ahonkhai, V. I., Cherubin, C. E., Sierra, M. F., Bokkenheuser, V. D., Shulman, M. A., and Mosenthal, A. C.**, In vitro susceptibility of campylobacter fetus subsp. jejuni to *N*-formimidoyl thienamycin, rosaramicin, cefoperazone and other antimicrobial agents, *Antimicrob. Agents Chemother.*, 20, 850, 1981.
2. **Brunton, W. A. T., Wilson, A. M. M., McRae, R. M.**, Erythromycin-resistant campylobacters, *Lancet*, 2, 1385, 1978.
3. **Buck, G. E. and Kelly, M. T.**, Susceptibility testing of *Campylobacter fetus* subsp. *jejuni*, using broth microdilution panels, *Antimicrob. Agents Chemother.*, 21, 274, 1982.
4. **Butzler, J. P., Dekeyser, P., and Lafontaine, T.**, Susceptibility of related vibrios and *Vibrio fetus* to twelve antibiotics, *Antimicrob. Agents Chemother.*, 5, 86, 1974.
5. **Butzler, J. P. and Skirrow, M. B.**, Campylobacter enteritis, *Clin. Gastroenterol.*, 8, 737, 1979.
6. **Chow, A. W., Patten, V., and Bednorz, D.**, Susceptibility of *Campylobacter fetus* to twenty-two antimicrobial agents, *Antimicrob. Agents Chemother.*, 13, 416, 1978.
7. **Dekeyser, P., Gossuin-Detrain, M., Butzler, J. P., and Sternon, J.**, Acute enteritis due to related Vibrio: first positive stool cultures, *J. Infect. Dis.*, 125, 390, 1972.
8. **Fleming, P. C., D'Amico, A., De Grandis, S., and Karmali, M. A.**, The detection and frequency of β-lactamase production in *C. jejuni*, in *Campylobacter: Epidemiology, Pathogenesis, and Biochemistry*, Newell, D. G., Ed., MTP Press, Lancaster, U.K., 1982, 214.
9. **Karmali, M. A., De Grandis, S., and Fleming, P. C.**, Antimicrobial susceptibility of *Campylobacter jejuni* and *Campylobacter fetus* subsp. *fetus* to eight cephalosporins with special reference to species differentiation, *Antimicrob. Agents Chemother.*, 18, 948, 1980.
10. **Karmali, M. A., De Grandis, S., and Fleming, P. C.**, Antimicrobial susceptibility of *Campylobacter jejuni* with special reference to resistance patterns of Canadian isolates, *Antimicrob. Agents Chemother.*, 19, 593, 1981.
11. **Metzler, C. M. and De Haan, R. M.**, Susceptibility tests of anaerobic bacteria: Statistical and clinical considerations, *J. Infect. Dis.*, 130, 588, 1974.
12. **Ringertz, S., Rockhill, R. C., Ringertz, O., and Sutomo, A.**, Susceptibility of *Campylobacter fetus* subsp. *jejuni*, isolated from patients in Jakarta, Indonesia to ten antimicrobial agents, *J. Antimicrob. Chemother.*, 8, 333, 1981.
13. **Skirrow, M. B.**, Campylobacter enteritis — a "new" disease, *Br. Med. J.*, 2, 9, 1977.
14. **Svedhem, A., Kaijser, B., and Sjogren, E.**, Antimicrobial susceptibility of *Campylobacter jejuni* isolated from humans with diarrhoea and from healthy chickens, *J. Antimicrob. Chemother.*, 7, 301, 1981.
15. **Taylor, D. E., De Grandis, S., Karmali, M. A. and Fleming, P. C.**, Transmissible plasmids from *Campylobacter jejuni*. *Antimicrob. Agents Chemother.*, 19, 831, 1981.
16. **Taylor, D. E., De Grandis, S., Karmali, M. A., Fleming, P. C., Van Hoof, R., and Butzler, J. P.**, Erythromycin resistance in *Campylobacter jejuni*, in *Campylobacter: Epidemiology, Pathogenesis, and Biochemistry*, Newell, D. G., Ed., MTP Press, Lancaster, U.K., 1982, 211.
17. **Vanhoof, R., Butzler, J. P., and Yourassowsky, E.**, In vitro activity of a new cephalosporin (HR756) and cefazolin, *Lancet*, 2, 209, 1978.
18. **Vanhoof, R., Coignau, Stas, G., and Butzler, J. P.**, Evaluation of the in vitro activity of nifuroxazide on enteropathogenic microorganisms: determination of bacteriostatic and bactericidal concentrations and disk susceptibility, *Acta Clin. Belg.*, 36, 126, 1982.
19. **Vanhoof, R., Coignau, H., Stas, G., Goossens, H., and Butzler, J. P.**, Activity of Bicozamycin (CGP 3543/E) on different enteropathogenic microorganisms: comparison with other antimicrobial agents, *J. Antimicrob. Chemother.*, 10, 343, 1982.
20. **Vanhoof, R., Dierickx, R., Coignau, H., and Butzler, J. P.**, Antibiotic disk susceptibility tests with *Campylobacter fetus* subsp. *jejuni*, abstr. C 228, 81st Annu. Meet. Am. Soc. Microbiology, Dallas, Texas, 1981.
21. **Vanhoof, R., Goossens, H., Coignau, H., Stas, G., and Butzler, J. P.**, Susceptibility pattern of *Campylobacter jejuni* from human and animal origin to different antimicrobial agents, *Antimicrob. Agents Chemother.*, 21, 990, 1982.
22. **Vanhoof, R., Gordts, B., Dierickx, R., Coignau, H., and Butzler, J. P.**, Bacteriostatic and bactericidal activities of 24 antimicrobial agents against *Campylobacter fetus* subsp. *jejuni*, *Antimicrob. Agents Chemother.*, 18, 118, 1980.
23. **Vanhoof, R., Vanderlinden, M. P., Dierickx, R., Lauwers, S., Yourassowsky, E., and Butzler, J. P.**, Susceptibility of *Campylobacter fetus* subsp. *jejuni* to twenty-nine antimicrobial agents, *Antimicrob. agents Chemother.*, 14, 553, 1978.
24. **Walder, M.**, Susceptibility of *Campylobacter fetus* subsp. *jejuni* to twenty antimicrobial agents, *Antimicrob. Agents Chemother.*, 16, 37, 1979.
25. **Walder, M. and Forsgren, A.**, Erythromycin-resistant Campylobacters, *Lancet*, 2, 1201, 1978.

Chapter 8

PLASMIDS FROM *CAMPYLOBACTER*

Diane E. Taylor

TABLE OF CONTENTS

I. Introduction 88

II. Procedures for the Isolation of Plasmids from *Campylobacter* 88
A. Growth of *Campylobacter* Cells Prior to Plasmid DNA Isolation 88
B. Triton® X-100 Lysis Procedure 88
C. Sodium Dodecyl Sulfate (SDS) Lysis Procedure 89
D. Agarose Gel Electrophoresis and Visualization of Plasmid DNA 89
E. Molecular Weight Determination of Plasmids 89

III. Conjugative Transfer of Tetracycline-Resistance Plasmids from *Campylobacter jejuni* 89

IV. Physical Characterization of Tetracycline-Resistance Plasmids from *C. jejuni* 91
A. Molecular Weights 91
B. Guanine + Cytosine (G+C)-Content 92
C. Restriction Endonuclease Analysis 92

V. Ampicillin Resistance in *C. jejuni* 93

VI. Erythromycin Resistance in *C. jejuni* 93

VII. Conclusions 95

Acknowledgments 95

References 95

I. INTRODUCTION

The relatively recent recognition of *Campylobacter jejuni* as a pathogen, as well as its microaerophilic nature, have delayed the study of plasmids in this organism. Austin and Trust[1] in 1980 were the first to report plasmids in *C. jejuni*. They found that approximately 19% of strains from various geographic locations contained plasmids. Only one isolate harbored more than one plasmid. Molecular weights of these plasmids varied from 5.0×10^6 to 77×10^6. Although some of the isolates were resistant to antibiotics, no direct correlation could be made between antibiotic resistance and plasmid-carriage. Since then the author and co-workers have identified transmissible plasmids which specify tetracycline resistance in a number of strains of *C. jejuni*.[2,3] These studies as well as the questions of erythromycin and ampicillin resistance in *C. jejuni* are discussed in this article.

II. PROCEDURES FOR THE ISOLATION OF PLASMIDS FROM *CAMPYLOBACTER*

Several methods developed for the isolation of plasmid DNA from *Escherichia coli* and *Pseudomonas aeruginosa* have been modified for isolation of plasmid DNA from *C. jejuni* and *C. fetus* subsp. *fetus*. Austin and Trust[1] used the method of Hansen and Olsen[4] for the isolation of plasmid DNA from clinical isolates of *C. jejuni*. Taylor and colleagues[3] have successfully isolated plasmid DNA from *Campylobacter* using a number of different procedures. A modification of method described by Meyers et al.[5] was used in our earlier studies.[3] Subsequent work showed that Portnoy and White's procedure for the isolation of large plasmids, detailed by Crosa and Falkow,[6] gave better yields of plasmid DNA. Most recently *Campylobacter* plasmid DNA has been isolated in a purified form suitable for restriction endonuclease digestion using cesium chloride-ethidium bromide density gradient centrifugation.[17]

A. Growth of *Campylobacter* Cells Prior to Plasmid DNA Isolation

Bacterial strains were grown on Columbia base agar (GIBCO Diagnostics or Oxoid Ltd.) containing 7% defibrinated horse blood (BA) for 48 hr. The agar plates were incubated at 37 or 42°C in an atmosphere in which the oxygen tension was reduced to approximately 7%. Such an atmosphere was obtained by evacuating two thirds of the air from an anaerobic jar (without catalyst) and replacing the evacuated air with a hydrogen-carbon dioxide mixture to give a final concentration of approximately 7% O_2, 7% CO_2, 26% N_2, and 60% H_2. Alternatively, agar plates were incubated at 37°C in a carbon dioxide incubator (CO_2 tension: 7%, humidity: 85%).

B. Triton® X-100 Lysis Procedure

The method of Meyers et al.[5] as modified by Taylor et al.[3] gave poor yields compared with plasmids from *E. coli*. Improved yields of *C. jejuni* plasmid DNA were obtained using a Triton® X-100 lysis procedure. The method of Quackenbush, as described by Crosa and Falkow,[6] was used with the following modifications. After the Triton® X-100 lytic mixture had been added, the preparations were heated at 65°C for 5 min when clearing of the samples was observed. The cleared lysate was treated with ribonuclease (50 μg/mℓ) for 20 min on ice and then with pronase (Boehringer-Mannheim Corporation) at a final concentration of 200 μg/mℓ for 30 min at 37°C. After centrifugation at 17,000 r/min at 4°C for 20 min, the DNA was subjected to cesium chloride-ethidium bromide density gradient centrifugation at 55,000 r/min for 24 hr in a Beckman 75 Ti rotor. The plasmid DNA was collected from the side of the tube using a syringe with an 18-gauge needle, and immediately rebanded by density gradient centrifugation as described above. Plasmid DNA was dialysed against extensively in TE buffer (0.01 *M* tris(hydroxymethyl) aminomethane (TRIS) — 0.002 *M* ethylene diaminetetraacetic acid [EDTA] pH 8.0) and stored at 4°C.

C. Sodium Dodecyl Sulfate (SDS) Lysis Procedure

A modification of the method devised by Portnoy and White[6] for the isolation of large plasmids usually gave good yields of *Campylobacter* plasmid DNA. Cells were collected from plates in TE buffer. Washed pellets were resuspended in 1.0 mℓ TE buffer and left on ice for 60 min to ensure complete suspension of cells prior to lysis. Lysis buffer, which was freshly prepared, contained (0.05 *M* TRIS, 0.01 *M* EDTA, and 3% SDS) and the pH was adjusted to 12.4. The cells were lysed by addition of 12 mℓ lysis buffer to the *C. jejuni* cells in test tubes, which were then inverted 20 times. The cell suspensions were incubated at 50°C for 20 min to achieve complete lysis of the cells and denaturation of DNA. Neutralization of the suspensions was achieved by addition of 2 mℓ of 2 *M* TRIS-HCl pH 7.0. This resulted in a final pH of 8.0 to 8.5 and allowed preferential renaturation of closed circular DNA molecules. Membrane-chromosomal complexes were precipitated by adding 5 *M* NaCl to a concentration of 1 *M*, followed by incubation at 4°C for 10 hr. Debris was pelleted at 13,500 × gravity for 45 min at 4°C. The plasmid DNA-enriched supernatant fraction was precipitated with isopropyl alcohol, treated with ribonuclease (50 μg/mℓ) and extracted with phenol. Traces of phenol were removed by extraction with chloroform and DNA was precipitated with ethanol and resuspended in 50 μℓ TE buffer. Aliquots of 5 to 20 μℓ were examined for plasmid DNA content by agarose gel electrophoresis.

D. Agarose Gel Electrophoresis and Visualization of Plasmid DNA

Samples were subjected to electrophoresis in horizontal 0.7% agarose gels in TRIS-borate buffer.[5] The gels were stained as described by Meyers et al.[5] in a solution of ethidium bromide in TRIS-borate buffer pH 8.3. The plasmid DNA bands were visualized on an Ultraviolet Products Transilluminator containing shortwave bulbs (254 nm). Photographs were taken on Ilford FP4 film using a Wratten 23A gelatin filter (see Figure 1).

E. Molecular Weight Determination of Plasmids

Molecular weights of plasmids from clinical isolates of *C. jejuni* were estimated using plasmids of known molecular weights as standards in each gel. These plasmids were present in an *E. coli* K-12 host and were as follows: pMB8 1.8 × 10^6, pBR322, 2.6 × 10^6, RSF1030 5.5 × 10^6; S-a, 23 × 10^6, RP1, 38 × 10^6, MP10 60 × 10^6, RA1, 86 × 10^6. Standard curve of the log of the molecular weight vs. the log of the distance migrated by the known plasmids were prepared for each gel. One known plasmid of appropriate size was used as an internal standard in each well containing a plasmid of unknown size.[7] The molecular weight determination was performed at least three times for DNA from each isolate.

III. CONJUGATIVE TRANSFER OF TETRACYCLINE-RESISTANCE PLASMIDS FROM *CAMPYLOBACTER JEJUNI*

Antibiotic susceptibilities of *C. jejuni* strains isolated at The Hospital for Sick Children in Toronto, Canada in 1978 and 1979 have been determined.[8] Approximately 20% of these strains were resistant to tetracycline (4 μg/μℓ) and about 12% were resistant to high levels of tetracycline, with minimal inhibitory concentration (MIC) of 64 to 256 μg/mℓ. In *C. jejuni*, high level tetracycline-resistance is plasmid-mediated and is transmissible within the genus *Campylobacter*. Taylor et al.[3] used a clinical isolate of *C. jejuni* MK22 with tetracycline MIC of 128 μg/mℓ as the donor strain. Tetracycline resistance transfer was demonstrated to a *C. jejuni* SD2, a spontaneous nalidixic acid-resistant mutant of *C. jejuni* MK

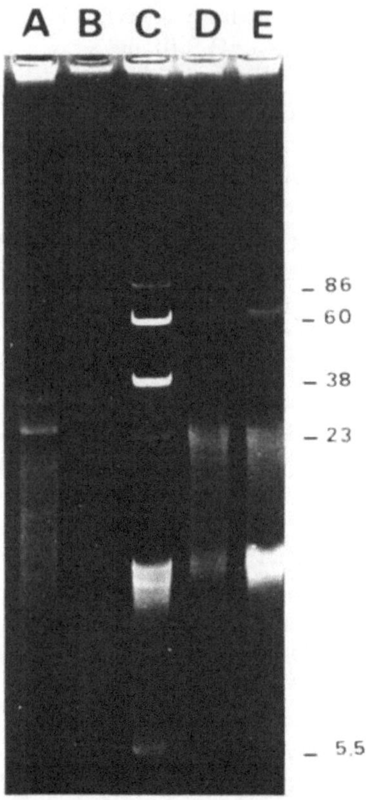

FIGURE 1. Agarose gel electrophoresis of plasmids from clinical isolates of *C. jejuni* resistant to erythromycin and tetracycline or to erythromycin. Plasmid-enriched DNA fractions were prepared by the modification of the Portnoy and White procedure as described in the text. Samples were subjected to electrophoresis on a 0.7% agarose gel at for 2 hr and gels were photographed under shortwave ultraviolet light. Track A: *C. jejuni* clinical isolate BA37 resistant to erythromycin and tetracycline; Track B: *C. jejuni* SD2 transconjugant harboring pBJA37, resistant to tetracycline; Track C: reference plasmids RA1 (molecular weight 86 × 10^6), MP10 (molecular weight 60 × 10^6) RP1 (molecular weight 38 × 10^6), S-a (molecular weight 23 × 10^6) and RSF1030 (molecular weight 5.5 × 10^6); Track D: *C. jejuni* BA48, erythromycin-resistant isolate from West Germany; Track E: *C. jejuni* BA49, erythromycin-resistant isolate from West Germany.

118, with a tetracycline MIC of ≤4 μg/mℓ and a nalidixic acid MIC of 256 μg/mℓ. Broth and filter-mating methods were used, in which the strains were grown in a liquid medium described by Taylor et al.[3] Transfer of tetracycline resistance was demonstrated by both broth and filter-mating methods, however, the filter-mating method gave approximately 100-fold higher frequency of transfer than the broth mating (5.0 × 10^{-4} transconjugants per recipient for the 48-hr filter-mating procedure compared with 2.4 × 10^{-6} for the broth mating).

The higher transfer frequency observed for the filter-mating procedure indicated that a solid surface facilitates intraspecies transfer of the *C. jejuni* tetracycline resistance determinant. As the filter-mating method is somewhat tedious, a plate-mating procedure was developed and was used for both intraspecies and interspecies transfer.

Interspecies transfer was demonstrated with two donor strains of *C. jejuni*, MK22, described above, and MK175, a clinical isolate resistant to both tetracycline (MIC = 64 μg/mℓ and ampicillin (MIC = 128 μg/mℓ). The recipient strain was *C. fetus* subsp. *fetus* ATCC 27374 (CIP 5396), which is naturally resistant to nalidixic acid (MIC = 256 μg/

Table 1
TRANSFER OF TETRACYCLINE RESISTANCE DETERMINANTS FROM *C. JEJUNI* CLINICAL ISOLATES TO *C. FETUS* SUBSP. *FETUS* ATCC27374

Strain[a]	Antibiotic resistance of *C. jejuni* clinical isolate[b]	Plasmid	Tetracycline resistance transfer frequency[c]
MK175	ApTc	pMAK175	4.0×10^{-5}
BA37	ErTc	pBJA37	1.3×10^{-3}
BA39	ErTc	pBJA39	1.0×10^{-4}

[a] Origin of strains: MK175 Toronto, Canada; BA37 and BA39 Brussels, Belgium.
[b] Ap, ampicillin; Er, erythromycin; Tc, tetracycline.
[c] Tetracycline-resistance determinants were transferred from *C. jejuni* clinical isolates to *C. fetus* subsp. *fetus* ATCC27374 in a 24-hr plate mating on BA at 37°C. The other resistance phenotypes, Ap and Er-resistance, were not transferred.

mℓ). The strains were grown on (BA) and incubated for 48 hr at 42°C for *C. jejuni* and at 37°C for *C. fetus* subsp. *fetus*. The cells from the plates were suspended in 0.05 *M* sodium phosphate buffer at 1×10^9 cells per milliliter. Aliquots of 0.15 mℓ of donor and recipient cell suspensions were mixed together and spread over BA plates. The plates were incubated at 37°C for 24 or 48 hr. To select for *C. fetus* subsp. *fetus* transconjugants, cells were washed off the plates, diluted and spread on Diagnostic Sensitivity Testing agar (DST, Oxoid) containing 5% lysed horse blood, 75 μg/mℓ nalidixic acid and 16 μg/mℓ tetracycline.

Interspecies transfer of tetracycline resistance from clinical isolates of *C. jejuni* to *C. fetus* subsp. *fetus* ATCC 27374 is shown in Table 1. The tetracycline resistance determinant transferred equally well to both the *C. fetus* subsp. *fetus* and *C. jejuni* recipients. Attempts were also made to transfer tetracycline resistance to *Escherichia coli*. Three strains were used: *E. coli* K12, J53-1 *(pro met gyrA)*, NM148 (Kr − Km + *gyrA)*, a restriction-deficient mutant of *E. coli* K12, and the *E. coli* C strain RG176[9] which does not possess the K restriction system. None of the three *E. coli* strains was able to act as a recipient.

Experiments with cell-free filtrates of the donor strains could not promote transfer of tetracycline resistance determinants to a plasmid-free recipient ($<1 \times 10^{-8}$ transconjugants per recipient after a 48-hr mating period). This result suggested that the transfer process was not bacteriophage-mediated. The transfer frequencies of *C. jejuni* plasmids were not affected by incorporation of deoxyribonuclease (100 μg/mℓ) into the agar used in the plate-mating experiments. This fact appeared to rule out transformation in the transfer of tetracycline resistance. The process, therefore, appears to involve conjugation via cell-to-cell contact and to be facilitated by a solid surface.

IV. PHYSICAL CHARACTERIZATION OF TETRACYCLINE-RESISTANCE PLASMIDS FROM *C. JEJUNI*

A. Molecular Weights

Taylor et al.[3] showed that four tetracycline-resistant Canadian isolates of *C. jejuni* harbored a plasmid with a molecular weight of approximately 38×10^6 as determined by agarose gel electrophoresis. A plasmid of the same molecular weight was observed in tetracycline-resistant transconjugants of *C. jejuni* and *C. fetus* subsp. fetus.[3] Two clinical isolates of *C. jejuni* from Belgium, resistant to both erythromycin and tetracycline, have been examined. The clinical isolates from Belgium, BA37 and BA39, contained two plasmids of approxi-

Table 2
CHARACTERIZATION OF TETRACYCLINE RESISTANCE PLASMIDS FROM *C. JEJUNI*

Serotype of *C. jejuni*[a]	Source	Antibiotic resistance of *C. jejuni* strain[b]	Plasmid designation	Buoyant density (g/cm^3)[c]	G+C content (mol %)	Number of restriction fragments after digestion with *Acc* I[d]
13, 16	Canada	Tc	pMK22	1.693	32.5	4
3	Canada	Tc	pMAK122	1.694	33.0	4
4	Canada	ApTc	pMAK175	1.693	32.5	4
51	Belgium	ErTc	pBJA37	1.693	32.5	3
51	Belgium	ErTc	pBJA39	1.692	32.0	3

[a] Determined by the passive hemagglutination technique, as described previously.[15]
[b] Ap, ampicillin; Er, erythromycin; Tc, tetracycline.
[c] Determined by centrifugation in a Beckman Model E analytical ultracentrifuge in CsCl at 44,000 r/min and 25°C for 24 hr as described previously.[10]
[d] Plasmid DNA was digested with the restriction enzyme *Acc* I (Bethesda Research Laboratories, Inc.). Enzyme digestions were carried out as described by Davis et al.[16] for 90 min using 0.3 μg DNA. Fragments were subjected to electrophoresis on a 0.5% agarose gel at 5 V/cm for 6 hr.[17]

mately 23 × 10^6 and 30 × 10^6. Plasmids from BA37 are shown in Figure 1, Track A. Tetracycline resistance was associated with the larger plasmid which was transferred to *C. fetus* subsp. *fetus* ATCC 27374 (Table 1). Plasmid DNA from the *C. fetus* subsp. *fetus* transconjugant is seen in Figure 1, Track B.

B. Guanine + Cytosine (G+C)-Content

The buoyant density of the tetracycline-resistance plasmids from *C. jejuni* strains in Canada and Belgium was determined by ultracentrifugation in CsCl in a Beckman Model E analytical ultracentrifuge at 44,000 r/min at 25°C for 24 hr as described by Szybalski and Szybalski.[10] The relative buoyant densities were determined based on their positions in relation to *Micrococcus lysodeikticus* DNA (buoyant density 1.731 g/cm^3). The base composition (G+C content in mol %) of various DNA species was determined graphically from published data.[10] The buoyant densities of the plasmids varied from 1.692 to 1.694 g/cm^3 corresponding to a G+C content of 32.0 to 33.0% (Table 2). These values are similar to those of the *C. jejuni* chromosome, buoyant density 1.693 g/cm^3 and corresponding G+C content 32.5%,[17] whereas *C. fetus* subsp. *fetus* DNA had a slightly higher buoyant density of 1.695 g/cm^3 or 35.0% G+C. As their G+C content is characteristic of *C. jejuni* DNA, the tetracycline resistance plasmids have most probably arisen in this species, rather than being acquired from elsewhere.

C. Restriction Endonuclease Analysis

More detailed comparisons of five tetracycline resistance plasmids from Canada and Belgium were made by restriction endonuclease digestion. One plasmid pMAK175 isolated from a Toronto clinical isolate was digested with 20 different enzymes and the results are shown in Table 3. Four enzymes *Acc* I, *Bcl* I, *Bgl* II, and *Pst* I yielded 4 to 10 fragments and were judged to be suitable to compare the five plasmids. With *Acc* I, *Bgl* II, *Pst* I, the two Belgian isolates gave identical fragmentation patterns and were uncut with *Bcl* II. These plasmids appeared to be identical. In contrast, the three Canadian isolates showed some variations in restriction patterns. Digestion of pMAK122 and pMAK175 with *Acc* I, *Bgl* II and *Pst* I yielded similar fragmentation patterns, whereas pMAK22 gave distinct patterns.

Table 3
RESTRICTION ENDONUCLEASE ANALYSIS OF pMAK175, A TETRACYCLINE-RESISTANCE PLASMID FROM *CAMPYLOBACTER JEJUNI*

Enzymes Generating One Fragment

Ava I	*Bsts*EII	*Puv*II	*Xba*I
*Ava*II	*Hpa*I	*Sau*961	*Xho*I
*Bam*HI	*Kpn*I	*Sma*I	

Enzymes Generating 2 to 10 Fragments

*Acc*I	*Bgl*II
*Bcl*I	*Pst*I

Enzymes Generating Numerous Fragments

*Alu*I	*Hin*fI	*Taq*I
*Hin*dIII	*Hpa*II	

[a] Conditions for restriction digestion of plasmid DNAs are as described in the legend to Table 2.

Data from Allan, B. J. and Taylor, D. E., *Abstr. Am. Soc. Microbiol. Intersci. Conf. Antimicrob. Agents Chemother.*, No. 877, 1982, 222.

Figure 2 illustrates the fragments generated by treatment of the plasmids with *Bgl* II. The data indicate that the two Belgian plasmids are identical and this is consistent with the fact that both were present in strains of the same serotype (Penner 51) and both strains were resistant to erythromycin as well as to tetracycline. The Canadian isolates, which were all isolated from patients at the Hospital for Sick Children in Toronto, were from three different serotypes (Table 2). Two of the plasmids pMAK 175 and pMAK122 appear to be very similar, by their restriction enzyme profiles after digestion with *Bgl* II (Figure 2). Further work is required to determine the degree of relatedness among these tetracycline-resistance plasmids.

V. AMPICILLIN RESISTANCE IN *C. JEJUNI*

Ampicillin resistance was observed in about 15% of clinical isolates of *C. jejuni*[8] and is associated with β-lactamase production in these strains.[11] *C. jejuni* MK175 is one such strain from which transfer of tetracycline resistance was associated with transfer of a 38 $\times$ 10^6 plasmid in both intra- and interspecies matings (Table 1). Ampicillin resistance was not cotransferred with tetracycline resistance from *C. jejuni* MK175 and was not located on the same plasmid. No other plasmid was visualized in DNA preparations of MK175, therefore, it is likely that ampicillin resistance in this strain is of chromosomal origin.

VI. ERYTHROMYCIN RESISTANCE IN *C. JEJUNI*

Approximately 1% of *C. jejuni* strains isolated at the Hospital for Sick Children, in Toronto, 1978 and 1979 were resistant to erythromycin[8] with MIC >1024 μg/mℓ. In Belgium[12] 8.4% and in Sweden[13] 10% of *C. jejuni* strains were reported to be erythromycin-resistant.

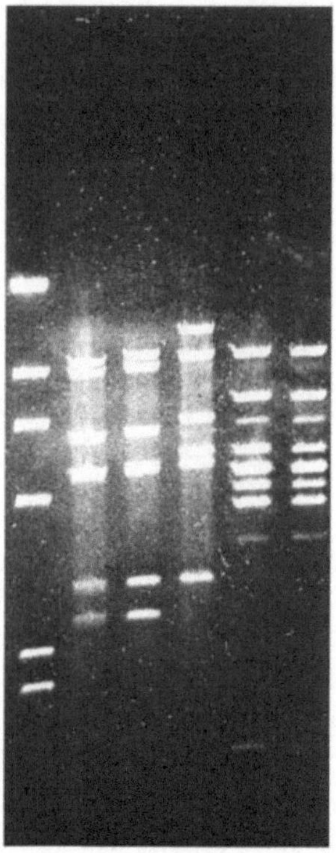

FIGURE 2. Restriction endonuclease *BglII* digestion of tetracycline resistance plasmids from *C. jejuni*. Plasmid DNA was isolated as described in the text, then subjected to digestion with the enzyme *BglII* for 90 min at 37°C. Restriction fragments were subjected to electrophoresis on a 0.5% agarose gel at 5 V/cm for 6 hr Track A: Lambda phage DNA reference fragments from digestion with *Hin*dIII; Track B: pMAK175, Track C: pMAK122; Track D: pMAK22; Track E: pBJA37; Track F: pBJA39.

Plasmid DNA has been isolated from a number of erythromycin-resistant strains.[14] Strains from Belgium and Canada contained plasmids which varied in size from 1.5×10^6 to 25×10^6. Strains of *C. jejuni* resistant to erythromycin were isolated in West Germany from pigs by Sticht-Groh.[18] Two of these erythromycin-resistant strains contained plasmids of approximately 70×10^6.[19] Plasmid DNA from these two strains (BA48 and BA49) is shown in Figure 1, Tracks D and E. Numerous attempts have been made to transfer erythromycin-resistance from strains containing plasmids, ranging in molecular weight from 20×10^6 to 70×10^6, to *C. jejuni* SD2 or to *C. fetus* subsp. *fetus* ATCC 27374.[20] In no experiment was transfer of the erythromycin-resistance phenotype demonstrated. Four other erythromycin-resistant strains contained small plasmids of molecular weight 1.5×10^6 to 5×10^6, which are too small to encode their own conjugative transfer. Two other strains of *C. jejuni* originating in Belgium were plasmid-free.

Three other strains of *C. jejuni* resistant to both erythromycin and tetracycline were tested for transfer of each of these resistance determinants. Although transfer of tetracycline resistance was observed in these isolates (Table 1), erythromycin resistance was neither transferred alone nor in combination with tetracycline resistance. Although a number of erythromycin-resistant strains of *C. jejuni* contain plasmid DNA, there is no evidence, as yet, that erythromycin resistance is plasmid-associated.

VII. CONCLUSIONS

Tetracycline resistance in *C. jejuni* is encoded by transmissible plasmids. Intraspecies transfer to *C. jejuni* was observed, moreover interspecies transfer to *C. fetus* subsp. *fetus* was also demonstrated. Transfer of plasmids in *Campylobacter* is facilitated in filter- or plate-matings, indicating that a solid surface is beneficial. Neither DNA transformation nor bacteriophage-mediated transduction appear to be involved in transfer of *C. jejuni* plasmids. The process probably involves conjugation, via cell to cell contact. *Campylobacter* plasmids could not be transferred to *E. coli*.

Tetracycline resistance plasmids from *C. jejuni* were characterized further by determining their molecular weight, buoyant density, and restriction endonuclease digestion profiles. From their restriction digestion patterns, plasmids from two Belgian isolates appeared to be very similar. Plasmids from three Canadian isolates, however, gave different patterns from the Belgian isolates and also showed differences among themselves. These studies should enable us ultimately to identify the degree of similarity among *C. jejuni* plasmids from different geographic locations.

Note added in proof — *C. jejuni* strains BA37 and BA39 have now been reclassified as *C. coli*.

ACKNOWLEDGMENTS

To my colleagues B. J. Allan and R. Garner at the University of Alberta, S. A. DeGrandis, J. G. Levine, M. A. Karmali, and P. C. Fleming at The Hospital for Sick Children, Toronto and J. L. Penner at the University of Toronto. Thanks to R. Vanhoof and V. Sticht-Groh for supplying *C. jejuni* strains.

This work was supported by Grants from the Medical Research Council of Canada, from the Alberta Heritage Foundation for Medical Research, and from The Hospital for Sick Children Foundation. The author was supported by Scholarships from MRC and from AHFMR.

REFERENCES

1. **Austin, R. A. and Trust, T. J.,** Detection of plasmids in the related group of the genus *Campylobacter*, *Fems Microbiol. Lett.*, 8, 201, 1980.
2. **Taylor, D. E., DeGrandis, S. A., Karmali, M. A., and Fleming, P. C.,** Transmissible tetracycline resistance in *Campylobacter jejuni*, *Lancet*, 2, 797, 1980.
3. **Taylor, D. E., DeGrandis, S. A., Karmali, M. A., and Fleming, P. C.,** Transmissible plasmids from *Campylobacter jejuni*, *Antimicrob. Agents Chemother.*, 19, 831, 1981.
4. **Hansen, J. B. and Olsen, R. H.,** Isolation of large bacterial plasmids and characterization of the P2 incompatibility group plasmids pMG1 and pMG5, *J. Bacteriol.*, 135, 227, 1978.
5. **Meyers, J. A., Sanchez, D., Elwell, L. P., and Falkow, S.,** Simple agarose gel electrophoresis of plasmid deoxyribonucleic acid, *J. Bacteriol.*, 127, 1592, 1976.
6. **Crosa, J. H. and Falkow, S.,** Plasmids, in *Manual of Methods for General Bacteriology*, Gerhardt, P., Murray, R. G. E., Costilow, R. N., Nester, E. W., Wood, W. A., Krieg, N. R., Phillips, G. B., Eds., American Society for Microbiology, Washington, D.C., 1981, 266.
7. **Taylor, D. E. and Levine, J. G.,** Studies of temperature-sensitive transfer and maintenance of H incompatibility group plasmids, *J. Gen. Microbiol.*, 116, 475, 1980.
8. **Karmali, M. A., DeGrandis, S., and Fleming, P. C.,** Antimicrobial susceptibility of *Campylobacter jejuni* with special references to resistance patterns of Canadian isolates, *Antimicrob. Agents Chemother.*, 19, 593, 1981.
9. **Taylor, D. E. and Grant, R. G.,** Incompatibility and Bacteriophage inhibition properties of N-1, a plasmid belonging to the H_2 incompatibility group, *Mol. Gen. Genet.*, 153, 5, 1977.

10. **Szybalski, W. and Szybalski, E.,** Equilibrium density gradient centrifugation, *Proc. Nucl. Acid Res.*, 2, 311, 1971.
11. **Fleming, P. C., DeGrandis, S., D'Amico, A., and Karmali, M. A.,** The detection and frequency of beta-lactamase production in *Campylobacter jejuni*, in *Campylobacter: Epidemiology, Pathogenesis, and Biochemistry*, Newell, D. G., Ed., MTP Press, Lancaster, U.K., 1982, 214.
12. **Vanhoof, R., Vanderlinden, M. P., Dierickx, R., Lauwers, S., Yourassowsky, E., and Butzler, J. P.,** Susceptibility of *Campylobacter fetus* subsp. *jejuni* to twenty-nine antimicrobial agents, *Antimicrob. Agents Chemother.*, 14, 553, 1978.
13. **Walder, M. and Forsgren, A.,** Erythromycin-resistant Campylobacters, *Lancet*, 2, 1201, 1978.
14. **Taylor, D. E., DeGrandis, S. A., Karmali, M. A., Fleming, P. C., Vanhoof, R., and Butzler, J. P.,** Erythromycin resistance in *Campylobacter jejuni*, in *Campylobacter: Epidemiology, Pathogenesis, and Biochemistry*, Newell, D. G., Ed., MTP Press, Lancaster, U.K., 1982, 211.
15. **Penner, J. L. and Hennessy, J. N.,** Passive hemagglutination technique for serotyping *Campylobacter fetus* subsp. *jejuni* on the basis of soluble heat-stable antigens, *J. Clin. Microbiol.*, 12, 73, 1980.
16. **Davis, R. W., Botstein, D., and Roth, J. R.,** *A Manual for Genetic Engineering Advanced Bacterial Genetics*, Cold Spring Harbor, New York, 1980.
17. **Allan, B. J. and Taylor, D. E.,** Physical characterization of tetracycline resistance plasmids from *Campylobacter jejuni*, *Abstr. Am. Soc. Microbiol. Intersci. Conf. Antimicrob. Agents Chemother.*, No. 877, 1982, 222.
18. **Sticht-Groh, V.,** Campylobacter in healthy slaughter pigs: a possible source of infection for man, *Vet. Rec.*, 110, 104, 1985.
19. **Taylor, D. E. and Garner, R. S.,** unpublished data, 1982.
20. **Taylor, D. E., Allan, B. J., DeGrandis, S. A., Garner, R. S., and Vanhoof, R.,** unpublished results, 1980—1982.

Chapter 9

SEROLOGICAL RESPONSES TO *CAMPYLOBACTER JEJUNI* INFECTION

D. M. Jones

TABLE OF CONTENTS

I. Introduction 98

II. Techniques 98
- A. Bacterial Agglutination 98
- B. Slide Agglutination 98
- C. Complement Fixation 98
- D. Immunofluorescence 99
- E. Bactericidal Assay 99
- F. Enzyme-Linked Immunosorbent Assay 99

III. Individual Patient Response 101

IV. Outbreaks 101

V. Conclusion 103

References 103

I. INTRODUCTION

Patients infected with *Campylobacter jejuni* acquire serum antibodies and these have been investigated by a variety of techniques including agglutination,[1] complement fixation,[1,2] bactericidal assay,[3,4] indirect immunofluorescence,[5] and enzyme-linked immunosorbent assay.[6] Because the nature and variation of the antigens present on the organism are still not well understood, these techniques have undefined degrees of specificity and the specificity may also vary with the way the antigen is prepared for a particular technique. Studies of the antibody response to infection lead in two directions. First, a method using a "universal" antigen, reactive with both IgM and IgG antibodies, would detect all those infections that are accompanied by an antibody response. This would, therefore, be useful and reliable for diagnosis when no culture was made and also applicable for serological surveys to determine the prevalence of infection in particular groups of individuals. Second, antibody detection systems that are specific for particular antigens on the organism would be useful for studying the homologous antibody response and also may throw light onto the antigens involved in immunity and pathogenesis. Most of the serological methods so far reported fall short of these requirements to some extent and it will be convenient to review some of the techniques.

II. TECHNIQUES

A. Bacterial Agglutination

Tube agglutination tests with formolized suspensions are suitable for detecting antibody in convalescent sera but the reaction is specific and the homologous organism is preferable.[1] Indeed this reaction is suitable for serotyping[7] and gives results similar to those obtained by haemagglutination using a soluble surface antigen. Heating the organisms results in the suspensions having greater cross reactivity and boiling for 15 min[2] or 1 hr[8] results in suspensions that are suitable for detecting antibody in convalescent sera. Because of antigenic similarities and differences that exist between strains, some suspensions have greater cross reactivity than others. Kosunen et al.[8] found that 64% of convalescent patient sera reacted with formolized suspensions prepared from four reference strains (two of which had previously been used by Watson),[9] 74% of the sera reacted with boiled antigens from the same strains, while only 28% of sera reacted with autoclaved suspensions. The proportion of sera reacting with homologous heated suspensions was 81% in a study of two outbreaks.[2] Sometimes homologous suspensions react less well than heterologous ones for some reason that is not yet clear.[10]

B. Slide Agglutination

The technique is used for serotyping[11] and for this purpose the antisera are absorbed and the use of saline-containing nuclease has been found to reduce autoagglutination. There are no reports of the method being used to examine convalescent human sera, but we have found it to be frequently positive using the homologous strain. The agglutination is fairly serotype-specific so that the homologous serotype (live or formolized) is usually necessary to detect antibody in this way. It appears that IgM antibody is involved in slide agglutination (Table 3). The method is very simple but not as sensitive as agglutination in tubes.

C. Complement Fixation

In early work with the complement fixation test (CFT), establishing that it was useful but of some specificity for detecting antibody,[12] the antigen used was probably whole organisms. The introduction of a sonicate as antigen[2] demonstrated that the CFT was commonly positive in convalescent sera and, further, a sonicate of a few strains served as a broadly reactive antigen able to detect most infections irrespective of the serotype involved. We have also observed that this test will detect antibody to *C. coli* in convalescent sera. The usefulness of the technique has been confirmed in studies on several outbreaks of *C. jejuni* infection[4,13,14] where between 85 to 100% of sera reacted in the test. Between 2 to 5% of normal sera may be positive.[4,12,15] The test is therefore useful for serological surveys to demonstrate possible exposure to infection. For example, in groups of workers in the poultry industry, between 30 to 60% were found to have complement-fixing antibody.[16]

D. Immunofluorescence

Blaser et al.[17] used boiled, homologous organisms as antigen and demonstrated greater than fourfold rises in antibody titer in convalescent sera from a small number of patients. This technique is readily adaptable for detecting IgM or IgG by use of the appropriate conjugate. Watson and Kerr[10] used washed, heat-fixed organisms of their two reference strains and were able to detect antibody in 82% of convalescent sera. Their conjugate reacted with IgG, IgM, and IgA. It is likely that many antigens present on *C. jejuni*, and not only the serotype antigen, react with various antibodies in the convalescent serum so that the reactivity of this technique would be expected to be broad and again the method is suitable for surveys.

E. Bactericidal Assay

C. jejuni is killed by antibody in the presence of complement. Utilizing this reaction for serotyping,[7] we showed that the effect was serotype-specific and the specificity was closely similar to that of hemagglutination using heat-soluble surface antigens to sensitize red cells,[18] and also similar in specificity to agglutination of formalized suspensions. It is likely, therefore, that outer somatic serotype antigens are involved in the bactericidal reaction. Bactericidal antibodies can be detected in most, but not all, convalescent sera and react with the infecting serotype only. The bactericidal antibody is predominantly IgM (Table 3). Because this is a serotype-specific reaction, this technique is inappropriate for prevalence studies, but the presence of bactericidal antibody in convalescent serum gives confirmatory evidence of infection with a particular serotype.

F. Enzyme-Linked Immunosorbent Assay

This technique has been used with different antigens and various modifications. A glycoprotein surface antigen prepared from cultures with glycine-HCl buffer[19] has been used on conventional fluid phase ELISA and also in the diffusion-in-gel enzyme-linked immunosorbent assay (DIG-ELISA).[6] In this mode the antigen seems to be broadly reactive and there is a low proportion of positive reactions in normal serum and a high proportion of reactions in convalescent serum.

Satisfactory results have also been obtained using a lightly centrifuged sonicate of six different strains of *C. jejuni* as antigen. Alternate rows of wells in microtiter plates are coated overnight with a 1/100 dilution of antigen in buffer pH 9.6; after washing, the plates can be stored at −20°C until required. A negative antigen is placed in the other rows of wells to avoid nonspecific attachment of the serum protein. Serum dilutions are allowed to react for 1 hr, the plates washed, and then anti-IgG or anti-IgM horseradish peroxidase conjugates are added. Substrate is then added and this may be ABTS 2,2′-azino-di-(3-ethyl benzthiazoline sulfonic acid) or OPD (orthophenyldiamine) in citric acid phosphate buffer pH 4.0. The color may be read by eye or preferably by reader and is then expressed in absorbence units. Care must be taken to establish the level of negative reaction and this may be done by examining a large number of normal sera. Positive reactions are then taken as those in excess of this predetermined negative value. A better method is to determine a positive/negative ratio taken from the readings with negative and positive sera put up and read at the same time as the tests. If this ratio is satisfactory the batch of tests is accepted

Table 1
DURATION OF THE ANTIBODY RESPONSE

Days after onset of symptoms	Complement fixation[a]	Agglutination[b]	Bactericidal
6	4[c]	80	8
20	64	160	32
40	16	160	16
48	8	80	16
63	8	40	<2
84	8	20	<2

[a] Antigen was a sonicate of three heterologous strains.
[b] Homologous heated suspensions.
[c] Reciprocal titer.

Table 2
SEROLOGICAL RESPONSE FOLLOWING EXPERIMENTAL INFECTION

Days after infection	Complement fixation[a]	Agglutination[b]	ELISA IgG[a]	ELISA IgM
0	2	10	300[c]	184
5	2	10	288	152
10	16	160	518	362
46	16	160	488	229

[a] Antigen was a sonicate of six heterologous strains.
[b] Heated suspension of homologous strains.
[c] ELISA results expressed in absorbence units: >200 is regarded as positive.

Table 3
FRACTIONATION OF A CONVALESCENT SERUM

	Fractions[a]									
	1	2	3	4	5	6	7	8	9	10
Tube agglutination[b]	—	—	—	—	20	80	160	160	20	—
Complement fixation[c]	4	8	—	—	2	16	16	2	—	—
Bactericidal	64	64	32	—	—	—	8	—	—	—
Slide agglutination	++	+++	—	—	—	—	—	—	—	—
ELISA IgG[d]	34	26	15	66	332	561	453	128	46	34
ELISA IgM[d]	283	351	49	24	0	9	9	13	19	15
Hemagglutination[e]	—	—	—	—	—	—	—	—	—	—

[a] Fractions 1 to 3 contain IgM, 5 to 9 IgG.
[b] Heated homologous suspension.
[c] Sonicate of heterologous strains.
[d] Expressed in absorbance units: >200 is regarded as positive.
[e] Red cells sensitized with heated extract of homologous strain.

as valid and any serum giving a reading at least twice that of the negative control serum is regarded as positive. This method may also be used with the glycoprotein antigen. The ELISA techniques are suitable for serological surveys and react with high proportions of convalescent sera.

III. INDIVIDUAL PATIENT RESPONSE

The different techniques described detect different antibodies in the patient and the responses disappear at different rates (Table 1). In this patient, antibody was detectable 6 days after the onset of symptoms and the bactericidal antibody was the first to disappear. In experimental infection,[20] symptoms were experienced on days 4, 5, and 6 after infection and antibody appeared between day 5 and day 10 (Table 2). The antibody response was, therefore, well developed by 6 days after the onset of symptoms. Initially, this subject had some preexisting antibody detectable by three techniques but IgM antibody was not detectable by ELISA. The antibody response did not appear to be accelerated, IgM as well as IgG was produced in response to the infection which was detectable in all four tests.

When convalescent human serum is fractionated it is possible to show how the different techniques react in the fractions containing IgM or IgG (Table 3). These results were obtained with a serum collected 14 days after the onset of symptoms. Tube agglutination is predominantly with the IgG fractions, whereas slide agglutination involves only IgM. The ELISA technique can be slanted to either (or both) immunoglobulin classes by choice of conjugate. This option would also be available for immunofluorescence, not tested in this serum. The CFT reacts with both IgG and IgM, whereas the bactericidal antibody is mainly, but not exclusively, IgM. Passive hemagglutination has not been found to be useful for detecting convalescent antibody, although the method is very satisfactory for serotyping.

IV. OUTBREAKS

The study of outbreaks of *C. jejuni* infection which are point-source outbreaks and where the individuals involved are infected with the same organism in terms of serotype and virulence, enables observations to be made not only on the prevalence of symptoms but also on the variation, extent, and rapidity of development of the antibody response together with rates of infection. The variable that remains in these circumstances is the dose of infecting organisms taken in by the individuals.

In a milk-borne outbreak in a village community,[21] symptoms occurred over a 4 to 5 day period following what was thought to be contamination of the milk at a single milking. In the herd of 85 cows that provided the milk for the villagers, 7 were fecal excreters of *C. jejuni* and 2 of these were excreting the same serotype that was isolated from the villagers. Serum collected from 21 patients all contained antibody, but sera from 10 individuals without symptoms and who had an alternative milk supply had no antibody detectable by any test. Sera were collected 10 to 12 days after the onset of symptoms. All 21 sera had complement fixing, agglutinating and ELISA IgG antibodies, 71% had bactericidal antibody, and 52% had ELISA IgM antibody. An abbreviated list of results is shown in Table 4. It is clear that the extent of the responses detected by each technique are unrelated and that the tests are therefore detecting different antibodies. There is a tendency for sera with high bactericidal titers also to be positive in the IgM ELISA test, but this is not invariably so. Also shown is the variation of response between individual patients. From observations of this and other outbreaks it is clear that a proportion of patients may not react in some tests while showing adequate antibody response detectable by other methods.

When there was the opportunity to examine serum from those at risk of infection and who did or did not develop gastrointestinal symptoms, serological tests enabled further observations of interest to be made. In another point-source outbreak of *C. jejuni* infection among students,[14] sera were collected from 131 individuals, 14 of whom had gastrointestinal symptoms. Observation of the antibody responses in these students (Table 5) again shows the variability of the response, but also that four individuals were positive in all three tests but had no symptoms, and that four more had both agglutinins and complement-fixing

Table 4
ANTIBODY RESPONSES IN SOME PATIENTS IN A MILK-BORNE OUTBREAK OF *C. JEJUNI* INFECTION

Bactericidal	Tube agglutination[a]	Complement fixation[b]	ELISA IgG	ELISA IgM
<2	1280	8	447[c]	197
<2	2560	8	400	158
<2	1280	8	458	110
4	2560	64	444	113
8	2560	4	378	299
16	1280	8	274	209
16	320	8	309	124
32	2560	64	413	553
128	2560	64	312	363
2048	1280	16	444	113
2048	2560	128	526	591

[a] Heated homologous suspension.
[b] Sonicate of heterologous strains.
[c] ELISA results in absorbence units: >200 regarded as positive.

Table 5
MILK-BORNE OUTBREAK OF *C. JEJUNI* INFECTION

Complement fixation[a]	Tube agglutination[b]	Bactericidal	Symptoms
32	160	256	+
16	40	32	+
8	20	256	+
16	320	256	+
8	320	256	+
4	40	64	
32	80	256	+
4	20	128	+
2	40	32	+
16	<10	256	+
4	320	256	+
4	20	256	+
4	320	256	+
8	320	256	+
32	320	256	
16	640	128	
8	10	256	
2	40	256	
32	640	<2	
8	160	<2	
4	160	<2	
4	40	<2	
32	<10	<2	
4	<10	<2	
4	<10	<2	
4	<10	<2	

Note: Results of tests on 26 sera that contained antibody; 105 sera were negative in the 3 tests.

[a] Sonicate of heterologous strains.
[b] Heated homologous suspension.

antibody and so were almost certainly infected, but again had no symptoms. Individuals with only low titers of complement-fixing antibody, or the 105 with no detectable antibody, were unlikely to have been infected recently and certainly had no relevant gastrointestinal symptoms. So of the 131 at risk, 23 (17%) had serological evidence of infection and 14 (11%) had symptoms. To put it another way, of the 23 infected, 9 (39%) did not have symptoms. Observations on several other outbreaks have led us to suppose that between 20 and 40% of individuals infected with *C. jejuni* may not develop symptoms.

In outbreaks where more than one serotype has caused infection some individuals develop serotype-specific antibody to one of the infecting strains only, whereas other individuals have antibody to more than one of the relevant serotypes. This finding suggests the possibility of simultaneous infection with more than one serotype. It has been observed that when more than one colony is picked from primary culture plates for serotyping, sometimes more than one serotype may be isolated from a patient.[22]

V. CONCLUSION

Following *C. jejuni* infection, antibodies are regularly produced and may be detected by a variety of techniques. Human antibody response to infection with this organism shows some variation and seems to bear no clear relationship to the presence of gastrointestinal symptoms except that those who have clinical illness almost invariably develop antibody detectable by some method. Those techniques that use broadly reacting antigens such as sonicates, glycoprotein extracts, or indirect immunofluorescence are most regularly positive and the choice of technique will often depend on the laboratory facilities available and, for population studies, the ease with which large numbers of sera can be examined.

REFERENCES

1. **Butzler, J. P. and Skirrow, M. B.,** Campylobacter enteritis, *Clin. Gastroenterol.*, 8, 737, 1979.
2. **Jones, D. M., Eldridge, J., and Dale, B.,** Serological response to *Campylobacter jejuni/coli* infection, *J. Clin. Pathol.*, 33, 767, 1980.
3. **Karmali, M. A. and Fleming, P. C.,** Campylobacter enteritis in children, *J. Paediatr.*, 94, 527, 1979.
4. **Jones, D. M., Robinson, D. A., and Eldridge, J.,** Sero-epidemiological studies of *C. jejuni* infection, in *Campylobacter: Epidemiology, Pathogenesis, and Biochemistry,* Newell, D. G., Ed., MTP Press, Lancaster, U.K., 1981, 209.
5. **Blaser, M. J., Berkowitz, I. D., LaForce, F. M., Cravens, J., Reller, L. B., and Wang, W. L.,** Campylobacter enteritis: clinical and epidemiological features, *Ann. Intern. Med.*, 91, 179, 1979.
6. **Svedhem, A., Gunnarsson, H., and Kaijser, B.,** Serological diagnosis of *Campylobacter jejuni* infections by using the enzyme-linked immunosorbent assay principle, in *Campylobacter: Epidemiology, Pathogenesis, and Biochemistry,* Newell, D. G., Ed., MTP Press, Lancaster, U.K., 1981, 118.
7. **Abbott, J. D., Dale, B., Eldridge, J., Jones, D. M., and Sutcliffe, E. M.,** Serotyping of *Campylobacter jejuni/coli, J. Clin. Pathol.*, 33, 762, 1980.
8. **Kosunen, T. U., Pitkänen, T., Pettersson, T., and Pönkä, A.,** Clinical and serological studies in patients with *Campylobacter fetus* ss *jejuni* infection. II. Serological findings, *J. Infect.*, 9, 279, 1981.
9. **Watson, K. C. Kerr, E. J. C., and McFadzean, S. M.,** Serology of human campylobacter infections, *J. Infect.*, 1, 151, 1979.
10. **Watson, K. C. and Kerr, E. J. C.,** Comparison of agglutination, complement fixation and immunofluorescence tests in *Campylobacter jejuni* infections, *J. Hyg. (Cambridge)*, 88, 165, 1982.
11. **Lior, H., Woodward, D. L., Edgar, J. A., Laroche, L. J., and Gill, P.,** Serotyping of *Campylobacter jejuni* by slide agglutination based on heat labile factors, *J. Clin. Microbiol.*, 15, 761, 1982.
12. **Butzler, J. P.,** Related vibrios in Africa, *Lancet*, 2, 858, 1973.
13. **Reid, T. M. S. and Porter, I. A.,** Serological epidemiology of campylobacter infection, in *Campylobacter: Epidemiology, Pathogenesis, and Biochemistry,* Newell, D. G., Ed., MTP Press, Lancaster, U.K., 1981, 293.

14. **Jones, D. M., Robinson, D. A., and Eldridge, J.**, Serological studies in two outbreaks of Campylobacter infection, *J. Hyg. (Cambridge)*, 87, 163, 1981.
15. **McCartney, A. and Ross, C. A. C.**, Serological studies of Campylobacter infection by complement fixation, *Commun. Dis. Scotland*, No. 51, 1981.
16. **Jones, D. M. and Robinson, D. A.**, Occupational exposure to *Campylobacter jejuni*, *Lancet*, 1, 440, 1981.
17. **Blaser, M., Cravens, J., Powers, B. W., and Wang, W. L.**, Campylobacter enteritis associated with canine infection, *Lancet*, 2, 979, 1978.
18. **Penner, J. L. and Hennessy, J. N.**, Passive haemagglutination technique for serotyping *Campylobacter fetus* subsp. *jejuni* on the basis of soluble heat stable antigens, *J. Clin. Microbiol.*, 12, 732, 1980.
19. **McCoy, E. C., Doyle, D., Burda, K. I., Corbeil, L. B., and Winter, A. J.**, Superficial antigens of *Campylobacter (vibrio) fetus:* characterisation of an anti-phagocytic component, *Infect. Immunol.*, 11, 517, 1975.
20. **Robinson, D. A.**, Infective dose of *Campylobacter jejuni* in milk, *Br. Med. J.*, 282, 1584, 1981.
21. **Robinson, D. A., Edgar, W. J., Gibson, G. L., Matchett, A. A., and Robertson, L.**, Campylobacter enteritis associated with consumption of unpasteurised milk, *Br. Med. J.*, 1, 1171, 1979.
22. **Lior, H.**, personal communication.

Chapter 10

DIAGNOSTIC SEROLOGY FOR *CAMPYLOBACTER JEJUNI* INFECTIONS USING THE DIG-ELISA PRINCIPLE AND A COMPARISON WITH A COMPLEMENT FIXATION TEST

Å. Svedhem, S. Lauwers, and B. Kaijser

TABLE OF CONTENTS

I. Introduction 106

II. Different Techniques for Antibody Determination against *Campylobacter* 106

III. DIG-ELISA for Diagnostic Serology of *Campylobacter* Infections 106
A. Comparison of DIG-ELISA and C.F. 109

IV. Conclusion 110

References 111

I. INTRODUCTION

Until recently the importance of *Campylobacter jejuni* (C.j.) in the etiology of acute diarrhea in humans was not recognized.[2,3,18,21] One reason for this might be the unusual conditions necessary for optimal growth — microaerophilic milieu and 42°C. Although the bacteria are relatively easy to grow under these conditions, culture-negative cases of campylobacter diarrhea do appear. In such cases, and in such not uncommon cases of nonenterocolitis appearance as, for example, reactive arthritis,[16] diagnostic serology should be useful.

Details of the incidence of C.j. in different social environments as well as of the modes of bacterial transmission are still largely unknown.[3] The main source of infection, however, is contaminated, unheated food — especially chicken and raw milk — though water, too, has been described as a vehicle.[12,14,17,19,22] In developing countries C.j. enteritis is particularly prevalent in the very young, while in developed countries it is primarily adults that become infected.[1] Antibody determination could be useful in epidemiological investigation.

Though reinfections have been described — generally with different serotypes[7,14] — not much is known at present about immunity against C.j. infection. To improve our knowledge in this respect, techniques in diagnostic serology are needed.

II. DIFFERENT TECHNIQUES FOR ANTIBODY DETERMINATION AGAINST *CAMPYLOBACTER*

Techniques like direct agglutination, indirect hemagglutination, immunofluorescence, immunodiffusion, complement fixation, and enzyme-linked immunosorbent assay (ELISA) have been used for C.j. antibody determination.[5,8,13,20] In most cases, whole bacteria or poorly defined or imperfectly purified antigen preparations have been used, contributing to the possibility of unspecific reactions. Furthermore, few of the techniques used hitherto permit determination of immunoglobulin class-specific antibody response.

In the typing of bacteria, antigens specific for different serotypes are necessary, but very little is known about C.j. antigens. Two different principles for serotyping C.j. have been published — one based on a thermolabile, and one based on a thermostable antigen.[9,10,15] There is great complexity in both systems, with up to at least 50 antigens involved, which renders the more serotype-specific antigens less suitable for diagnostic serology.

III. DIG-ELISA FOR DIAGNOSTIC SEROLOGY OF *CAMPYLOBACTER* INFECTIONS

A few years ago, McCoy et al. described a surface antigen that was possibly part of a microcapsule of *Campylobacter fetus*.[11] Using their technique, we recently prepared a corresponding antigen from C.j.[20] This antigen was found to be present in all tested strains, but the amount released from different strains could vary, as could the antigenicity of the various preparations. By cross-reaction tests using immunological techniques, we were able to show that a pool of the antigens from two different patient strains presented wide cross reactivity within the C.j. species. A study of the occurrence or development of C.j. antibodies in different groups of individuals was performed, using a modification of the diffusion-in-gel enzyme-linked immunosorbent assay (DIG-ELISA).[4,20] This technique is based on a solid-phase ELISA and immunodiffusion in gel. In the first step the antigen is adsorbed to polystyrene dishes. In the second step the test serum is allowed to diffuse from wells in an agarose gel applied to the coated surface. Third, the agarose is removed and a solution of enzyme-conjugated antihuman immunoglobulin added, and finally substrate in agarose is used to develop the reaction. The diameter of the circular colored reaction zones is correlated to the concentration of antibodies (Figure 1).

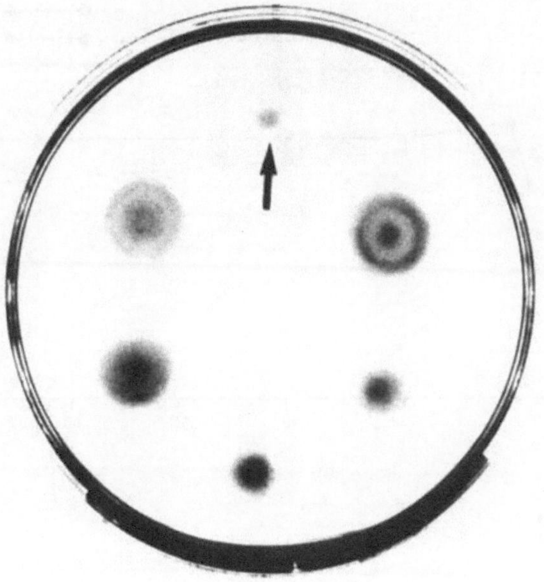

FIGURE 1. C.j. DIG-ELISA dishes with the circular color reactions developed. The arrow indicates the negative control serum.

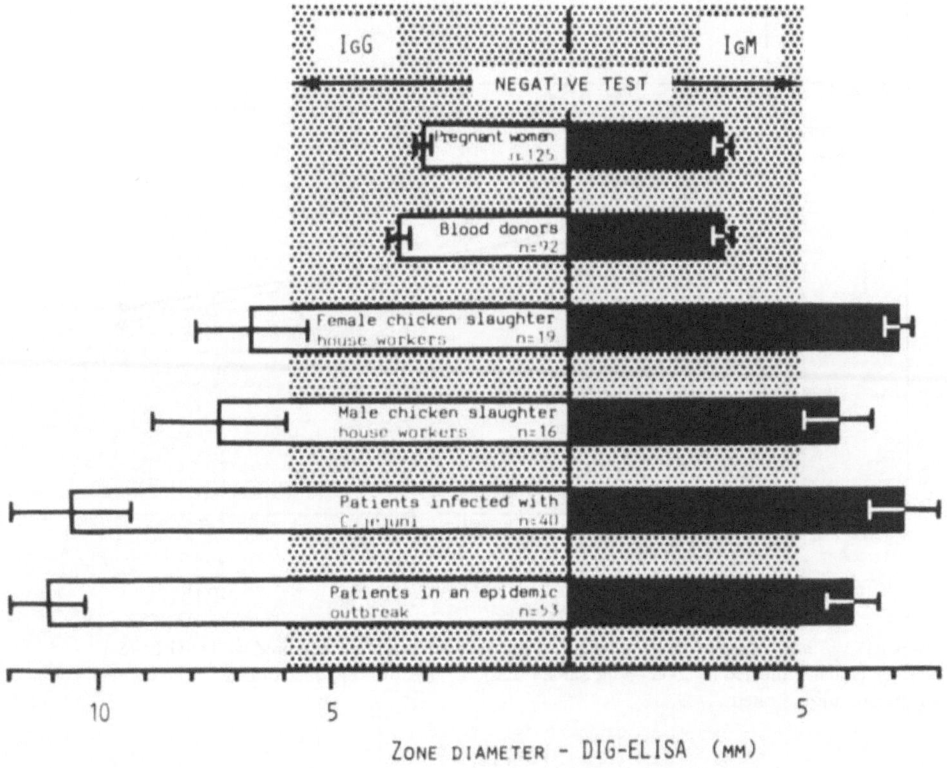

FIGURE 2. Monitoring of C.j. antibodies in different groups of individuals by DIG-ELISA. Means and ± SEM are indicated (p = 0.05).

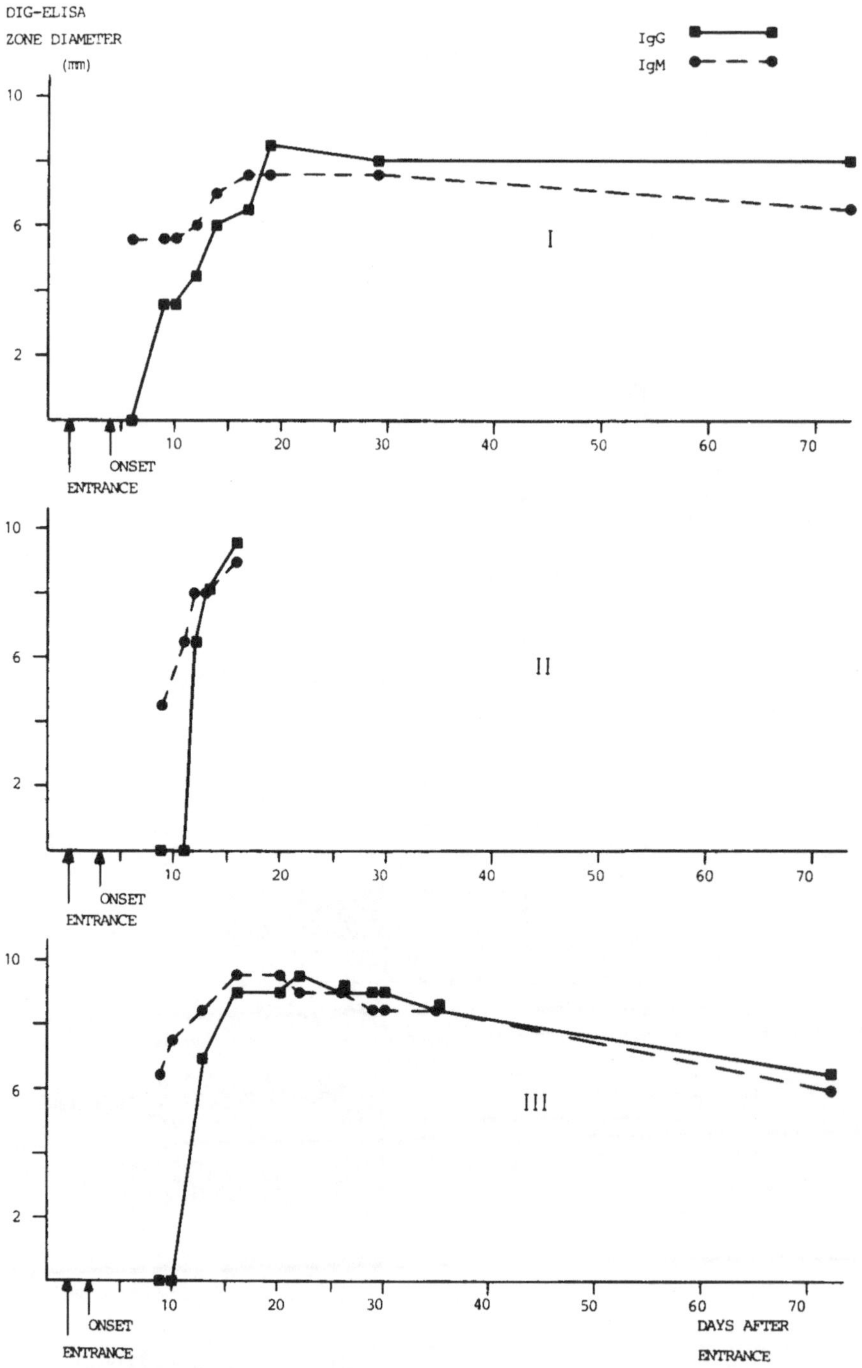

FIGURE 3. The course of IgG and IgM C.j. antibody formation as measured with DIG-ELISA in three patients infected by food on the same occasion. Times for entrance of bacteria and onset of disease are indicated.

With the C.j. DIG-ELISA, no antibody values, or only very low ones, were registered in most healthy people examined in Sweden[20] (Figure 2). In some cases, however, raised values were seen in these individuals, but this could usually be correlated to recent travel in countries where diarrheal diseases are more common. Not all the individuals with raised values had had overt diarrhea, indicating that — even where clinical symptoms are lacking — contact with C.j. may be enough to initiate antibody production. This might well be an anamnestic booster reaction to earlier contact with the bacteria. C.j. is almost regularly present in the gut of chickens.[19,22] The finding of increased levels of C.j. antibodies in occupationally exposed healthy chicken slaughterhouse workers lends further support to the idea that mere contact may be sufficient for antibody induction.[6,20] This was strikingly apparent in the levels of IgM antibodies in female workers, as compared to those in their male colleagues who were considerably less involved in actually handling the animals.[20]

C.j. IgG and IgM antibodies made their appearance after specific periods following the entrance of the bacteria into the body, as illustrated by sequences of serum samples from three patients in Figure 3. Thus IgM antibodies were detected on day 6 after entrance, while no IgG antibodies were to be seen before day 12. Both IgG and IgM reached their maximum levels around day 20 and then slowly declined. In larger patient materials we have shown that IgM antibodies remain detectable for at least 3 months and IgG from 6 to 12 months, depending on the initial strength of antibody response. This explains why raised values may sometimes be noted in healthy individuals. The best way of using DIG-ELISA as a diagnostic tool is to obtain two serum samples, the first as soon as possible after onset of the disease, and the second 2 to 3 weeks later. However, one serum sample is often sufficient to establish the diagnosis, if the IgG/IgM ratio is considered together with the clinical course.

In a special study of a waterborne outbreak of C.j. diarrhea, an antibody response was registered in all cases of diarrhea when serum samples were obtained.[12,20] In that study, as well as in other patients with diarrhea, we were also able to show C.j. antibody response where C.j. was strongly suspected as being the cause from clinical appearance or for epidemiological reasons, but where feces cultures were negative.

A. Comparison of DIG-ELISA and C.F.

In 1981, Mosimann et al.[13] (Institut Virion AG, Rüschlikon, Switzerland) described a complement-fixation test for serodiagnosis of infections caused by C.j. and *C. fetus* subsp. *fetus*. The two antigen preparations are now commercially available. Before the comparison of DIG-ELISA and C.F. test was started, the efficacy of the C.F. antigen was investigated, using 33 hyperimmune rabbit antisera, prepared against C.j. reference strains corresponding to 33 different Brussels serotypes.[9] All reacted with the C.j. antigen, in titers varying from 1/40 to 1/640. Normal rabbit sera were negative.

For the testing of human sera, the manufacturer's technical instructions were followed without modification. Control of the specificity of the C.j. antigen was satisfactory. Indeed, patients with C.j. enteritis developed only complement-fixing antibodies to the C.j. antigen, and a few patients with *C. fetus* subsp. *fetus* infections had only a positive titer in the *C. fetus* subsp. *fetus* C.F. test.[13]

In healthy Belgian blood donors (n = 139), C.F. antibodies against C.j. were found in 8% and against *C. fetus* subsp. *fetus* in less than 4% — which is in accordance with the Mosimann et al. observations.[13]

In our interpretation of the C.F. test, a titer of 1/10 or greater was considered positive, since the reproducibility of even such a low titer was very good.

For comparison of DIG-ELISA and C.F. the following were analyzed: 3 sera (S_I-S_{III}) from each of 21 patients involved in a Swedish waterborne outbreak of C.j.;[12,20] 10 Swedish reference sera; 8 sera from healthy pregnant women; 7 sera from healthy workers at a chicken slaughterhouse; 10 sera from Swedish patients with C.j. enteritis; and 31 sera from Belgian patients with a positive C.j. stool culture (Table 1).

Table 1
DIAGNOSTIC SEROLOGY OF *CAMPYLOBACTER JEJUNI* ANTIBODIES—COMPARISON BETWEEN DIG-ELISA AND C.F.

Group	No. of individuals investigated	No. of individuals with positive test for C.j. serum antibodies		
		DIG-ELISA		
		IgG	IgM	C.F.
Waterborne outbreak				
Serum I (2—3 weeks)[a]	21	21	17	16
Serum II (3.5 months)[a]	21	21	15	8
Serum III (6.5 months)[a]	21	14	9	5
Swedish reference sera	10	3	3	1
Healthy Swedish pregnant women	8	0	0	0
Healthy Swedish chicken slaughterhouse workers	7	5	7	5
Swedish patients with C.j. enteritis	10	10	9	7
Belgian children with C.j. in stool culture	31	16	18	16

[a] Time between onset of disease and collection of serum samples.

With the DIG-ELISA IgG, antibodies were detected for 3.5 months in all 21 patients from the waterborne outbreak, and for 6.5 months in 14 of them (Table 1). The C.F., on the other hand, failed to detect antibodies in all three sera from five patients. In most cases with detectable C.F. antibodies, titers between 1/30 and 1/60 were found. C.F. antibodies were still present in only 5 patients 6.5 months after the acute infection, whereas the DIG-ELISA still detected IgG antibodies in 14 patients and even IgM antibodies in 9 patients.

Ten Swedish reference sera from blood donors were selected because of different antibody activities as measured with DIG-ELISA. C.F. failed to reveal antibody activity in two of the three positive sera (Table 1). Sera from healthy pregnant women were negative by both methods (Table 1).

The sera from healthy chicken slaughterhouse workers gave similar results with both techniques (Table 1).

Of ten serum samples from Swedish C.j. enteritis patients tested with DIG-ELISA, a positive antibody response in both immunoglobulin classes was shown by nine, and in IgG by the tenth. Low titers of 1/10 were found in three patients in the C.F. Titers in the remaining four patients were 1/10, 1/30, 1/40, and 1/120 (Table 1).

In sera from Belgian children with a positive C.j. stool culture, no antibodies were detectable by either method in ten patients (Table 1). Two of these patients were symptomatic C.j. carriers; the others were very young infants. In 13 patients, antibodies were detectable by both methods. In eight patients only one test was positive. Five patients with a borderline DIG-ELISA IgM had a negative C.F. result, and in three patients with detectable C.F. antibodies the DIG-ELISA was negative.

IV. CONCLUSION

The overall correlation between DIG-ELISA and C.F. was good. More C.j. infections were missed by C.F. than by DIG-ELISA. This might be due to a suboptimal selection of strains used for the antigen preparation, and the addition of a few more representative strains to the C.F.-antigen might improve the results. Furthermore, as noted above, the antibodies are detectable for a longer period with DIG-ELISA (even IgM) than with C.F.

Besides diagnostic and epidemiological studies, immunoglobulin class-specific antibody techniques are important in studies of C.j. immunity. The high incidence of diarrhea in chicken slaughterhouse workers, occurring only in the first weeks of their employment, and antibody levels that persist for years indicate an immunity to C.j.[20] This also bears out the observation that reinfections often seem to be caused by different serotypes.[7,14] Additional support for immunity is that in developing countries, where the bacteria are possibly more common in the food due to lower hygiene standards, C.j. enteritis is especially prevalent in children, in contrast to the situation in the developed countries.[1] In studies on immunity, information is needed on different immunoglobulin classes — not only in serum, but also in secretions (e.g., breast milk). Techniques like the C.j. DIG-ELISA would be very useful in this respect.

In the determination of antibodies in C.j. infections, the DIG-ELISA presented here is a specific, sensitive, and handy method using an antigen that is common to most of our analyzed strains of C.j.

DIG-ELISA has shown itself to be more sensitive and capable of detecting antibodies over longer periods of time and permits the separate determination of different immunoglobulin classes. The traditional, commercially available complement fixation test is cheap and easy to perform by diagnostic laboratories more familiar with C.F. reaction. Both C.F. and DIG-ELISA can be used for the diagnosis of campylobacter infection, but DIG-ELISA has to be preferred in epidemiological and immunity studies.

REFERENCES

1. **Blaser, M. J., Glass, R. I., Hut, M. I., Stoll, B., Kibriya, G. M., and Alim, A. R. M. A.,** Isolation of *Campylobacter fetus* subsp. *jejuni* from Bangladeshi children, *J. Clin. Microbiol.*, 12, 744, 1980.
2. **Butzler, J. P., Dekeyser, P., Detrain, M., and Dehaen, F.,** Related *Vibrio* in stools, *J. Pediatr.*, 82, 493, 1973.
3. **Butzler, J. P. and Skirrow, M. B.,** *Campylobacter enteritis, Clin. Gastroenterol.*, 8, 737, 1979.
4. **Elwing, H. and Nygren, H.,** Diffusion-in-gel enzyme-linked immunosorbent assay (DIG-ELISA): A simple method for quantitation of class-specific antibodies, *J. Immunol. Methods*, 31, 101, 1979.
5. **Jones, D. M., Eldridge, J., and Dale, B.,** Serological response to *Campylobacter jejuni/coli* infection, *J. Clin. Pathol.*, 33, 767, 1980.
6. **Jones, D. M. and Robinson, D. A.,** Occupational exposure to *Campylobacter jejuni* infection, *Lancet*, 1, 440, 1981.
7. **Karmali, M. A., Kosoy, M., Newman, A., Tischler, M., and Penner, J. L.,** Reinfection with *Campylobacter jejuni*, *Lancet*, 2, 1104, 1981.
8. **Kosunen, T. U., Pettersson, I., and Pönkä, A.,** Clinical and serological studies in patients with *Campylobacter fetus* ssp. *jejuni* infection. II. Serological findings, *Infection*, 9, 279, 1981.
9. **Lauwers, S., Vlaes, L., and Butzler, J. P.,** *Campylobacter* serotyping and epidemiology, *Lancet*, 1, 158, 1981.
10. **Lior, H., Woodward, D. L., Edgar, J. A., Laroche, L. J., and Gill, P.,** Serotyping of *Campylobacter jejuni* by slide agglutination based on heat-labile antigenic factors, *J. Clin. Microbiol.*, 15, 761, 1982.
11. **McCoy, E. C., Doyle, D., Burda, K., Corbeil, L., and Winter, A. J.,** Superficial antigens of *Campylobacter (Vibrio) fetus:* Characterization of an antiphagocytic component, *Infect. Immun.*, 11, 517,
12. **Mentzing, L. O.,** Waterborne outbreaks of *Campylobacter enteritis in central Sweden, Lancet*, 2, 352, 1981.
13. **Mosimann, J., Jung, M., Schär, G., Bonifas, V., Heinzer, I., Brunner, S., Hermann, G., and Lambert, R. A.,** Serologische Diagnose menschlicher *Campylobacter*-Infektionen, *Schweiz Med. Wschr.*, 111, 846, 1981.

14. **Norkrans, G. and Svedhem, Å.,** Epidemiological aspects of *Campylobacter jejuni* enteritis, *J. Hyg.*, 89, 163, 1981.
15. **Penner, J. L. and Hennessy, J. N.,** Passive hemagglutination technique for serotyping *Campylobacter fetus* subsp. *jejuni* on the basis of soluble heat-stable antigens, *J. Clin. Microbiol.*, 12, 732, 1980.
16. **van de Putte, L. V. A., Berden, J. H. M., Boerbooms, A. M. T., Müller, W. H., Rasker, J. J., Reynvaan-Groendijk, A., and van der Linden, S. M. J. P.,** Reactive arthritis after *Campylobacter jejuni* enteritis, *J. Rheumatol.*, 7, 531, 1980.
17. **Robinson, D. A. and Jones, D. M.,** Milk-borne *Campylobacter* infection, *Br. Med. J.*, 1, 1374, 1981.
18. **Skirrow, M. B.,** *Campylobacter* enteritis: a "new" disease, *Br. Med. J.*, 2, 9, 1977.
19. **Smith, M. V. and Muldoon, P. J.,** *Campylobacter fetus* subspecies *jejuni (Vibrio fetus)* from commercially processed poultry, *Appl. Microbiol.*, 27, 995, 1974.
20. **Svedhem, Å., Gunnarsson, H., and Kaijser, B.,** Diffusion-ion-gel enzyme-linked immunosorbent assay for routine detection of IgG and IgM antibodies, *to Campylobacter jejuni, J. Infect. Dis.*, 148, 82, 1983.
21. **Svedhem, Å. and Kaijser, B.,** *Campylobacter fetus* subspecies *jejuni:* A common cause of diarrhoea in Sweden, *J. Infect. Dis.*, 142, 353, 1980.
22. **Svedhem, Å. and Kaijser, B.,** Isolation of *Campylobacter jejuni* from domestic animals and pets: probable origin of human infection, *J. Infect.*, 3, 37, 1981.

Chapter 11

EXPERIMENTAL STUDIES OF *CAMPYLOBACTER* ENTERITIS

D. G. Newell

TABLE OF CONTENTS

I. Introduction 114

II. Evidence for the Pathogenicity of *C. jejuni* 114

III. Animal Models of *Campylobacter* Enteritis 115
- A. Primates 115
- B. Cattle 116
- C. Sheep 116
- D. Pigs 116
- E. Dogs and Cats 117
- F. Small Mammals 117
- G. Birds 122

IV. Pathogenesis 122
- A. Attachment and Colonization 122
- B. Toxins 124
  - 1. Endotoxins 124
  - 2. Enterotoxins 124
  - 3. Cytotoxins 124
- C. Invasion 124
- D. Virulence Factors 125

V. Conclusion 128

References 129

I. INTRODUCTION

*Campylobacter jejuni** infection in man produces a wide spectrum of symptoms but is generally manifest as gastroenteritis beginning with a prodromal fever of 1 to 2 days duration, followed by severe abdominal pain and watery diarrhea with stools containing blood, leukocytes, and mucus. The disease is usually self-limiting with termination of diarrhea by day 7 of the infection although persistent or relapsing diarrhea is not infrequent.[1,2,3]

Enteric disease associated with campylobacter infections has been widely recognized as a veterinary problem for many years and the identification of cases of human infections with an apparent animal source have implicated domestic and wild animals as environmental reservoirs,[3] although there is no direct evidence that campylobacters of environmental origin are pathogenic to man.

Catalase-positive campylobacters other than *C. jejuni* have been implicated as causative agents in a number of other human and veterinary infections. In contrast to *C. jejuni, C. fetus fetus*** rarely causes diarrheal disease in man but is associated with systemic infection, as a bacteremia without localized infection, and frequently occurs in immunologically compromised patients.[2,5]

Although this review will not attempt to encompass all manifestations of campylobacteriosis much of the experimental work on campylobacter pathogenic mechanisms has been with *C. fetus fetus* and *C. fetus venerealis*. Where relevant these references will be included as an indication of the pathogenic characteristics of the genus.

II. EVIDENCE FOR THE PATHOGENICITY OF *C. JEJUNI*

To a limited extent *C. jejuni* can fulfill the required criteria of Koch postulates for differentiating an adventitious organism from a pathogen. The correlation between gastroenteritis and the presence of *C. jejuni* in feces is significant but not conclusive. Approximately 3 to 14% of patients with diarrhea who seek medical attention in developed countries have campylobacters isolated from their stool.[2,5] In these cases, *C. jejuni* is generally isolated as the sole presumptive pathogen although multiple infections have been observed with other potential enteric pathogens isolated such as *Salmonellae, Shigellae*, and enteropathogenic *E. coli*.[3,7] Isolation of *C. jejuni* from asymptomatic controls in developed countries is usually less than 2%;[5] however, asymptomatic carriers may be considerably more frequent in underdeveloped countries.[8-11]

In animal populations, however, the situation is less rigorously defined. In cattle the isolation rate from diarrheic calves was up to 40%,[12,13] but *C. jejuni* can also be isolated from the feces and intestines of normal calves.[13,14] The isolation of *C. jejuni* from apparently healthy animals has been similarly recorded in a number of species including primates, various birds, rodents, dogs, and pigs.[14-19]

In *C. fetus fetus* infections the debilitated patient is most at risk but previously healthy individuals are predominantly affected with campylobacter enteritis. The isolation of *C. jejuni* from the blood and other extragastrointestinal sites, as well as the stool, of these patients indicates the pathogenic rather than opportunistic nature of the organism.[3]

Similarly, the production of high titers of specific circulating IgG and IgM antibodies during and after the enteritis is indicative of a pathogenic organism.

The treatment of diarrheic patients with erythromycin results in a rapid remission of

* Although *C. coli* and *C. jejuni* are separate species, as defined by DNA hybridization studies,[4] for the purposes of this review they will be grouped together as *C. jejuni*. This is justified by their similarity in biochemical characteristics and antigenicity.

** The classification of species of the *Campylobacter* genus is that of the *Approved List of Bacterial Names*.[6]

symptoms and clearance of campylobacters from the stools. As *C. jejuni* is particularly sensitive to erythromycin in vitro, while most other enteric pathogens are resistant, this would suggest that *C. jejuni* is the causative agent of the diarrhea. However as the period of diarrhea and excretion of *C. jejuni* is variable and may be intermittent, it is difficult to assess the importance of this information.

C. jejuni has been isolated in pure culture, on artificial media, and used for experimental inoculations of man and animals in attempts to reproduce enteric disease, with varying degrees of success. In two instances of oral self-administration and one case of accidental inoculation with laboratory cultured strains of *C. jejuni* from human fecal isolates, disease symptoms similar to naturally acquired campylobacter enteritis were developed, accompanied by rising antibody titers specific against the inoculated strain and recovery of a serologically identical strain from the stool.[1,20,21]

The experimental oral or intragastric inoculation of susceptible animals generally results in colonization with rapid intragastric multiplication and excretion of the inoculated organism for variable time periods. Occasionally a systemic spread has been observed, but the bacteremia is transient. However, few animal inoculations result in a diarrheic disease of a similar nature to that of naturally acquired infections.

The evidence for the pathogenic nature of *C. jejuni* is, therefore, inconclusive. Differences in pathogenic mechanism or virulence of strains of *C. jejuni* could account for the inconsistencies in the results of experimental infections, but until a convenient animal model is developed the pathogenic nature of *C. jejuni* cannot be determined.

III. ANIMAL MODELS OF *CAMPYLOBACTER* ENTERITIS

Campylobacters have been implicated as the causative agent of swine dysentery, winter scours in cattle, bluecomb disease in turkeys, abortion in sheep, abortion and infertility in cattle, avian vibrionic hepatitis of chickens, and enteritis in many species including primates, dogs, and cats. For some of these diseases, such as winter scours in cattle, *C. jejuni* is no longer considered the pathologic agent. Convincing evidence of pathogenicity seems only to have been established for *C. jejuni*-induced abortion in sheep and *C. fetus venerealis*-induced infertility in cattle. Obtaining conclusive experimental evidence appears to be hampered by difficulties in reproducing disease symptoms during experimental animal infections. However, as the establishment of a suitable animal model of campylobacter enteritis is considered to be essential to an understanding of the pathogenesis of the disease, and is necessary for quantitative and qualitative assessments of virulence in epidemiologically important isolates, many attempts have been made to infect various animals with *C. jejuni*, some of which are reviewed below. Where possible, the pathological and clinical comparisons are made with the naturally acquired infections. The list is not intended to be comprehensive but is confined to those species with experimental or epidemiological relevance or where information is available on the pathogenicity of the disease.

A. Primates

Although *C. jejuni* can be isolated from apparently healthy primates,[15] it was also isolated from the digestive tract, gall bladder, and urinary bladder of monkeys with intermittent or persistent diarrhea.[22] However, few monkeys orally inoculated with *C. jejuni* develop diarrhea even though they are colonized with and excrete the organism for up to 6 weeks postinfection.[3,22,23] A transient bacteremia can occur with consistent colonization of the gall bladder and liver, but development of a positive specific serum antibody titer is rare and no histopathological lesions attributable to the campylobacter infection have been identified in the gastrointestinal tract, although foci of inflammatory cell infiltration in the liver are reported.[23] The experimental infection of monkeys is, however, protective against challenge by the same organism.[23]

B. Cattle

C. jejuni, C. fetus fetus, and *C. faecalis* can all be isolated from cattle with enteric disease and occasionally from multiple infections of the same site.[24] *C. jejuni* can also be isolated from the intestines and feces of cattle without diarrhea and is assumed to be part of the normal gut flora.[13,14,25]

In the naturally acquired infections the major area of involvement is the distal portion of the ileum, with histological changes including the stunting of villi, crypt abscess formation, and inflammatory response in the lamina propria with hyperplasia of submucosal lymphoid tissue.[18]

Experimental infection of calves with *C. jejuni* can produce fever, diarrhea, and sporadic dysentery[12,26] with a transient bacteremia 1 to 2 days post infection, and the appearance of similar histopathological changes to those observed in the naturally occurring enteritis.[13] *C. faecalis* also causes the production of soft feces with blood and mucus in experimentally infected calves, but the disease is milder than that caused by *C. jejuni.*[27]

Because of the epidemiological evidence implicating raw milk as a vehicle for transmission of human campylobacter enteritis, the question of campylobacter infections of cow udders is epidemiologically important. Lander and Gill[28] argue that the degree of fecal contamination required to produce an infectious dose in milk would result in milk spoilage, and they were able to induce mastitis by experimental infection of udders which seems to be consistent with the recovery of campylobacters from milk socks.[29]

C. Sheep

C. jejuni but not *C. fetus fetus* may be isolated from the feces of healthy sheep.[14] Both *C. jejuni* and *C. fetus fetus* have been implicated as the causative agents of ovine abortion but the infection, unlike that in cattle, is not venerally transmitted. Ewes orally infected with *C. jejuni* in the later stages of pregnancy may abort[14] and the placental tissue and expelled fetus are infected,[30] presumably via the systemic infection of the ewe originating from invasion of the intestinal tract.[12] Fetopathogenic effects may also be induced by the infection.[31]

The gastrointestinal tract of pregnant ewes and young lambs become colonized with *C. jejuni* after experimental inoculation and a bacteremia is developed, but there is no frank diarrhea although mucoid feces flecked with blood may be observed.[12,14,31]

Lamb intestinal loops inoculated with cultures of *C. jejuni* showed no distention, but an infiltration of leukocytes into the lumen and defoliation of epithelial cells was observed with accumulations of large amounts of mucus. Histologically necrosis of the villus tips, accumulations of polymorphonuclear cells, and some increased mitotic activity adjacent to the crypts were detected.[12]

D. Pigs

Swine dysentery is no longer believed to be due to infection with the *Vibrio coli (C. coli)* described by Doyle,[32] but *C. jejuni* is a member of the normal porcine intestinal flora, particularly of young pigs,[14] and has been isolated, in the absence of other potential enteric pathogens, from the feces of piglets with diarrhea and weaned pigs with colitis induced by drug treatment.[33] The major site of involvement in these naturally acquired infections is the small intestine with stunting of the villi, thickening of the terminal ileum, and infiltration of the lamina propria by polymorphonuclear cells and monocytes.[33]

Pigs may be experimentally colonized by *C. jejuni*[14] but suckling piglets from herds with endemic campylobacter infections exhibit only mild clinical signs of infection while colostrum-deprived piglets develop creamy mucoid diarrhea flecked with blood by 48 hr postinfection.[24,33] The organism can be recovered from lesions in the small intestine which are histologically similar to those seen in naturally acquired infections, and a specific serological response was observed in all infected piglets.[33]

In contrast, gnotobiotic piglets infected with human or porcine isolates of *C. jejuni* developed only a transient, moderate inflammation of the colonic mucosa 2 to 3 days postinfection, with an increased fluidity of intestinal contents accompanied by diarrhea at 5 to 6 days postinfection in the absence of histopathological changes.[34]

E. Dogs and Cats

The incidence of isolation of *C. jejuni* from dogs with and without diarrhea appears to be approximately the same.[35,36] Although cats and dogs are susceptible to campylobacter enteritis and can transmit the disease to humans[3] there seems to be some doubt as to the significance of *C. jejuni* carriage in healthy and diarrheic dogs.[25]

Campylobacter enteritis in dogs is characterized by vomiting with diarrhea and occasional dysentery[37] and cases of fatal septicemia have been recorded.[31] *C. jejuni* is isolated from the spleen, liver, gall bladder, and mesenteric lymph nodes as well as the gastrointestinal tract of infected dogs, but the main site affected appears to be the small intestine. The gross lesions observed include mild hyperemia, thickening of the jejunum and ileum, and enlarged mesenteric lymph nodes while the stunting of villi, inflammatory cell infiltration, and enlarged Peyer's patches were seen on histological investigations.[37]

Despite the obvious susceptibility of dogs and cats to campylobacter infections, experimental inoculation produced inconsistent results. Puppies and kittens orally inoculated with human isolates of *C. jejuni* did not develop significant diarrhea even though the isolate was sufficiently virulent to infect the experimenter,[21] but adult dogs infected with canine isolates developed watery diarrhea or mucoid feces within 7 days and histopathological changes compatible with those observed in natural infections.[37]

Gnotobiotic puppies, however, developed mild diarrhea within 72 hr of challenge with human and canine isolates of *C. jejuni*, but the histopathological changes were observed in the colon rather than the small intestine.[34,38]

F. Small Mammals

The experimental animal infections mentioned so far, although relevant, would be outside the facilities of most laboratories as well as being expensive and difficult to handle. Several smaller mammal and chicken models have been proposed which may provide more suitable animal models of campylobacter infections. Natural infections in small mammals do not appear to be frequent. Campylobacter species have been isolated from the nondiarrheic stools of rats, bank voles,[16] hamsters,[39] rabbits,[25] and ferrets although diarrhea could be induced by stress in these asymptomatic ferrets.[39]

Proliferative colitis in ferrets[40] and transmissible ileal hyperplasia in hamsters[41] are both enteric disease in which *C. jejuni* is implicated as the causative agent. Experimental infection of ferrets with a ferret isolate results in long-term colonization and induces diarrhea,[40] but hamsters appear to be more resistant to colonization than other small mammals[42] and when colonized did not develop diarrhea.[39,41] Neither experimentally colonized ferrets nor hamsters develop any histologically abnormal features indicative of disease.[39,41]

Short-term colonization of neonatal rats and rabbits can be established, but again without signs of enteritis.[42] *C. jejuni* is not normally excreted by mice nor is its excretion induced in tobramycin-treated mice.[43] Normal mice do not have specific anticampylobacter antibodies detectable with ELISA techniques[44] and there is no published evidence of naturally occurring murine campylobacter enteritis. Despite this most of the in vivo attempts to define campylobacter pathogenesis and establish a laboratory model of the disease have been with mice.

Weanling and adult mice are relatively refractory to colonization with *C. jejuni*,[42] even when treated with a gut motility-reducing drug.[43] Adult mice are, however, consistently colonized when the competitive gut microflora is eliminated using tobramycin,[43] but colo-

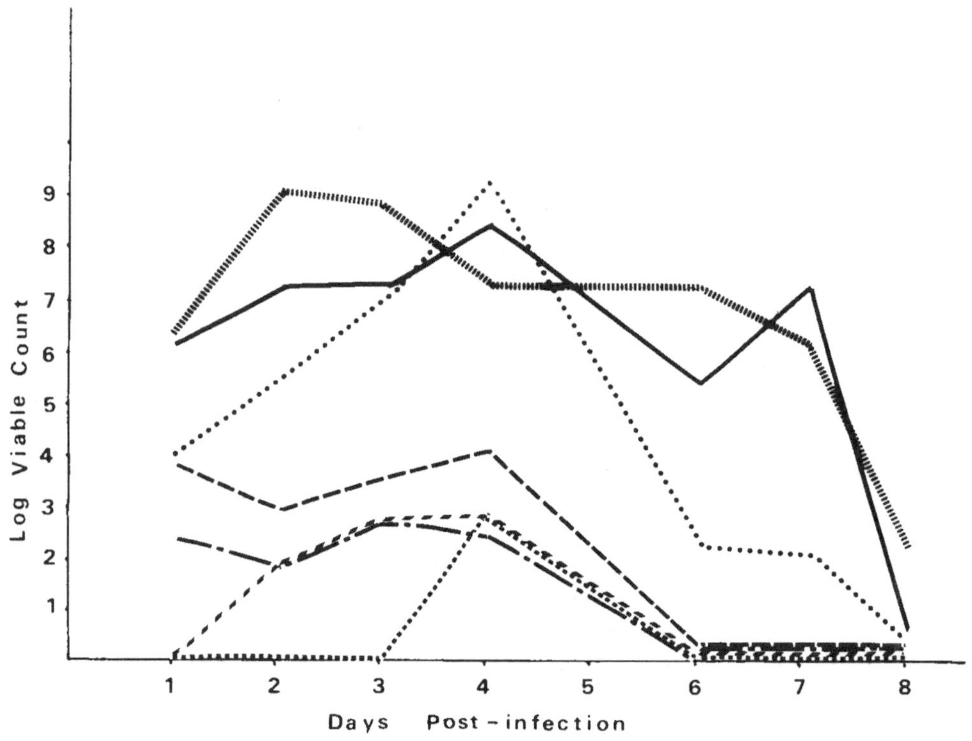

FIGURE 1. The recovery of viable *C. jejuni* (Strain 3571) from infant mice. Five-day old infant mice were inoculated intragastrically with 1 × 10^7 organisms in 0.05 mℓ of phosphate buffered saline. Duplicate mice were killed at various times postinoculation. The stomach, duodenum, and the upper, middle, and lower small intestine, cecum, and colon were homogenized in 1 mℓ of phosphate buffered saline and viable counts performed using Skirrow's medium. Organisms could be recovered from the colon up to 14 days postinoculation. The results are the mean of two experiments.

Stomach
Duodenum
Upper small intestine
Middle small intestine
Lower small intestine
Cecum
Colon

nized mice do not exhibit diarrhea[43,45] even when inoculated directly into the ileum and colon[46] although excretion of *C. jejuni* may persist for up to 48 days. The period of colonization is probably dependent on the susceptibility of the strain of mouse and the virulence of the *C. jejuni* isolate.

Consistent colonization with human isolates of *C. jejuni* is achievable using 5- to 6-day-old neonatal mice inoculated intragastrically.[42] Colonization persists for variable lengths of time, dependent on the *C. jejuni* isolate, but the gastrointestinal tract is usually cleared by 3 weeks postinoculation. In the infant mouse, colonization is greatest in the cecum and colon with up to 100-fold increase in recoverable organisms, but clearance from the stomach, duodenum, and the upper and middle small intestine is rapid (Figure 1). Overt enteric disease is not apparent in infected infants but there is occasionally evidence of runting and a low incidence of mortality.[42] The susceptibility of the mouse strain may be an important factor in the incidence of mortality.

Although adult mice are not naturally susceptible to infection with *C. jejuni*, some dams

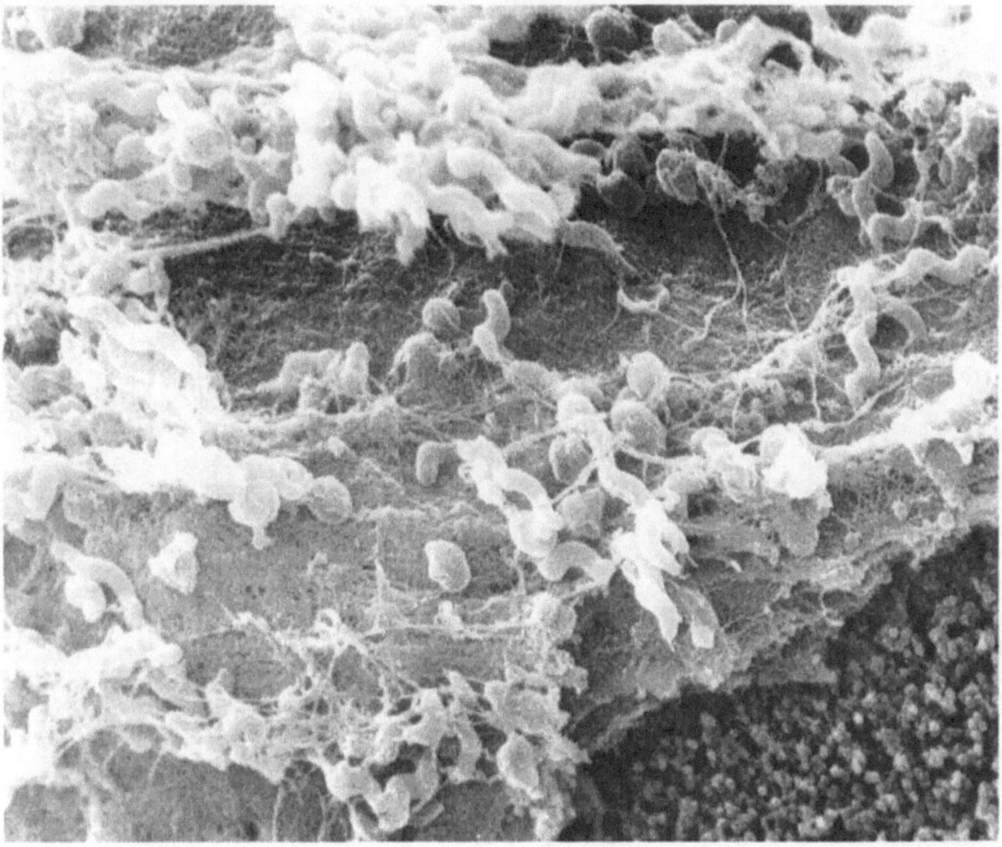

FIGURE 2. Scanning electron micrograph of the lower ileum of an infant mouse inoculated with *C. jejuni* (Strain ACI) with S-shaped organisms in, on, and beneath the mucus layer (× 135,000). (From Field, L. H., Underwood, J. L., Pope, L. M., and Berry, L. J., *Infection and Immunity*, 33, 884, 1981. With permission.)

of infected infants excrete soft stools which may contain blood. This excretion may continue for up to 29 weeks[42] and may reflect enhanced virulence of the organism associated with passage through the infant and transmitted via the coprophagic activity of the dam. Uninfected infants caged with infected infants also become colonized, which again suggests that animal passage may enhance virulence.[44] Field et al.[43] have shown that weanling mice, normally resistant to colonization, exhibit increased colonization when intragastrically inoculated with fecal suspensions of soft stools from a nursing dam.

A transient systemic infection in inoculated mice occurs at 1 to 4 hr post infection with recovery of organism from the spleen, kidneys, liver, lungs, and heart, but clearance from extragastrointestinal sites occurs by 12 hr postinfection.[43] Systemic spread seems to occur via invasion of the intestinal epithelium. The ultrastructural investigation of the cecum and colon of infected neonatal mice shows S-shaped organisms on, in, and below the mucus layer[42] (Figure 2) and closely associated with the microvilli of intestinal epithelial cells.[44] Similar organisms are frequently found deep within the crypts of the colon (Figure 3).

At 3 to 4 days post infection, organisms with the morphology of campylobacters are observed within the cytoplasm of cells of the epithelium and lamina propria of the cecum and colon (Figure 4). The cytoplasmic integrity of these invaded cells is frequently disrupted with swollen mitochondria and necrotic nuclei, although the epithelial layer appears intact and the only histological change observed was mild accumulations of polymorphonuclear cells in the lamina propria. Circulating antibody specific against *C. jejuni* has been detected in infected neonatal mice using the ELISA technique.[44]

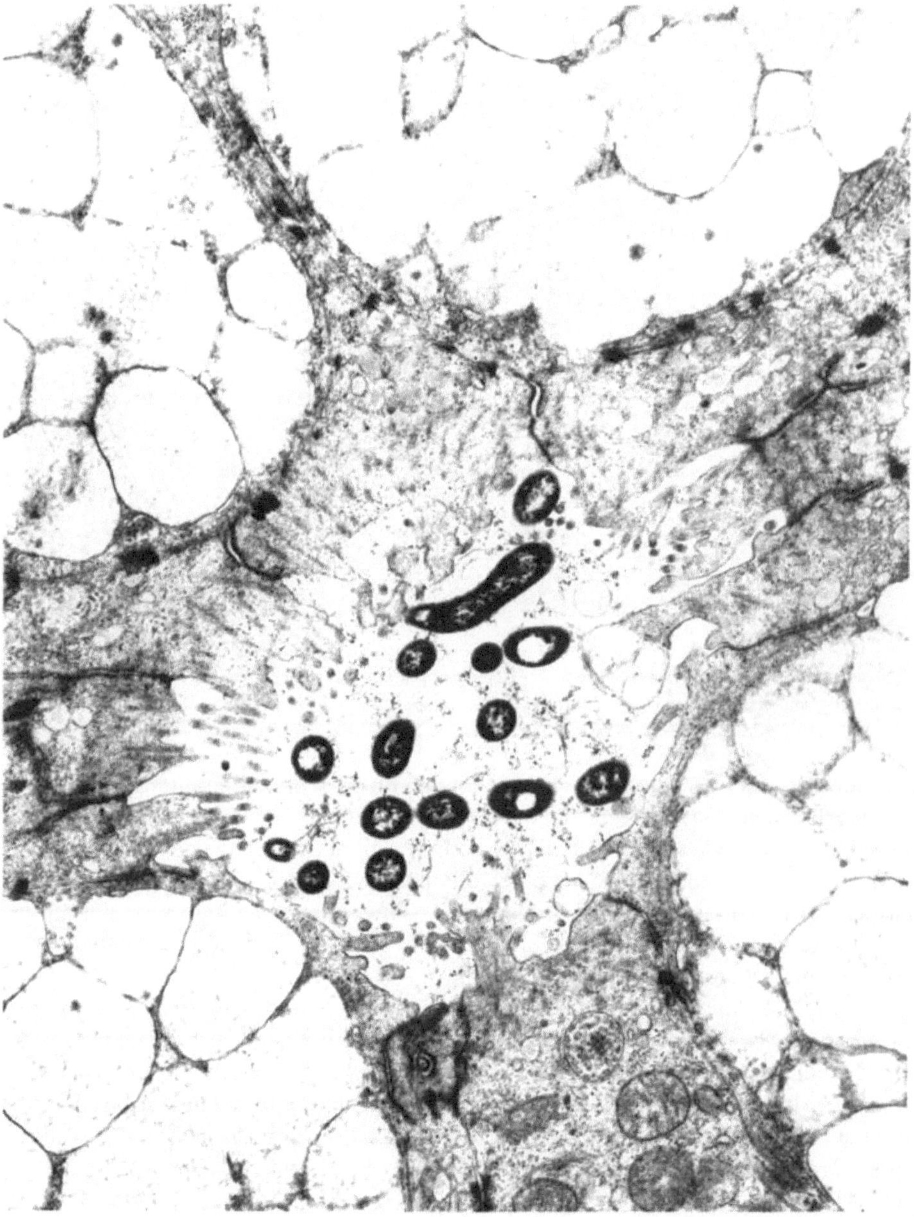

FIGURE 3. Association of S-shaped organisms with the microvilli of intestinal epithelial cells in crypts of the colon of an infant mouse infected with *C. jejuni*. (Strain 3571; × 17390.)

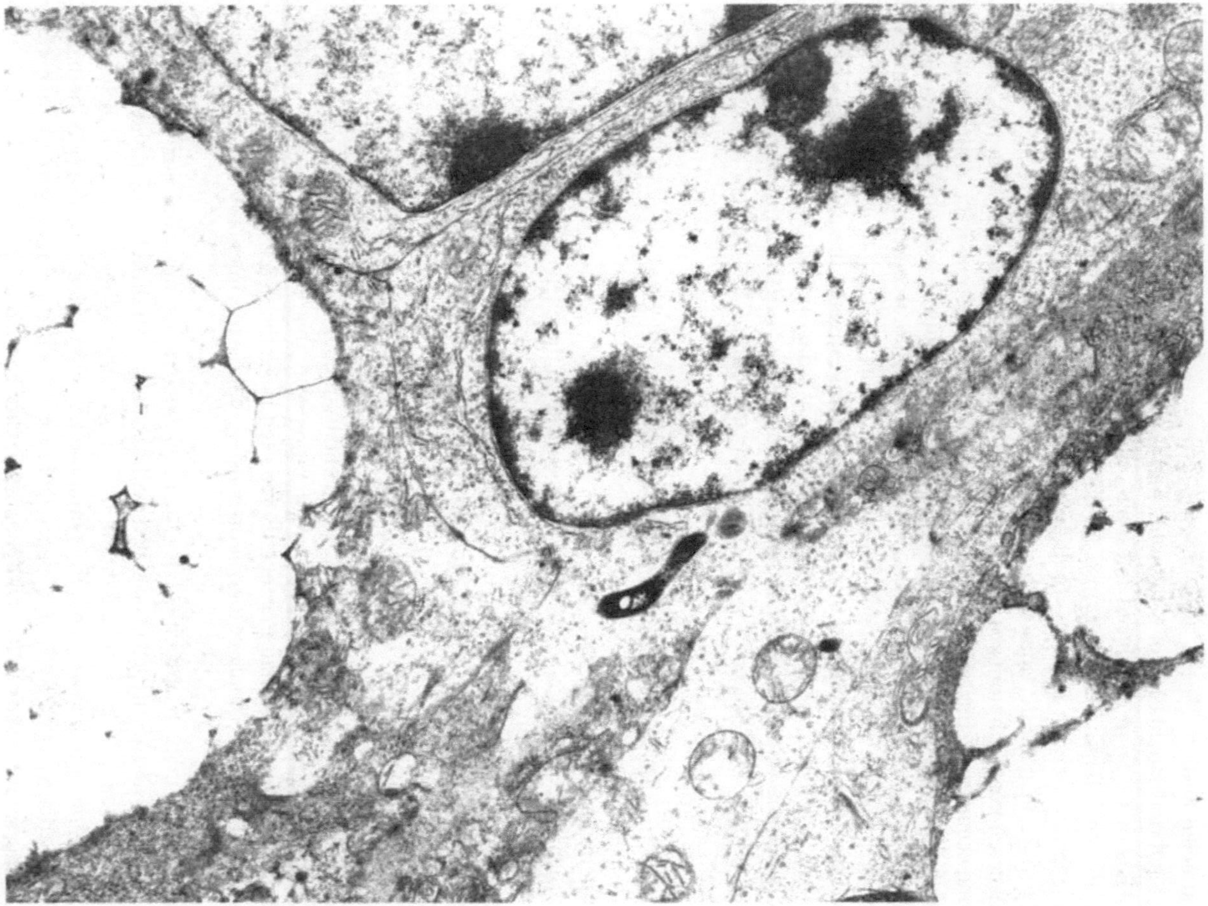

FIGURE 4. Transmission electron micrograph of S-shaped organism within the cytoplasm of an intestinal epithelial cell of the cecum of an infant mouse infected with *C. jejuni*. (Strain 3571; × 11,200.)

G. Birds

Intestinal carriage of *C. jejuni* by poultry and wild birds is common but not obligatory.[14,15,25,47] In the confined conditions of broiler houses all of the chickens of flocks with endemic *C. jejuni* may be infected, but there is no mortality or difference in average weights compared with unaffected flocks.[17]

Both bluecomb disease of turkeys and infectious hepatitis of chickens are associated with campylobacter infections, with *C. jejuni* isolated from the liver, small intestine, bile, and cecum of affected birds.[14]

Experimental oral inoculation of 8-day-old chicks[3] or 5-day-old gnotobiotic chicks[34] with human isolates of *C. jejuni* produced intestinal colonization, with isolation from the liver and blood and an inflammatory response in the cecal lamina propria and submucosa, but did not induce diarrhea. In contrast, orally inoculated 3-day-old chicks developed severe diarrhea 24 to 72 hr postinfection, with a mortality rate of 32%. As few as 90 organisms could produce clinical symptoms of enteritis in 90% of infected chickens.[48] In this model, the ileum and cecum were the most affected areas of the gastrointestinal tract, with organisms detected in the phagocytic cells of the lamina propria in the lower portion of the jejunum, the ileum, and the upper portion of the colon by immunofluorescence techniques, and S-shaped organisms were observed penetrating the epithelial layer by transmission electron microscopy. However, Manninen et al.[49] have recently investigated the pathogenicity of 15 *C. jejuni* and *C. coli* isolates from animals and man in 3-day-old chicks and observed no evidence of diarrhea or mortality. The absence of indications of infection in gnotobiotic chicks suggests that such differences in results are not due to the immune status of the animal.

IV. PATHOGENESIS

A. Attachment and Colonization

The establishment of an enteric infection is dependent on the bacterial (virulence) factors, instrumental in colonization, being more efficient than the host defense mechanisms directed at prevention of colonization.[50] Because enteropathogenic bacteria must resist intestinal motility and mucus flow, attachment to intestinal mucosa is a prerequisite for colonization and establishment of disease.

The recovery of vast numbers of organisms from diarrheic stools indicates that *C. jejuni* is highly successful at colonization and multiplication in the human gut. Up to 100-fold increases in recoverable organisms from experimentally inoculated infant mice confirm the efficiency of colonization and subsequent proliferation of *C. jejuni* in the intestinal milieu.

The site of colonization in man is generally reported to be the distal ileum, but the colon may also be involved and colonic colonization is usual in the experimentally infected animal models.

In cattle,[24] infant mice,[42] and chickens,[48] campylobacters are observed adjacent to the mucosal epithelium and in the crypts of the small and large intestine by light and electron microscopic techniques. The mechanism of this close association of *C. jejuni* with the intestinal epithelium is unknown. There is no morphological evidence of pili and the hemagglutination mediated by *C. jejuni* appears to be nonspecific on the basis of mannose resistance, temperature insensitivity, and chemical extraction data.[51]

Most strains of *C. jejuni* demonstrate poor spontaneous adhesion to epithelial cells or brush borders isolated from the jejunum of piglets[51] and to the epithelial cell lines HeLa and INT407.[49,52] However, attachment of *C. jejuni, C. fetus fetus,* and *C. faecalis* to isolated brush borders has been reported.[24]

Centrifugation of *C. jejuni* on to monolayers of HeLa and INT407 cells enhances attachment of sufficient strength to resist the shearing forces inherent in electron microscopic preparation techniques and is a preliminary to internalization of the organism by the cell. By scanning electron microscopy (Figure 5) the attached organisms are frequently associated with cell surface microvilli. Attachment involves a close association between the bipolar flagella and the cell surface (Figure 5), but aflagellate variants appear to attach to the cell surface just as efficiently as the flagellate wild type and their rate of internalization is similar.

The necessity for centrifugation of the organisms on to the cell surface probably reflects either a lack of initial attachment receptor sites on the cells or an insufficient collision rate between organisms and cells to allow the observation of spontaneous attachment by electron microscopic techniques.

The importance of motility for virulence in *C. jejuni* has yet to be assessed. Such studies require the three-dimensional structures of the intestine in vivo using aflagellate and non-motile mutants.

Once the intestinal epithelium has been colonized the mechanisms of bacterial diarrheal disease involve production of a toxin and/or invasion of epithelial tissue.[50]

B. Toxins

1. Endotoxins

Like other Gram-negative organisms, campylobacters possess an LPS with endotoxic properties.[2] The molecular weight of this LPS has been reported as 18,600 daltons[53] and 26,300 daltons.[54] Endotoxin-induced anaphylactic shock was initiated in cattle on intravenous inoculation of broth cultures and supernatants as early as 1962,[55] and an endotoxin-like activity associated with heat-killed *C. sputorum bubulus, C. coli* and *C. jejuni* was detected by the limulus gelation assay and Schwartzman skin reaction in rabbits.[56] The significance of endotoxin in the pathogenesis of campylobacter infections has yet to be established but the inflammatory involvement of major blood vessels, often with associated thrombosis, is not infrequent in *C. fetus* infections and may be the result of local endotoxin activity.[57]

2. Enterotoxins

Many strains of *C. jejuni* have been investigated for the production of enterotoxins using in vitro and in vivo techniques.

Gubina et al.[58] have described a heat-labile enterotoxin activity, detectable by changes in Y-1 cell morphology in cell-free dialysates of broth cultures of strains of *C. jejuni* and *C. fetus fetus*. However, neither heat-labile nor heat-stable enterotoxic activity have been found by other workers using Y-1, INT 407, MRC-5, and CHO cells,[44,58,59] the suckling mice test,[12,49,58,60,61] or intestinal loops of pigs,[62,49] lambs,[12] calves,[49,62] and rabbits.[61]

3. Cytotoxins

A cytotoxic activity in campylobacter culture supernatants could account for the confusion in the investigation of enterotoxic activity. Although Drake et al.[63] could not find any cytopathic effect of *C. jejuni* on human lung cells, cytotoxic activity of filtrates has been detected using CHO cells[61] and of cell-free sonicates detected by bovine kidney and Vero cells.[64] This latter cytotoxic factor has been partially characterized and is nondialyzable, precipitated by ammonium sulfate, excluded by G 100 Sephadex,® and inactivated at 60°C and by extremes of pH.

The significance of the cytotoxic activity in the pathogenicity of *C. jejuni* requires further investigation, but a cytotoxic enzyme may be involved in the destruction of cell membranes allowing penetration through the intestinal mucosa.[38,48]

C. Invasion

The clinical presentation of campylobacter enteritis suggests that invasion is the most likely mechanism by which *C. jejuni* causes diarrhea. In humans there is a prodromal fever,

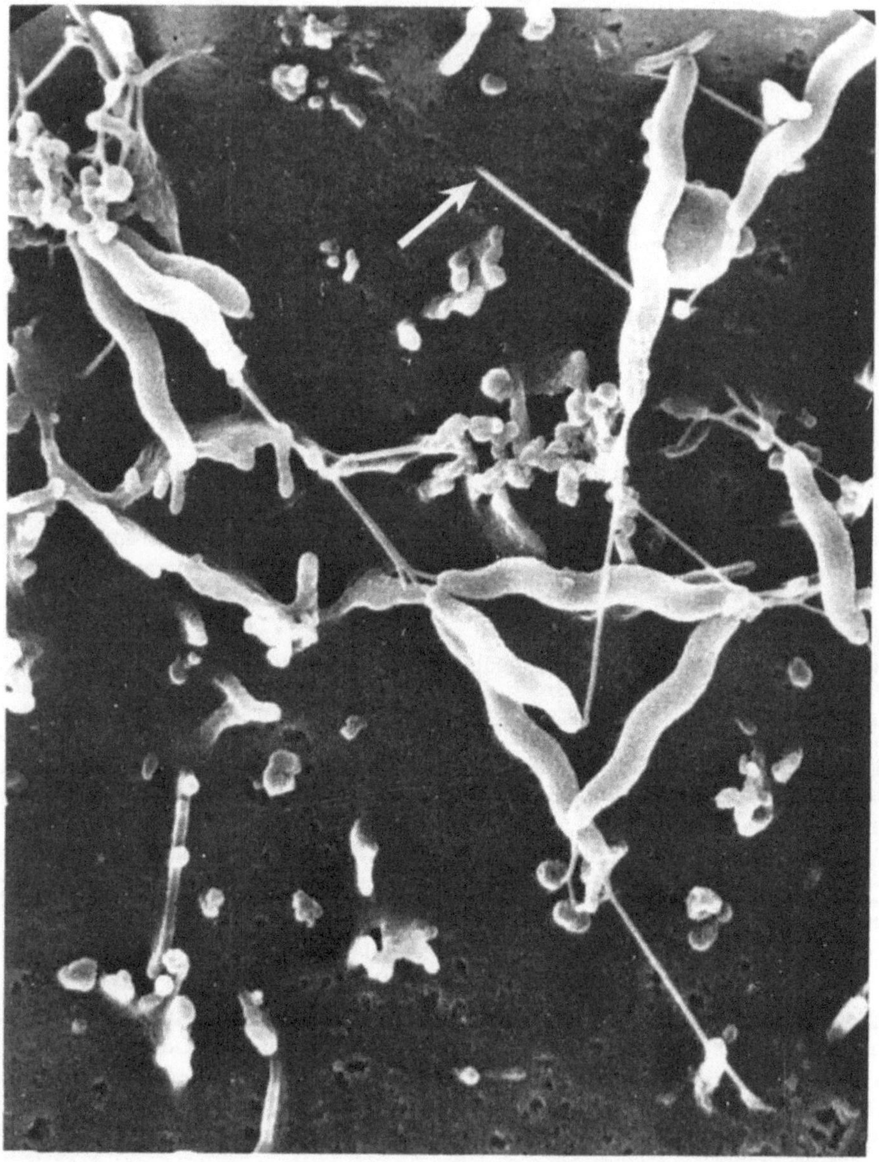

FIGURE 5. *C. jejuni* (Strain 68869) attached to the surface of HeLa cells after centrifugation for 30 min at 1100g. Note the association of organisms with the cell microvilli and the close proximity of the flagella to the cell surface (↑). (× 22,500.)

blood and leukocytes in the stool, frequent dissemination to extragastrointestinal sites, bacteremia, and hemorrhagic necrosis of the small intestine with numerous hyperplastic mesenteric lymph nodes.[65]

In the experimentally infected infant mouse,[43] calves,[26] and chickens[3] there is an early and transient systemic spread of organisms indicative of invasion.

Using indirect immunofluorescence, Ruiz-Palacios et al.[48] have demonstrated organisms within the phagocytic cells in the lamina propria of infected chickens. Additionally, organisms with the characteristic S-shaped morphology of campylobacters have been observed at the ultrastructural level, apparently penetrating the intestinal epithelium[48] and the cecal wall[3] of infected chickens, and within the cells of the intestinal epithelium and lamina propria of infected infant mice. [44]

The penetration of chicken embryonic chorioallantoic membrane by human isolates of *C. jejuni* provides further evidence of an invasive capacity.[66] Organisms are recovered from the embryonic hearts and livers of 11-day-old embryos but 17-day-old embryos are more resistant to invasion, which was considered to be related to the development of immunocompetence. Interestingly, the two strains of *C. jejuni* tested apparently possessed different invasive capacities.

Despite the in vivo evidence, neither the Sereny guinea pig conjunctival sac or the rabbit ileal loop tests indicate that *C. jejuni* has invasive potential.[49,58,60,61]

Cell cultures have been used in the past to investigate the invasive ability of several enteropathogenic organisms including strains of shigella, salmonella, and yersina. Butzler and Skirrow[3] showed that *C. jejuni* strains appeared to penetrate primary chicken embryo cell cultures, however they did not distinguish organisms contained in surface invaginations from organisms surrounded by membrane-bound vacuoles.

C. jejuni is rapidly internalized after centrifugation onto HeLa or INT 407 monolayers and completely internalized organisms can be distinguished from the surface-attached organisms by the use of thorium to label surface acidic glycoproteins (Figure 6). The process appears to involve a progressive movement of the organism into the cell from a polar position. The observation of elongated organisms embedded in the cell surface seems to confirm this mechanism of penetration (Figure 7). Although several organisms may be found in a vacuole there is no direct evidence of intracellular multiplication. Internalization requires metabolically active organisms (organisms prefixed in 0.1% glutaraldehyde[44] and heat-killed organisms[49] do not penetrate cells) and is concomitant with cell death. Cell death involves rounding-up of the cell and production of surface blebs, pits, and crevices, which is usually indicative of a cytotoxic phenomena.

Quantitative assessment of the invasion of HeLa cells by strains of campylobacter, using indirect fluorescence, indicates that differences in invasive capacity between strains can be determined although the number of organisms internalized without centrifugation is very low.[49]

D. Virulence Factors

A microcapsule has been identified on *C. fetus fetus* which prevents ingestion by macrophages in the absence of specific antiserum.[67,68] The presence of the capsular material has been identified immunologically and ultrastructurally using a variant without capsular antigen. The in vivo significance of this capsule to the pathogenicity of the organism was not demonstrated, but the antiphagocytic activity is presumably critical in the establishment of systemic infections by this organism.

The evidence for similar capsular antigens on the surface of *C. jejuni* is sparse. There is no conventional ultrastructural evidence of a capsule, but Merrell et al.[46] identified a capsule-like material on the surface of *C. jejuni* by ruthenium red staining, and a glycoprotein material with capsule-like properties can be isolated from *C. jejuni* by saline extraction procedures.[69]

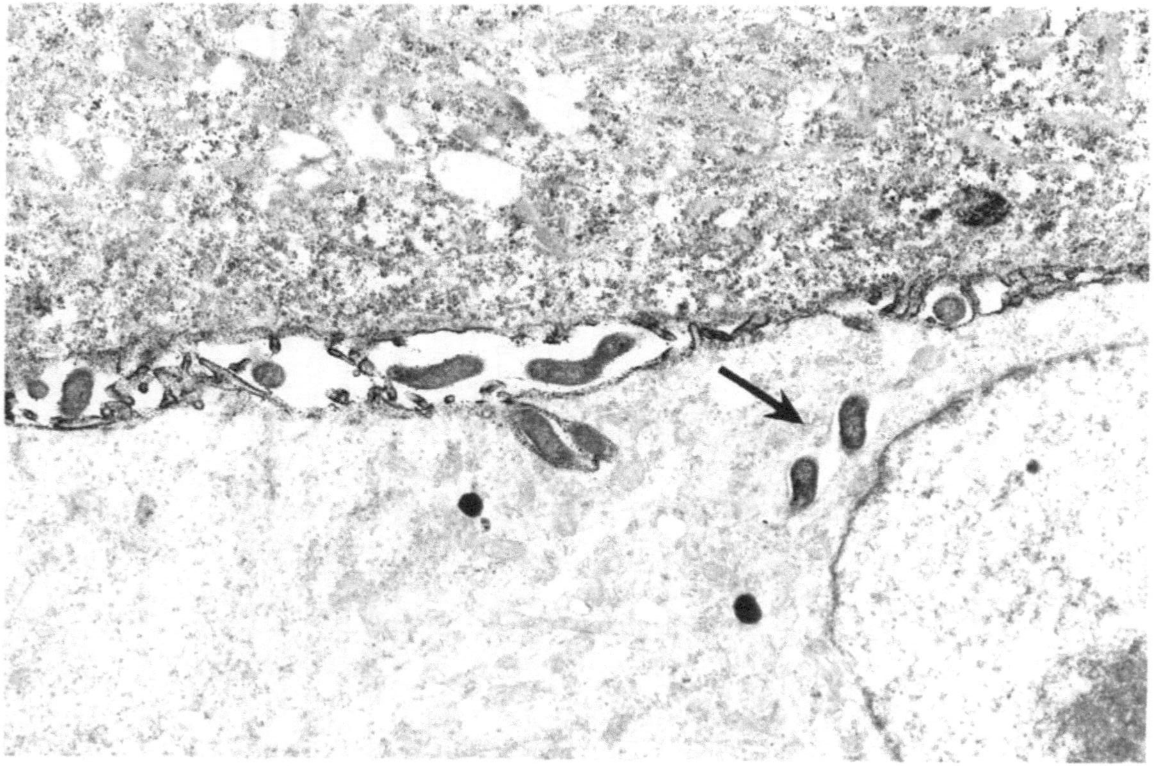

FIGURE 6. Transmission electron micrograph of the penetration of HeLa cells by *C. jejuni* (Strain 68869). The organisms were centrifuged on to the cell monolayer, cultured on plastic coverslips, and incubated for 1 hr prior to fixation. The surface glycoproteins of the fixed cells were labeled with thorium (1% thorium in 3% acetic acid) and orientated thin sections were cut after embedding the coverslip on Spurr resin blank blocks. Organisms situated in invaginations of the cell surface are surrounded by thorium-labeled surface membranes while organisms fully enclosed with the cell cytoplasm are surrounded by an unlabeled vacuolar membrane (↑). Note the diffuse cytoplasmic thorium staining of the adjacent cell due to the cytotoxicity of the *C. jejuni*. (× 10,760.)

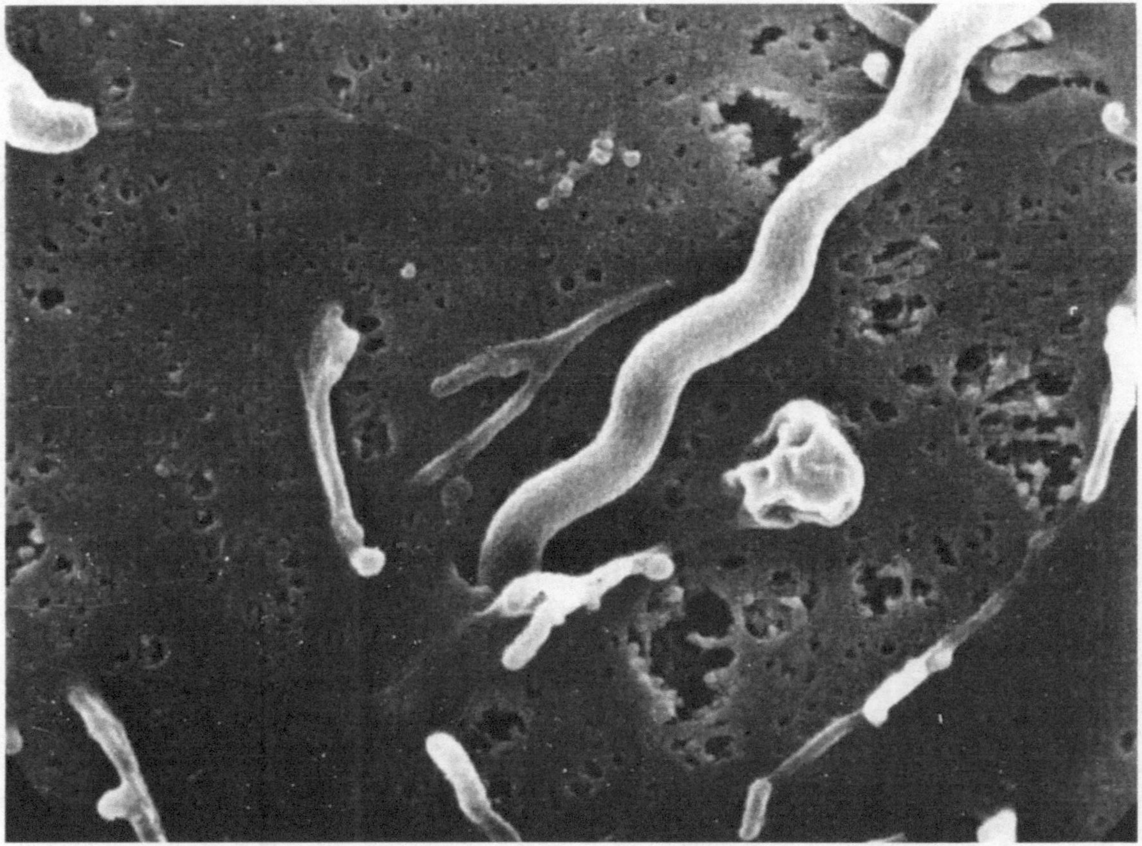

FIGURE 7. Scanning electron micrograph of an elongated campylobacter penetrating the surface of a HeLa cell. Note the absence of phagocytic activity around the site of penetration. The damage to the cell surface membrane is a combination of cytotoxicity of *C. jejuni* and artifactual shrinkage during the preparation for scanning electron microscopy. (× 37,500.) (From Newell, D. G. and Pearson, A. D., in *Campylobacter: Epidemiology, Pathogenesis and Biochemistry*, Newell, D. G., Ed., MTP Press, Lancaster, 1982, 196. With permission.)

V. CONCLUSION

Without doubt, campylobacter enteritis is a considerable world health problem contributing significantly to the mortality and morbidity due to diarrheal diseases, particularly in underdeveloped countries.[70]

Although a number of potential sources of infected material have been identified, including milk, water, poultry, and domestic animals, the ubiquitous nature of *C. jejuni* begs questions as to the relevance of many of these environmental isolates. Techniques for differentiating the pathogenic from the nonpathogenic strains are required to answer these questions. The problem is further complicated by the wide spectrum of symptoms presented by the patient and asymptomatic carriage, which may be a reflection of host immune status.

To date no markers of pathogenicity in *C. jejuni* have been identified and several major technical problems are hindering the investigation of such markers. There is a need for a simple and reproducible classification scheme for the thermophilic campylobacter group of organisms which is relevant to epidemiological and pathogenic investigations. The recent DNA hybridization studies confirm that *C. jejuni* and *C. coli* are separate species but appear to contribute nothing further.[4,71] The serotyping schemes developed are complicated and not readily available to most laboratories, but suitable for epidemiological studies,[72,73] and the biotyping scheme[74] is, as yet, poorly defined.

There is a lack of suitable in vivo or in vitro models of pathogenesis. Even the mechanism(s) of pathogenesis have yet to be established. Although the current evidence appears to indicate an invasive capacity the possibility of enterotoxin or cytotoxic production by some strains cannot be eliminated and could account for the spectrum of symptoms experienced with *C. jejuni* infections.

Differences in invasive potential between strains are suggested by several techniques (the chicken chorioallantoic membrane and cell monolayers), but development of quantitative assays of invasion are necessary to confirm these results.

Although much significant information can be gained from in vitro models, an animal model is essential for the assessment of certain virulence factors; for example, the relevance of motility and for the investigation of host immune responses to the infection. With, perhaps, the exception of the primates, no suitable animal model of campylobacter enteritis has yet been established. Few, if any, of the small mammals develop overt disease despite evidence of colonization and invasion of the intestinal mucosa. Manipulation of immune status using immunosuppressed or gnotobiotic animals and removal of competitive gut flora should be investigated as a means of enhancing the infection. Age and species susceptibility should also be investigated further with a view to producing a suitable model. The obvious susceptibility of man, compared with dogs, to the same strain of *C. jejuni*[21] indicates a species specificity which may be significant in the establishment of an animal model.

The selective pressures placed on strains of *C. jejuni* during isolation and purification procedures are considerable. The incubation temperature of 42°C enhances the growth rate of thermophilic campylobacters but suppresses the remainder of the enteric microflora, as does the selective antibiotic medium. These selective pressures, however, may be incompatible with retention of virulence factors. Nonetheless, laboratory-cultured strains of *C. jejuni* can induce enteritis in humans.[1,20,21]

Finally, the relevance of all the in vitro and in vivo models needs to be established using well-characterized mutants and variants of *C. jejuni* with pathogenic and nonpathogenic potentials. The biochemical investigation of these mutants will then establish those outer-membrane components which are virulence factors. The development of experimental models of campylobacter enteritis should, therefore, be concomitant with characterization of the surface structure of the organism.

REFERENCES

1. **Robinson, D. A.,** Infective dose of *Campylobacter jejuni* in milk, *Br. J. Med.*, 282, 1584, 1981.
2. **Blaser, M. J. and Peller, L. B.,** Campylobacter enteritis, *N. Engl. J. Med.*, 305, 1444, 1981.
3. **Butzler, J. P. and Skirrow, M. B.,** Campylobacter enteritis, *Clin. Gastroenterol.*, 8, 737, 1979.
4. **Owen, R. J. and Leaper, S.,** Base composition, size and nucleotide sequence simularities of genome deoxyribonucleic acids from specimens of the genus *Campylobacter*, *FEMS Microbiol. Lett.*, 12, 395, 1981.
5. **Rettig, P. J.,** Campylobacter infections in human beings, *J. Pediatr.*, 94, 855, 1979.
6. **Skerman, V. B. D., McGowan, V., and Sneath, P. H. A.,** Approved list of bacterial names, *Int. J. Syst. Bacteriol.*, 30, 270, 1980.
7. **McGechie, D. B., Teoh, T. B. and Bamford, V. W.,** Campylobacter enteritis in Hong Kong and Western Australia, in *Campylobacter: Epidemiology, Pathogenesis, and Biochemistry*, Newell, D. G., Ed., MTP Press, Lancaster, U.K., 1982, 19.
8. **De Mol, P.,** Campylobacter enteritis in developing countries, in *Campylobacter: Epidemiology, Pathogenesis, and Biochemistry*, Newell, D. G., Ed., MTP Press, Lancaster, U.K., 1982, 26.
9. **Glass, R. J., Hug, I., Stoll, P. J., Kibriya, G., and Blaser, M. J.,** Epidemiological features of campylobacter enteritis in Bangladesh, in *Campylobacter: Epidemiology, Pathogenesis, and Biochemistry*, Newell, D. G., Ed., MTP Press, Lancaster, U.K., 1982, 28.
10. **Richardson, N. J., Koornhof, H. J., and Bokkenheuser, V. D.,** The isolation of *C. jejuni* from babies with or without gastroenteritis, in *Campylobacter: Epidemiology, Pathogenesis, and Biochemistry*, Newell, D. G., Ed., MTP Press, Lancaster, U.K., 1982, 135.
11. **Rajan, D. P. and Mathan, V. I.,** Prevalence of *Campylobacter fetus* subsp. *jejuni* in healthy populations in Southern India, *J. Clin. Microbiol.*, 15, 749, 1982.
12. **Firehammer, B. D. and Myers, L. L.,** Experimental *Campylobacter jejuni* infections in calves and lambs, in *Campylobacter: Epidemiology, Pathogenesis, and Biochemistry*, Newell, D. G., Ed., MTP Press, Lancaster, U.K., 1982, 168.
13. **Al-Mashat, R. R. and Taylor, E. J.,** *Campylobacter* spp. in enteric lesions in cattle, *Vet. Rec.*, 107, 31, 1980.
14. **Smibert, R. M.,** The genus *Campylobacter*, *Annu. Rev. Microbiol.*, 32, 673, 1978.
15. **Leuchtefeld, N. W. and Wang, W. L. L.,** Animal reservoirs of *Campylobacter jejuni* in *Campylobacter: Epidemiology, Pathogenesis, and Biochemistry*, Newell, D. G., Ed., MTP Press, Lancaster, U.K., 1982, 249.
16. **Fernie, D. S. and Park, R. W. A.,** The isolation and nature of Campylobacters (Microaerophilic vibrios) from laboratory and wild rodents, *J. Med. Microbiol.*, 10, 325, 1977.
17. **Cruickshank, J. G., Egglestone, S. I., Gawler, A. H. L., and Lanning, D. G.,** *Campylobacter jejuni* and the broiler chicken process, in *Campylobacter: Epidemiology, Pathogenesis, and Biochemistry*, Newell, D. G., Ed., MTP Press, Lancaster, U.K., 1982, 263.
18. **Fox, J. G., Zanotti, S., and Jordon, H. V.,** The hamster as a reservoir of *Campylobacter fetus* subspecies *jejuni*. *J. Infect. Dis.*, 143, 856, 1981.
19. **Bruce, D., Zochowski, W., and Fleming, G. A.,** Campylobacter species in cats and dogs, *Vet. Rec.*, 107, 200, 1980.
20. **Steale, T. W. and McDermott, S.,** Campylobacter enteritis in South Australia, *Med. J. Aust.*, 2, 404, 1978.
21. **Prescott, J. F. and Karmali, M. A.,** Attempts to transmit campylobacter enteritis to dogs and cats, *Can. Med. Assoc. J.*, 119, 1001, 1978.
22. **Tribe, G. W. and Frank, A.,** Campylobacters in monkeys, *Vet. Rec.*, 106, 365, 1980.
23. **Fitzgeorge, R. B., Baskerville, A., and Lander, K. P.,** Experimental infection of Rhesus monkeys with a human strain of *Campylobacter jejuni*, *J. Hyg.*, 86, 343, 1981.
24. **Taylor, D. J.,** Natural and experimental enteric infections with catalase-positive campylobacters in cattle and pigs, in *Campylobacter: Epidemiology, Pathogenesis, and Biochemistry*, Newell, D. G., Ed., MTP Press, Lancaster, U.K., 1982, 163.
25. **Prescott, J. F. and Bruin-Mosch, C. W.,** Carriage of *Campylobacter jejuni* in healthy and diarrhoeic animals, *Am. J. Vet. Res.*, 42, 164, 1981.
26. **Al-Mashat, R. R. and Taylor, D. J.,** Production of Diarrhoea and dysentery in experimental calves by feeding pure cultures of *Campylobacter fetus* subspecies *jejuni*, *Vet. Rec.*, 107, 459, 1980.
27. **Al-Mashat, R. R. and Taylor, D. J.,** Production of enteritis in calves by the oral inoculation of pure cultures of *Campylobacter faecalis*, *Vet. Rec.*, 109, 97, 1981.
28. **Lander, K. P. and Gill, K. P. W.,** Experimental infection of the bovine udder with *Campylobacter coli/jejuni*, *J. Hyg.*, 84, 421, 1980.
29. **Patrick, S.,** Salisbury Public Health Laboratory, personal communication.
30. **Duffell, S. J. and Skirrow, M. B.,** Shepherd's scours and ovine campylobacter abortion, a 'new' zoonosis, *Vet. Rec.*, 103, 144, 1978.

31. **Shaw, I. G. and Ansfield, M.,** Fetopathogenicity of *Campylobacter jejuni* in sheep, in *Campylobacter: Epidemiology, Pathogenesis, and Biochemistry*, Newell, D. G., Ed., MTP Press, Lancaster, U.K., 1982, 177.
32. **Doyle, L. P.,** The etiology of swine dysentery, *Am. J. Vet. Res.*, 9, 50, 1948.
33. **Taylor, D. J. and Olubunmi, P. A.,** A re-examination of the role of *Campylobacter fetus* subspecies *coli* in the enteric disease of the pig, *Vet. Rec.*, 109, 112, 1981.
34. **Prescott, J. F., Manninen, K. I., and Barker, I. K.,** Experimental pathogenesis of *Campylobacter jejuni* enteritis — Studies in gnotobiotic dogs, pigs and chickens, in *Campylobacter: Epidemiology, Pathogenesis, and Biochemistry*, Newell, D. G., Ed., MTP Press, Lancaster, U.K., 1982, 1970.
35. **Bruce, D.,** Campylobacters in cats and dogs, in *Campylobacter: Epidemiology, Pathogenesis and Biochemistry*, Newell, D. G., Ed., MTP Press, Lancaster, U.K., 1982, 253.
36. **Hosie, B. D., Nicholson, T. B., and Henderson, D. G.,** Campylobacter infections in normal and diarrhoeic dogs, *Vet. Rec.*, 105, 80, 1979.
37. **Macartney, L., McCandlish, I. A. P., Al-Mashat, R. R., and Taylor, D. J.,** Natural and experimental enteric infections with *Campylobacter jejuni* in dogs, in *Campylobacter: Epidemiology, Pathogenesis, and Biochemistry*, Newell, D. G., Ed., MTP Press, Lancaster, U.K., 1982, 172.
38. **Prescott, J. F., Barker, I. K., Manninen, K. I., and Miniats, O. P.,** *Campylobacter jejuni* colitis in gnotobiotic dogs, *Can. J. Comp. Med.*, 45, 377, 1981.
39. **Fox, J.,** Massachusetts Institute of Technology, personal communication.
40. **Ackerman, J., Fox, J., and Murphy, J.,** Intestinal colonization of ferrets with *Campylobacter fetus* subspecies *jejuni*, presented at the 82nd Annu. Meet. Am. Soc. Microbiology, Atlanta, March 7-12, 1982.
41. **Lentsch, R. H., McLaughlin, R. M., Wagner, J. E., and Day. T. J.,** *Campylobacter fetus* subspecies *jejuni* as a possible etiologic agent of transmissible ileal hyperplasia in Syrian hamsters, presented at the 82nd Annu. Meet. Am. Soc. Microbiology, Atlanta, March 7-12, 1982.
42. **Field, L. H., Underwood, J. L., Pope, L. M., and Berry, L. J.,** Intestinal colonisation of neonatal animals by *Campylobacter fetus* subspecies *jejuni*, *Infect. Immunity*, 33, 884, 1981.
43. **Field, L. H., Underwood, J. L., and Berry, L. J.,** Colonization of the digestive tract of adult mice with *Campylobacter fetus* subspecies *jejuni*, *Med. Microbiol.*, in press.
44. **Newell, D. G.,** unpublished observation.
45. **Welkos, S.,** Pathogenesis of campylobacter enteritis — animal model, in *Campylobacter: Epidemiology, Pathogenesis, and Biochemistry*, Newell, D. G., Ed., MTP Press, Lancaster, U.K., 1982, 182.
46. **Merrell, B. R., Walker, R. I., and Coolbaugh, J. C.,** Experimental morphology and colonization of the mouse intestine by *Campylobacter jejuni*, in *Campylobacter: Epidemiology, Pathogenesis, and Biochemistry*, Newell, D. G., Ed., MTP Press, Lancaster, U.K., 1982, 183.
47. **Fenlon, D. R., Reid, T. M. S., and Porter, I. A.,** Birds as a source of campylobacter infections, in *Campylobacter: Epidemiology, Pathogenesis, and Biochemistry*, Newell, D. G., Ed., MTP Press, Lancaster, U.K., 1982, 261.
48. **Ruiz-Palacios, G. M., Escamilla, E., and Torres, N.,** Experimental campylobacter diarrhoea in chickens, *Infect. Immunity*, 34, 250, 1981.
49. **Manninen, K. I., Prescott, J. F., and Dohoo, I. R.,** Studies on the pathogenicity of *Campylobacter jejuni* isolates from animals and man, *Infect. Immunity*, 38, 46, 1982.
50. **Giannella, R. A.,** Pathogenesis of acute bacterial diarrheal disorders, *Annu. Rev. Med.*, 32, 341, 1981.
51. **Dijs, F. and de Graaf, F. K.,** In search of adhesive antigens on *Campylobacter jejuni*, in *Campylobacter: Epidemiology, Pathogenesis, and Biochemistry*, Newell, D. G., Ed., MTP Press, Lancaster, U.K., 1982, 243.
52. **Newell, D. G. and Pearson, A. D.,** Pathogenicity of *Campylobacter jejuni* — an *in vitro* model of adhesion and invasion ?, in *Campylobacter: Epidemiology, Pathogenesis, and Biochemistry*, Newell, D. G., Ed., MTP Press, Lancaster, U.K., 1982, 196.
53. **Austen, R. A. and Trust, T. J.,** Outer membrane protein composition of campylobacters, in *Campylobacter: Epidemiology, Pathogenesis, and Biochemistry*, Newell, D. G., Ed., MTP Press, Lancaster, U.K., 1982, 225.
54. **McBride, H. and Newell, D. G.,** unpublished observation.
55. **Osbourne, J. C. and Smibert, R. M.,** *Vibrio fetus* endotoxin, *Nature (London)*, 195, 1106, 1962.
56. **Fumarola, D., Miragliotta, G., and Jirillo, E.,** Activity associated with heat-killed organisms of the genus *Campylobacter* in *Campylobacter: Epidemiology, Pathogenesis, and Biochemistry*, Newell, D. G., Ed., MTP Press, Lancaster, U.K., 1982, 185.
57. **Musher, D. M.,** Cutaneous and soft tissue manifestations of sepsis due to Gram-negative enteric bacilli, *Rev. Infect. Dis.*, 2, 854, 1980.
58. **Gubina, M., Zajc-Satler, J., Dragas, A. Z., Zeleznik, Z., and Mehle, J.,** Enterotoxin activity of *Campylobacter species, in Campylobacter: Epidemiology, Pathogenesis, and Biochemistry*, Newell, D. G., Ed., MTP Press, Lancaster, U.K., 1982, 188.

59. **Campbell, W. F., Lambe, D. W., Tobe-Meyer, B., and Maybery-Carson, K. J.**, Assays for the presence of enterotoxin(s) in strains of *Campylobacter fetus* subspecies *jejuni* and strains of *Clostridium difficile*, presented at the 82nd Annu. Meet. Am. Soc. Microbiology, Atlanta, March 7-12, 1982.
60. **Butzler, J. P.**, Discussion, in *Campylobacter: Epidemiology, Pathogenesis, and Biochemistry*, Newell, D. G., Ed., MTP Press, Lancaster, U.K., 1982, 195.
61. **Guerrant, R. L., Lahita, R. G., Winn, W. C., and Roberts, R. B.**, Campylobacteriosis in Man: Pathogenic mechanisms and review of 91 blood stream infections, *Am. J. Med.*, 65, 584, 1978.
62. **Prescott, J. F.**, Discussion, in *Campylobacter: Epidemiology, Pathogenesis, and Biochemistry*, Newell, D. G., Ed., MTP Press, Lancaster, U.K., 1982, 195.
63. **Drake, A. A., Gilchrist, M. J. R., Washington, J. A., Huizenga, K. A., and van Scoy, R. F.**, Diarrhoea due to *Campylobacter fetus* subspecies *jejuni*. A clinical review of 63 cases, *Mayo Clin. Proc.*, 56, 414, 1981.
64. **Blankenship, L. C., Crawen, S. E., and Hopkins, S. R.**, Preliminary characterisation of a factor(s) from cell extracts of *Campylobacter fetus* subspecies *jejuni* that produces cytotoxic effects in cell cultures, presented at the 82nd Annu. Meet. Am. Soc. Microbiology, 1982.
65. **King, E. O.**, Human infections with *Vibrio fetus* and a closely related vibrio, *J. Infect. Dis.*, 101, 119, 1957.
66. **Davidson, J. A. and Solomon, J. B.**, Onset of resistance to pathogenic strains of *Campylobacter fetus* subspecies *jejuni* in chicken embryos, in *Aspects of Developmental and Comparative Immunology*, Solomon, J. B., Ed., Pergamon Press, Oxford, 1980, 289.
67. **McCoy, E. C., Doyle, D., Burda, K., Corbeil, L. B., and Winter, A. J.**, Superficial antigens of *Campylobacter (Vibrio) fetus* Characterisation of an antiphagacytic component, *Infect. Immunity*, 11, 517, 1975.
68. **Winter, A. J., McCoy, E. C., Fullmer, C. S., Burde, K., and Bier, P. J.**, Microcapsule of *Campylobacter fetus*: chemical and physical characterization, *Infect. Immunity*, 22, 963, 1978.
69. **Buck, G. E. and Parshall, K. A.**, Isolation of capsule material from *Campylobacter fetus* subsp. *jejuni*, presented at the 82nd Annu. Meet. Am. Soc. Microbiology, Atlanta, March 7-12, 1982.
70. **Evans, N.**, Pathogenic mechanisms in bacterial diarrhoea, *Clin. Gastroenterol.*, 8, 599, 1979.
71. **Belland, R. J. and Trust, T. J.**, Deoxyribonucleic acid sequence relatedness between thermophilic members of the genus *Campylobacter*, *J. Gen. Microbiol.*, 128, 2515, 1982.
72. **Lior, H., Woodward, D. L., Edgar, J. A., Laroche, L. J., and Gill, P.**, Serotyping of *Campylobacter jejuni* by slide agglutination based on heat labile antigenic factors, *J. Clin. Microbiol.*, 15, 761, 1982.
73. **Penner, J. L. and Hennessy, J. N.**, Passive haemagglutination technique for serotyping *Campylobacter fetus* subspecies *jejuni* in the basis of soluble heat-stable antigens, *J. Clin. Microbiol.*, 12, 732, 1980.
74. **Skirrow, M. B. and Benjamin, J.**, Differentiation of enteropathogenic campylobacter, *J. Clin. Pathol.*, 33, 1122, 1980.

Chapter 12

OUTER MEMBRANE AND SURFACE STRUCTURE OF *CAMPYLOBACTER JEJUNI*

Trevor J. Trust and Susan M. Logan

TABLE OF CONTENTS

I. Introduction 134

II. Structure 134

III. Involvement in Virulence 137

IV. Surface Antigens 139

V. Prospects 140

References 140

I. INTRODUCTION

The outer membrane of a Gram-negative cell and other cellular structures that are located in contact with its external surface serve as the interface between the bacterium and its surrounding milieu.[1] In the case of pathogens such as *Campylobacter jejuni* and its thermophilic relatives, this environment can be an animal, bird or human gut, human tissue cells and fluids, or, indeed, it can be fresh water in a stream, salt water in an estuary, milk, or even the skin on a chicken carcass.[2] In each of these widely differing environments, the outer membrane serves to protect the bacterium, and at the same time in certain of these niches it must also participate in bacterial growth. The outer membrane totally surrounds the Gram-negative cell, and in the case of *Campylobacter* appears only to be penetrated by the flagella. During cell growth this membrane must allow the passage of nutrients and waste products. It must expand as the cell grows larger, and in *Campylobacter* allow for a transition in cell shape from a vibrioid rod, through a spiral form, to the final coccoid form. The outer membrane must also serve as a selective barrier to the cell exterior. Indeed the outer membrane allows Gram-negative cells to be more resistant than Gram-positive cells to the actions of certain dyes, chemicals, enzymes, and antibiotics.[1] Also, in the case of enteric pathogens such as *C. jejuni*, the outer membrane clearly protects the cytoplasmic membrane from direct exposure to bile salts which would lyse the cells.[1]

In pathogenic bacteria the outer membrane also plays an important role in the outcome of host-parasite relationships. Studies with a variety of pathogens have shown that components of the outer membrane can participate in the adherence of the pathogen to host cells, the invasion of host cells, resistance to the bactericidal activities of serum, resistance to phagocytosis and to phagocytic killing mechanisms, and in the sequestering of iron.[3-7] The lipopolysaccharide (LPS) can also contribute directly to the toxicity of a Gram-negative cell.[8] Moreover, the surface exposure of the LPS and certain proteins allow contact with the immune system, and so they also serve as antigens.[1,9,10] This means that the surface-exposed constituents of the outer membrane are important in determining the specificity of the host's immune response, and become key determinants in serotyping schemes. Clearly then, the outer membrane is a structure of enormous importance.

II. STRUCTURE

The outer membrane is a highly asymmetric lipid bilayer containing LPS which is located in the outer leaflet, and phospholipid, much of which is located in the inner leaflet.[1,10] Interspersed in this bilayer is a set of outer membrane proteins.[1,10] The presence of the LPS allows the outer membrane to be separated from the cytoplasmic membrane by virtue of an increased density. The outer membrane can also be isolated after selective solubilization of the inner membrane by the detergent sodium lauryl sarcosinate.[11] The protein complement and LPS content can then be analyzed by sodium dodecyl sulfate-polyacrylamide gel electrophoresis (SDS-PAGE).[9] When sarcosinate-extracted outer membrane from a variety of type strains and wild type isolates of *C. jejuni* and *C. coli* were analyzed by SDS-PAGE they were shown to possess quite similar outer membrane protein profiles and the two species could not be differentiated on the basis of this profile.[12] This was somewhat unexpected since Belland and Trust[13] have recently demonstrated by DNA:DNA hybridization that *C. jejuni* and *C. coli* are separate species. The outer membrane of a typical human diarrheal isolate of *C. jejuni* is shown in Figure 1. It can be seen that there are seven major polypeptide bands in this preparation. The sarcosinate extraction procedure does, however, allow for coisolation of flagella. Since flagella are dissociated at low pH, outer membranes prepared from cells extracted with a glycine buffer at pH 2 should not display the flagellin monomer. When this was done with the strain shown in Figure 1, the glycine-extracted membrane

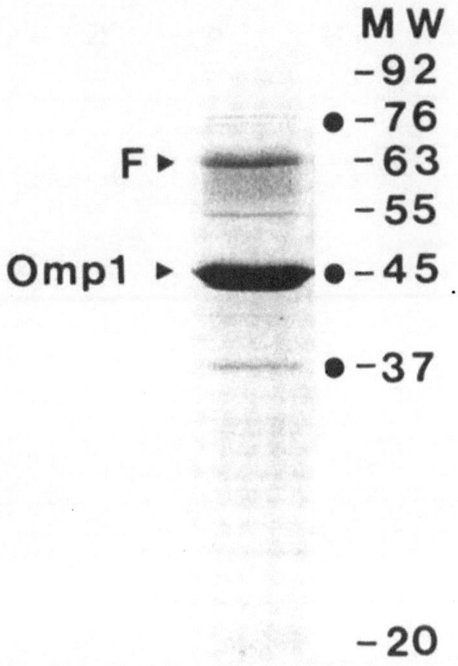

FIGURE 1. SDS-PAGE of isolated outer membrane of a typical fecal isolate of *C. jejuni* stained by Coomassie blue. The major outer membrane protein is labeled Omp 1 while the flagellin band which copurified with the outer membrane is labeled F. Other outer membrane proteins are identified by their apparent molecular weight (mol wt $\times$ 10^3), and surface-exposed proteins are also identified (●).

profile was distinguished by the absence of a 63-kdalton band. Subsequent neutralization of the glycine extract and ammonium sulfate precipitation allowed this 63-kdalton protein to be isolated. Electron microscopy clearly demonstrated the presence of partially reassociated flagella filaments, confirming this 63-kdalton band as flagellin.[14]

The SDS-PAGE profile of *C. jejuni* and *C. coli* is dominated by a single protein band, the apparent monomer molecular weight varying between 43 and 45 K depending on the strain examined.[12] This protein, *Campylobacter* Omp 1, is clearly the most important structural protein of the *C. jejuni* outer membrane, comprising well over 70% of the membrane protein content. Omp 1 is probably the porin or matrix protein of *C. jejuni*, the protein responsible for maintaining the hydrophilic size-dependent diffusion channels through the outer membrane. Indeed, subsequent studies may reveal the presence of several isoproteins within the porin complex. The protein is heat-modifiable, associates into oligomers, and spans the membrane with domains which are surface-exposed and peptidoglycan-associated. This dominance of the SDS-PAGE profile by a single polypeptide band readily distinguishes *C. jejuni* and *C. coli* from two other enteric organisms, *Escherichia coli* and *Salmonella typhimurium*, as well as the type species of the genus, *C. fetus* subsp. *fetus* CIP 5396. These species all display two or more major bands. Another significant difference shown by all species of *Campylobacter* is the absence of Braun's lipoprotein from outer membrane protein profiles.[12,15] In *E. coli* and *S. typhimurium* this protein is the most abundant in the cell in terms of number of molecules, with about one third being bound to the peptidoglycan.[1,9] This lipoprotein helps anchor the outer membrane to the cell and plays an important role in maintaining the integrity of the outer membrane structure. In the thermophilic campylobacters, the only peptidoglycan-associated protein appears to be *Campylobacter* Omp 1, emphasizing its important structural role.

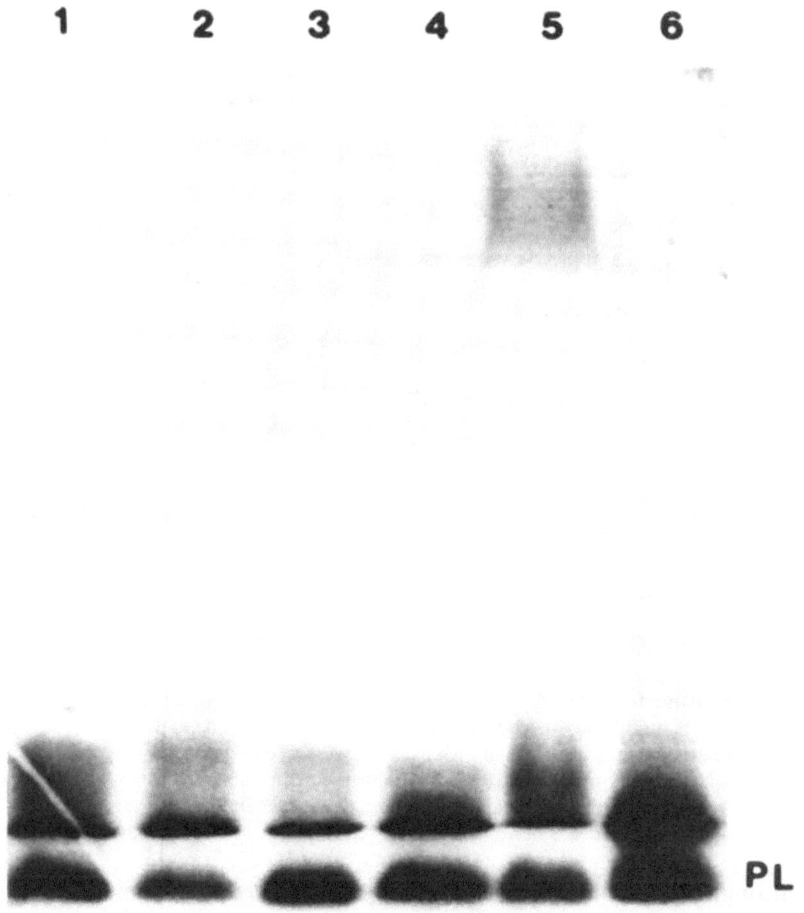

FIGURE 2. Autoradiogram of SDS-PAGE of isolated envelopes of bacteria grown on ^{32}P showing radiolabeled LPS. The lower band in each lane is phospholipid (PL). The thermophilic campylobacters are Lane 1, VC 117; Lane 2, VC 107; Lane 3, VC 108; Lane 4, VC 109. Control strains are Lane 5, *A. salmonicida* strain 449; Lane 6, *S. typhimurium* rough mutant strain SA22. In Lanes 5 and 6, core-lipid A LPS is visible as the lower-molecular-weight band of radiolabel migrating just behind the radiolabeled phospholipid. The O-polysaccharide-core-lipid A LPS migrates with a high apparent molecular weight and is clearly visible in Lane 5.

In the case of *C. jejuni* strain VC 74 studied by Logan and Trust,[12] two other outer membrane proteins were shown to be surface-exposed. These were a 76-kdalton (doublet), and a 37-kdalton band. Indeed, based on the relative amounts of proteins in Coomassie blue stained profiles, considerably more of the 76-kdalton protein appeared to be exposed than either of the other proteins. Preliminary experiments testing the effects of proteolytic enzymes on whole cells confirm the surface exposure of the 37-kdalton band.[14] The outer membrane of *C. jejuni* strain VC 74 also contains proteins of approximate molecular weight 92, 55, and 20 kdaltons, but their location in the outer membrane is unclear at this time.

The LPS content of the outer membrane of thermophilic campylobacters has also been studied. The lipid A component of the LPS is embedded in the lipid of the outer leaflet of the bilayer while the carbohydrate end is surface-exposed.[9] In SDS-PAGE, LPS migrates with lipid A present, and separates on the basis of size.[16-18] Migration distance is inversely related to polysaccharide content, and so the core LPS with its short carbohydrate chain migrates most rapidly, while LPS with long O polysaccharides migrates slowest. The technique is quite discriminating and has been shown to separate LPS molecules differing by as few as two or three saccharides. LPS can be visualized in gels by the periodic acid-Schiff stain, or better still, LPS can be radiolabeled by growing cells on ^{32}P and visualized in gels by autoradiography,[7] or by silver staining. When LPS in the outer membrane of wild-type *Aeromonas salmonicida*[7] is examined by this technique, the LPS runs in two major fractions. The fraction with high apparent molecular weight corresponds to core LPS attached to the long repeating polysaccharide subunits which make up the O chains. This is clearly shown with the *Aeromonas salmonicida* strain 449 in Figure 2 (lane 5). The other fraction with low apparent molecular weight corresponds to core lacking O polysaccharide chains, as shown by both *A. salmonicida* and the *S. typhimurium* rough mutant strain SA22 in Figure 2 (lane 6). When five strains of *C. jejuni* and *C. coli* were examined using this technique the LPS appeared to be of a low-molecular-weight variety and lacking in long repeating polysaccharide O antigen side chains.[12] In contrast Logan and Trust[12] showed that the type species of the genus, *C. fetus* subsp. *fetus* CIP 5396, did have O polysaccharide chains of intermediate length.

The LPS content in the OM fraction of another four strains of the thermophilic campylobacters is shown in Figure 2 (lanes 1 to 4). The LPS of these four strains also appears to be of a low-molecular-weight variety and lacking in long repeating O polysaccharide chains. This apparent absence of long repeating O polysaccharide chains in other strains of *C. jejuni* had also been reported by Naess and Hofstad.[19] These workers extracted LPS from three strains of *C. jejuni* with aqueous phenol. The sugar components of the LPS examined were typical constituents of the polysaccharide core structure of *Escherichia*, *Salmonella*, and *Neisseria* species.[9,20] Further, the water-soluble polysaccharide fraction isolated from the LPS of one of the three strains behaved like a core structure in gel filtration experiments.[19] The LPS from all three strains contained L-glycero-D-manno-heptose, glucose, galactose, and glucosamine, but in different molar ratios. One strain also contained galactosamine. *O*-acetyl groups were not present, but 3-deoxy-D-manno-octulosonic acid (KDO) was present in each LPS, as was phosphorus. The fatty acids present were mainly 3-hydroxy-tetradecanoid acid and *n*-hexadecanoic acid.[18] However, when LPS isolated by phenol-extraction was examined by SDS-PAGE, the LPS of several strains of *C. jejuni* clearly exhibited O polysaccharide chains.[14] These gave a ladder-like profile on SDS-PAGE gels similar to the LPS of wildtype smooth *E. coli* and *S. typhimurium*,[16-18] indicating that the O polysaccharides of *C. jejuni* are of heterogeneous chain length. Dilution titration of the purified LPS further revealed that the longer O polysaccharide chains were in a minority, with the majority of the LPS being of lower molecular weight. This accounted for the apparent absence of long O polysaccharide chains in the SDS-PAGE profiles presented by Logan and Trust,[14] and seen in Figure 2. That is, while present on the LPS of the stains examined, they were quantitatively below the limits of detection of the techniques as used. Subsequent examination of additional strains has indicated that in some cases, O polysaccharides are present in amounts sufficient to allow them to be seen in SDS-PAGE-processed outer membrane[14] and, indeed, in whole cell lysates. Preston et al.[21] have also reported that O polysaccharides appear to be present on some stains and do contribute to serotypic specificity.[21] The chemical nature of these O polysaccharides awaits elucidation.

III. INVOLVEMENT IN VIRULENCE

During growth, campylobacters release sizeable quantities of outer membrane. These blebs arise when the outer membrane expands at a greater rate than the underlying peptidoglycan layer, and so areas not attached to the underlying peptidoglycan will tend to bulge and grow until released.[22-25] This blebbing is readily seen in electron micrographs[26] (Figure 3). Indeed organisms such as *Campylobacter* and *Neisseria* which lack the anchoring effect

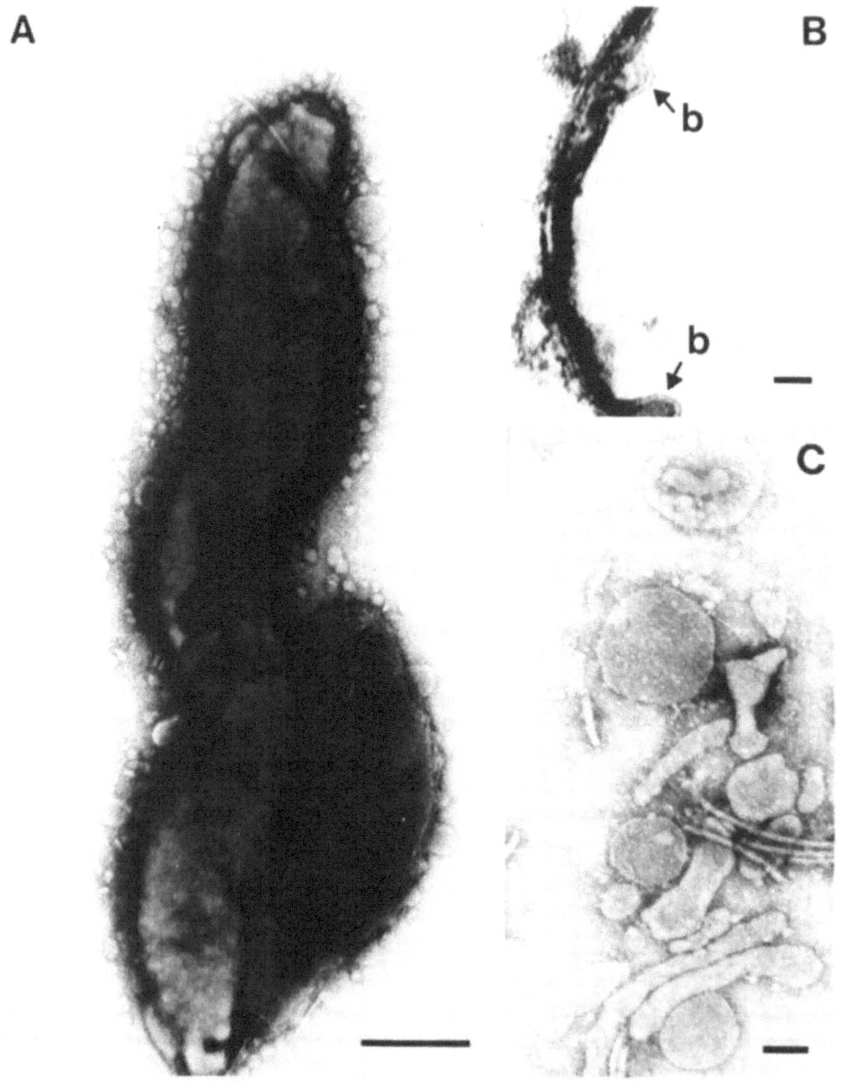

FIGURE 3. Release of outer membrane. A. *C. jejuni* cells surrounded by a large amount of released outer membrane. B. Formation of outer membrane blebs (b) by a Gram-negative cell prior to release of outer membrane. C. The released outer membrane fraction collected by differential centrifugation. The preparation is contaminated by fragments of flagella. The bar in A represents 0.3 μm, in B is 0.03 μm, and in C represents 0.1 μm.

produced by Braun's lipoprotein may release much more outer membrane than strains which have this lipoprotein.[27] The released membrane can be collected by differential centrifugation and its protein complement analyzed by SDS-PAGE. In the case of *C. jejuni* strain VC 74, the protein profile of the released outer membrane differed from that of outer membrane isolated by sarcosinate extraction.[12] Several of the outer membrane proteins were considerably enriched in the released membrane fraction. Studies with *E. coli* have shown that the released outer membrane fraction is enriched for newly synthesized LPS. Initial analyses suggest that the released outer membrane fraction of *C. jejuni* is also enriched for LPS.[14] The released outer membrane material should be endotoxic by virtue of the lipid A of the LPS, and indeed this material comprises the free endotoxin of Gram-negative bacteria.[9,27,28] Certainly at this point in our understanding of the pathogenesis of campylobacteriosis, endotoxicity is the only virulence property that can be unequivocally ascribed to the outer membrane of *C. jejuni*. Released outer membrane blebs were first reported for *C. fetus*,[3] and the LPS of this species was shown to produce a biphasic fever response and a generalized Schwartzman reaction in rabbits and to be lethal for mice. Endotoxin in spent culture medium caused an endotoxic shock reaction in cattle, and also aborted cows when injected intravenously.[29] In the case of *C. jejuni*, Kee Peng Ng et al.[30] have suggested that mouse lethality is due to endotoxin, since the majority of mice injected with 1.8×10^9 heat-killed cells died within 24 hr. Furthermore Fumarola et al.[31] produced a positive dermal Schwartzman reaction in rabbits with heat-killed cells of *C. jejuni* and *C. coli*. In an experiment with the released outer membrane fraction from two strains of *C. jejuni* we have also demonstrated endotoxicity as measured by mouse lethality. Intravenous injection of membrane fractions containing 136 μg LPS killed mice within 24 hr, with the mice displaying features typical of endotoxic shock.[14]

In the case of the animal pathogen *C. fetus*, a major surface component has been shown to be crucial to virulence.[3,32] Virulent strains produce a 97,000-dalton (glyco)protein which forms a microcapsule and protects the cell from phagocytic engulfment. Loss of this protein coincides with a loss of virulence. The protein is generally visible as a major component in outer membrane preparations prepared by sarcosinate extraction, and can also be extracted from the cell surface with glycine buffer at pH 2.[3,14,32] A corresponding antiphagocytic surface protein has not yet been identified on *C. jejuni* or *C. coli*. Certainly, it was not identified on the nine strains reported on by Logan and Trust,[12] nor was it seen on the type strain of the species *C. fetus* subsp. *fetus* CIP 5396.[12]

There is, however, some evidence for at least one loosely associated surface protein on *C. jejuni* which is not found in sarcosyl-extracted outer membrane preparations.[14] When intact cells of *C. jejuni* were radioiodinated by an impermeant labeling technique, and whole-cell SDS-PAGE protein profiles examined, a protein not present in OM preparations was clearly radiolabeled. This protein has a molecular weight of approximately 30 kdaltons and appears to be extractable by low-pH buffer, suggesting it is loosely surface associated. Further studies will clearly reveal whether this protein is a component of a microcapsule. Indeed the contribution of this or the other surface exposed proteins to virulence must await the development of suitable test systems, for although campylobacters have been reported to adhere to brush borders and other cells and to invade cells,[35-37] neither activity has been studied quantitatively. Until this occurs, the role of surface proteins in these virulence activities cannot be adequately gauged.

IV. SURFACE ANTIGENS

The antigenicity of components of the surface of the *C. fetus* cell has received much more attention than *C. jejuni* and its thermophilic relatives.[38-42] Winter and colleagues[38-42] have shown that the surface antigens of *C. fetus* include the major (glyco)protein component, two other superficial protein antigens readily released from the cell upon brief treatment in acid buffer, the matrix protein of the outer membrane, two antigens located on the flagella, and the LPS. The (glyco)protein microcapsular antigen contributes to the inagglutinability of wild-type cells in antisera directed at LPS. Indeed this prevents effective serogrouping based on heat-stable antigens until the microcapsule is lost upon subculture. Neither IgG nor IgA antibodies directed against the various surface antigens appear to be bactericidal to *C. fetus* in the presence of bovine complement. The antiphagocytic activity of the microcapsule is, however, overcome by antibodies of class IgG. Antibodies of this class directed at the microcapsule are opsonic, and after phagocytic engulfment by both neutrophils and macrophages, *C. fetus* cells are killed.

When we consider the cell surface of *C. jejuni* and *C. coli*, there is an important difference from *C. fetus* in the apparent absence of a high-molecular-weight microcapsule. Although glycine extraction has been used by at least one group to prepare a so-called glycoprotein antigen fraction from *C. jejuni*, no biochemical evidence has been presented for the actual presence of a glycoprotein, or indeed any protein of a molecular weight corresponding to that of the *C. fetus* protein.[42] Experiments in this laboratory have shown that the glycine-extractable fraction contains flagella and a variety of low-molecular-weight proteins, some of which appear to derive from the cell surface, while others derive from the periplasm.[14] At least one of the low-molecular-weight surface proteins may correspond to the acid-extractable surface antigens reported by McCoy et al.[3,42] for *C. fetus*.

Like the *C. fetus* cell, the *C. jejuni* cell has flagella, a major surface-exposed outer membrane protein, and LPS which, along with several other surface exposed outer membrane proteins are likely to contribute to the antigenic mosaic presented by the campylobacter cell to its animal host. Indeed, in experiments in this laboratory using the technique of immunoblotting, we have confirmed the immunogenicity of the LPS, the flagella, the matrix protein, and an acid-extractable surface protein of *C. jejuni*.[14] All but flagella are likely to participate in typing schemes based on bacteriophage binding.[44] Moreover the surface-exposed proteins will be important participants in serotyping schemes such as those proposed by Lior et al.,[45] which are based on heat-labile antigens, while the LPS will be the sero-determinant in schemes such as that of Penner and Hennessy[45] based on heat-stable antigens.

V. PROSPECTS

Clearly, much is still to be learned of the structural organization of the outer membrane and surface-associated proteins of *C. jejuni*. It is likely that some will play important roles in such virulence properties as attachment and invasion, and these are clearly areas that warrant attention. The major outer membrane protein also deserves greater attention. Its role as a pore protein needs to be confirmed, and the contribution of isoproteins to the putative porin complex determined. Perhaps of most immediate interest and attention is the antigenicity of the surface-exposed components of the *Campylobacter* cell. This necessitates a consideration of both LPS and proteins. Clearly, composition and arrangement of the sugars of the LPS needs to be determined in order to define their role as serogroup determinants. Similarly, the contribution of the various proteins to serotyping schemes based on heat-labile antigens needs to be elucidated. In this regard, monoclonal antibodies may be most useful because they should allow the antigenic structure to be defined at the molecular level. Some of these monoclonal antibodies will also have the practical benefit of allowing for highly specific serotyping.

REFERENCES

1. **Inouye, M.,** What is the outer membrane? in *Bacterial Outer Membranes*, Inouye, M., Ed., John Wiley & Sons, New York, 1979, 1.
2. **Karmali, M. A. and Fleming, P. C.,** Campylobacter enteritis, *Can. Med. Assoc. J.*, 120, 1525, 1979.
3. **McCoy, E. C., Doyle, D., Burda, K., Corbeil, L. B., and Winter, A. J.,** Superficial antigens of *Campylobacter (Vibrio) fetus*: characterization of an antiphagocytic component, *Infect. Immunity*, 11, 517, 1975.
4. **Smith, H. W.,** Microbial surfaces in relation to pathogenicity, *Bacteriol. Rev.*, 41, 475, 1977.
5. **Buchanan, T. M. and Pearce, W. A.,** Pathogenic aspects of outer membrane components of Gram-negative bacteria, in *Bacterial Outer Membranes*, Inouye, M., Ed., John Wiley & Sons, New York, 1979, 475.

6. **Lambden, P. R., Heckels, J. E., James, L. T., and Watt, P. J.,** Variations in surface protein composition associated with virulence properties in opacity types of *Neisseria gonorrhoeae, J. Gen. Microbiol.*, 114, 305, 1979.
7. **Munn, C. B., Ishiguro, E. E., Kay, W. W., and Trust, T. J.,** Role of surface components in serum resistance of *Aeromonas salmonicida, Infect. Immunity,* 36, 1069, 1982.
8. **Shands, J. W.,** Endotoxin as a pathogenetic mediator of Gram-negative infection, in *Microbiology 1975,* Schlessinger, D., Ed., American Society for Microbiology, Washington, 1975, 330.
9. **Lüderitz, O., Freudenberg, M. A., Galanos, C., Lehmann, V., Rietschel, E. T., and Shaw, D. H.,** Lipopolysaccharides of Gram-negative bacteria, *Curr. Top. Membrane Transp.*, 17, 79, 1982.
10. **Di Rienzo, J. M., Nakamura, K., and Inouye, M.,** The outer membrane proteins of Gram-negative bacteria: biosynthesis, assembly, and functions, *Ann. Rev. Biochem.*, 47, 481, 1978.
11. **Filip, C., Fletcher, G., Wulff, J. L., and Earhart, C. F.,** Solubilization of the cytoplasmic membrane of *Escherichia coli* by the ionic detergent sodium-lauryl sarcosinate, *J. Bacteriol.*, 115, 717, 1973.
12. **Logan, S. M. and Trust, T. J.,** Outer membrane characteristics of *Campylobacter jejuni, Infect. Immunity,* 38, 898, 1982.
13. **Belland, R. J. and Trust, T. J.,** Deoxyribonucleic acid sequence relatedness between thermophilic members of the genus *Campylobacter, J. Gen. Microbiol.*, 128, 2515, 1982.
14. **Logan, S. M. and Trust, T. J.,** unpublished data, 1982.
15. **Winter, A. J., Katz, W., and Martin, H. H.,** Murein (peptidoglycan) structure of *Vibrio fetus*. Comparison of a venereal and an intestinal strain, *Biochim. Biophys. Acta,* 244, 58, 1971.
16. **Palva, E. T. and Mäkelä, P. H.,** Lipopolysaccharide heterogeneity in *Salmonella typhimurium* analyzed by sodium dodecyl sulfate-polyacrylamide gel electrophoresis, *Eur. J. Biochem.*, 107, 137, 1980.
17. **Goldman, R. C. and Lieve, L.,** Heterogeneity of antigenic-side-chain length in lipopolysaccharide from *Escherichia coli,* 0111 and *Salmonella typhimurium,* LT2, *Eur. J. Biochem.*, 107, 145, 1980.
18. **Munford, R. S., Hall, C. L., and Rick, P. D.,** Size heterogeneity of *Salmonella typhimurium* lipopolysaccharides in outer membranes and culture supernatant membrane fragments, *J. Bacteriol.*, 144, 630, 1980.
19. **Naess, V. and Hofstad, T.,** Isolation and chemical composition of lipopolysaccharide from *Campylobacter jejuni, Acta Pathol. Microbiol. Immunol. Scand. Sect. B,* 90, 135, 1982.
20. **Adams, G. A., Kates, M., Shaw, D. H., and Yaguchi, M.,** Studies of the chemical constitution of cell wall lipopolysaccharides from *Neisseria perflava, Can. J. Biochem.*, 46, 1175, 1968.
21. **Preston, M. A., Bradbury, W. C., Barton, L. J., and Penner, J. L.,** Characterization of the somatic antigens of *Campylobacter jejuni, Abs. Annu. Mtg. Am. Soc. Microbiol.*, p. 358, C278, 1983.
22. **Mug-Opstelten, D. and Witholt, B.,** Preferential release of new outer membrane fragments by exponentially growing *Escherichia coli, Biochim. Biophys. Acta,* 508, 287, 1978.
23. **MacIntyre, S., Trust, T. J., and Buckley, J. T.,** Identification and characterization of outer membrane fragments released by *Aeromonas* sp., *Can. J. Biochem.*, 58, 1018, 1980.
24. **Gankema, H., Wensink, J., Guinée, P. A. M., Jensen, W. H., and Witholt, B.,** Some characteristics of the outer membrane material released by growing enterotoxigenic *Escherichia coli, Infect. Immunity,* 29, 704, 1980.
25. **Wensink, J. and Witholt, B.,** Outer membrane vesicles released by normally growing *Escherichia coli* contain very little lipoprotein, *Eur. J. Biochem.*, 116, 331, 1981.
26. **Pead, P. J.,** Electron microscopy of *Campylobacter jejuni, J. Med. Microbiol.*, 12, 383, 1979.
27. **De Voe, I. W.,** The meningococcus and mechanisms of pathogenicity, *Microbiol. Rev.*, 46, 162, 1982.
28. **Russell, R. R. B.,** Free endotoxin — a review, *Microbios Lett.*, 2, 125, 1976.
29. **Smibert, R. M.,** The genus *Campylobacter, Ann. Rev. Microbiol.*, 32, 673, 1978.
30. **Kee Peng, Ng, F., Wardlaw, A. C., and Stewart-Tull, D. E. S.,** Enhancement of the lethal effect of *Campylobacter fetus* s.s. *jejuni* in seven-day-old mice by ferric ammonium citrate, *Soc. Gen. Microbiol. Q.*, 8, 12, 1980.
31. **Fumarola, D., Meragliotta, G., and Jirillo, E.,** Endotoxin-like activity associated with heat-killed organisms of the genus *Campylobacter,* in *Campylobacter: Epidemiology, Pathogenesis, and Biochemistry,* Newell, D. G., Ed., MTP Press, Lancaster, 1982, 185.
32. **Winter, A. J., McCoy, E. C., Fullmer, C. S., Burda, K., and Bier, P. J.,** Microcapsule of *Campylobacter fetus:* chemical and physical characterization, *Infect. Immunity,* 22, 963, 1978.
33. **Taylor, D. J.,** Natural and experimental enteric infections with catalase-positive campylobacters in cattle and pigs, in *Campylobacter: Epidemiology, Pathogenesis, and Biochemistry,* Newell, D. G., Ed., MTP Press, Lancaster, U.K., 1982, 163.
34. **Field, L. H., Underwood, J. L., Pope, L. M., and Berry, L. J.,** Intestinal colonization of neonatal animals by *Campylobacter fetus* subsp. *jejuni, Infect. Immunity,* 33, 884, 1981.
35. **Butzler, J. P.,** Infection with campylobacters, in *Modern Topics in Infection,* Williams, J. D., Ed., William Heideman Medical Books, London, 1978, 214.

36. **Ruiz-Palacios, M., Escamilla, E., and Torres, N.,** Experimental *Campylobacter* diarrhea in chickens, *Infect. Immunity,* 34, 250, 1981.
37. **Newell, D. G. and Pearson, A. D.,** Pathogenicity of *Campylobacter jejuni*, an *in vitro* model of adhesion and invasion?, *Campylobacter: Epidemiology, Pathogenesis, and Biochemistry,* Newell, D. G., Ed., MTP Press, Lancaster, U.K., 1982, 196.
38. **Winter, A. J.,** An antigenic analysis of *Vibrio fetus*. III. Chemical, biologic, and antigenic properties of the endotoxin, *Am. J. Vet. Res.,* 27, 653, 1966.
39. **Myers, L. L.,** Purification and partial characterization of a *Vibrio fetus* immunogen, *Infect. Immunity,* 3, 562, 1971.
40. **McCoy, E. C., Wiltberger, H. A., and Winter, A. J.,** Antibody-mediated immobilization of *Campylobacter fetus:* inhibition by a somatic antigen, *Infect. Immunity,* 13, 1266, 1976.
41. **McCoy, E. C., Wiltberger, H. A., and Winter, A. J.,** Major outer membrane protein of *Campylobacter fetus:* physical and immunological characterization, *Infect. Immunity,* 13, 1258, 1976.
42. **Corbeil, L. B. and Winter, A. J.,** Animal model for the study of genital secretory immune mechanisms: Venereal vibriosis in cattle, in *Immunobiology of Neisseria gonorrhoeae,* Brooks, G. F., Gotschlich, E. C., Holmes, K. K., Sawyer, W. D., and Young, F. E., Eds., American Society for Microbiology Washington, 1978, 293.
43. **Svedhem, Å, Gunnarsson, H., and Kaijser, B.,** Serological diagnosis of *Campylobacter jejuni* infections by using the enzyme-linked immunosorbent assay principle, in *Campylobacter: Epidemiology, Pathogenesis, and Biochemistry,* Newell, D. G., Ed., MTP Press, Lancaster, U.K., 1982, 118.
44. **Bryner, J. H., Ritchie, A. E., and Foley, J. W.,** Techniques for phage typing *Campylobacter jejuni*, in *Campylobacter: Epidemiology, Pathogenesis, and Biochemistry,* Newell, D. G., Ed., MTP Press, Lancaster, U.K., 1982, 52.
45. **Lior, H., Woodward, D. L., Edgar, J. A., Laroche, L. J., and Gill, P.,** Serotyping of *Campylobacter jejuni* by slide agglutination based on heat-labile antigenic factors, *J. Clin. Microbiol.,* 15, 761, 1982.
46. **Penner, J. L. and Hennessy, J. N.,** Passive hemagglutination technique for serotyping *Campylobacter fetus* subsp. *jejuni* on the basis of soluble heat stable antigens, *J. Clin. Microbiol.,* 12, 732, 1980.

Chapter 13

EPIDEMIOLOGY OF *CAMPYLOBACTER* INFECTIONS

Martin J. Blaser, David N. Taylor, and Roger A. Feldman

TABLE OF CONTENTS

I. Introduction 144

II. Reservoirs 144
A. Animal Reservoirs 144
1. Poultry 144
2. Cattle 144
3. Swine 145
4. Sheep 145
5. Dogs 146
6. Cats 146
7. Rodents 146
8. Other Animals 146
B. Humans as Reservoirs 147
C. Inanimate Reservoirs 147
1. Water 147
2. Soil 147

III. Modes of Transmission 147
A. Transmission Following Direct Animal Contact 147
B. Ingestion of Contaminated Foods 148
C. Milk-Borne Transmission 149
D. Transmission by Other Foods 149
E. Ingestion of Contaminated Water 150
F. Person-to-Person Transmission 151
G. Perinatal Transmission 151
H. Transmission during Childhood 152

IV. Incidence and Prevalence 152
A. By Location 152
1. Developed Countries 152
2. Developing Countries 154
B. By Age and Sex 154
C. Seasonality 157
D. Occupation 157

V. Conclusions 157

References 158

I. INTRODUCTION

With new knowledge of the importance of *Campylobacter jejuni* as a pathogen of humans has come significant advances in our knowledge of the epidemiologic characteristics of the infection and illness associated with this infection. Although many important questions about the transmission of this agent are still unanswered, considerable progress has been made in our understanding of the reservoirs and prevalence of infection. The purpose of this chapter will be to review the progress in these areas. Because reliable methods for differentiating *C. jejuni* and *C. coli* have not been in common use until recently, in this discussion we shall use the former term to describe both species unless a distinction is necessary.

II. RESERVOIRS

A. Animal Reservoirs

C. jejuni or *C. coli* organisms may exist as commensals in the intestinal tracts of a wide variety of wild and domestic animals. Whether or not campylobacters are pathogenic in their animal hosts is beyond the scope of this chapter. That some of the organisms associated with animals cause infection and disease in humans is now known from data obtained in a few outbreaks in which an animal or animal product was ultimately identified as a source. Furthermore, recently some of the same serotypes of *Campylobacter* that cause disease in humans have been isolated from animals.[1] Our understanding of the relative importance of various reservoirs for human infection will be enhanced when improved typing schemes are developed and utilized. Nonetheless there are considerable data concerning *Campylobacter* infection in domestic animals.

1. Poultry

Much information is now available about *C. jejuni* isolated from poultry (Table 1). Most commercially raised poultry have campylobacters in their intestinal flora. Isolations have been made from poultry early in the growing process although some flocks apparently completely escape infection.[13] Potential sources for entry of organisms into a flock include infection of newborn chicks from older birds, contaminated feed (including partially pasteurized bird feather, offal, and blood) or contaminated water. For the most part, infection appears to be without obvious signs of illness. During the process of slaughtering, *C. jejuni* spreads from the intestinal contents to the carcasses. In those flocks in which intestinal carriage was not detected, carcasses were not contaminated.[13] Chlorine washing does not necessarily remove all campylobacters from giblets and carcasses.[10,13] Although freezing and storage may diminish the magnitude of the contamination, results of several studies have been variable.[5,14] In the U.S., the majority of chicken carcasses sold at retail markets are contaminated with *C. jejuni*.[6,8,9] Feces from infected poultry may also contaminate the surface of eggs.

Wild birds including ducks and geese may be excretors of *Campylobacter*.[15] Thus, attempts to eradicate the organism from commercial operations may fail. Some of the serotypes common in wild animals are those associated with infections in humans.[1]

2. Cattle

C. jejuni is a normal commensal of cows. Stool positivity rates peak in the summer months and decline in the winter. Individual cows excrete the same serotype for at least several months, if not for life.[16] Transmission occurs to calves but has not been shown between adult cows. Several different serotypes may be present in a herd at a given time.[17] Reported isolation frequencies have varied from herd to herd (Table 2). In part, these differences may be artifacts due to the use of different methods for isolation of campylobacters; methods

Table 1
ISOLATION OF *CAMPYLOBACTER* FROM COMMERCIAL POULTRY

Author—year	Ref.	Country	Animal	Stage in preparation	No. sampled	% Positive
Smith—1974	2	U.S.	Chicken	Parts after freezing 3 weeks	165	2
Bruce—1977	3	England	Chicken	Cecal contents	167	68
Ribeiro—1978	4	England	Chicken	Intestinal contents	34	91
Simmons—1979	5	England	Chicken	Eviscerated	50	72
				Eviscerated, water chilled	25	80
				Eviscerated, air chilled	10	80
			Turkey	Eviscerated, water chilled	6	83
				Eviscerated, air chilled	5	100
Grant—1980	6	U.S.	Chicken	Intestinal contents, retail	46	83
Goren—1980	7	Netherlands	Chicken	Intestinal contents	239	29
				Carcasses after freezing	750	0
Eiden—1980	8	U.S.	Chicken	Intestinal contents	62	50
				Carcasses after freezing	23	100
Park—1981	9	U.S.	Chicken	Whole chickens, retail	50	54
Luechtefeld—1981	10	U.S.	Turkey	Freshly killed, cecal contents	600	100
				Eviscerated carcass	33	94
				Eviscerated carcass, water chilled	83	94
				Viscera	24	33
Prescott—1981	11	Canada	Duck	Feces	94	88
Svedhem—1981	12	Sweden	Chicken	Feces	50	36

using enrichment techniques have shown a significantly higher isolation rate. Thus, adequate information on the levels of infection in herds awaits widespread use of enrichment techniques. As with the other animals discussed, carcasses may become contaminated with intestinal contents; however, contamination of carcasses is infrequent, and when it occurs the level of contamination is low (<1 organism per square centimeter).[22] Unpasteurized milk has been implicated as a vehicle for numerous outbreaks of *Campylobacter* enteritis.[25] Either contamination with fecal contents or mastitis are the likely conditions leading to introduction of *C. jejuni* into the milk. In an experimental model of *Campylobacter* mastitis, milk became contaminated with large numbers of organisms,[26] but whether this phenomenon occurs in nature is unknown. The presence of *C. jejuni* in the herd does not necessarily lead to contamination of milk.[27] Pasteurization is an effective method for eradicating *Campylobacter* from milk.[28]

3. *Swine*

Swine commonly carry *C. coli* and occasionally *C. jejuni* as intestinal commensals.[12,19,23,14] Studies in the U.S., The Netherlands, and Germany have all shown that more than half of commercially raised pigs excrete the organisms (Table 2). Washing and treating the intestines with salt diminish but do not eliminate contamination; thus, sausage may be contaminated.[24] Contamination of swine carcasses is more common that that of sheep and cattle carcasses and presumably occurs as a result of intestinal spillage.

4. *Sheep*

C. jejuni is an important cause of epizootic infectious abortion in sheep.[29] In many flocks it exists as a long-term intestinal commensal without apparent morbidity. Surveys of sheep intestinal contents have shown that isolation of *C. jejuni* is common; contamination of carcasses occurs less often (Table 2).

Table 2
ISOLATION OF *CAMPYLOBACTER JEJUNI* FROM INTESTINAL FLORA OF HEALTHY ANIMALS USED FOR FOOD PRODUCTION (EXCLUDING COMMERCIAL POULTRY)

Animal	Population sampled	Source of sample	Number sampled	% +	Author	Ref.
Cattle	Calves and steers	Feces	202	3	Prescott	18
	At slaughterhouse	Bile	525	12	Bryner	11
		Cecal contents	130	43	Luechtefeld	19
		Feces	90	19	Svedhem	12
		Feces	31	0	Stern	20
		Carcasses	58	2	Stern	20
		Carcasses	100	0	Hudson	21
	At retail store	Minced beef	2,015	1	Turnbull	22
Sheep	From several farms	Feces	35	23	Luechtefled	19
	At slaughterhouse	Bile	186	9	Bryner	18
		Feces	15	73	Stern	20
		Carcasses	54	24	Stern	20
		Carcasses	100	9	Hudson	21
Swine[a]	2 Farms	Feces	71	66	Luechtefeld	19
	At slaughterhouse	Intestinal contents	300	61	Oosterom	23
		Feces	116	66	Sticht-Groh	24
		Feces	138	95	Svedhem	12
		Feces	38	87	Stern	20
		Carcasses	58	22	Stern	20
		Carcasses	100	59	Hudson	21
		Carcasses—wet chill	50	26	Hudson	21
		Carcasses—dry chill	50	2	Hudson	21
Ducks	Wild	Cecal contents	445	35	Luechtefeld	15

[a] Most isolates from swine have phenotypic characteristics resembling those of *C. coli*.

5. Dogs

More than 30 years ago, "spirochetal organisms" were found to be present in the stools of both healthy dogs and those with diarrhea. More recently, these have been identified as *C. jejuni* and other *Campylobacter* species. Several points are now clear: (1) *C. jejuni* may be isolated from both healthy and diarrheal dogs;[30,31] (2) isolation rates are higher in puppies than in mature dogs;[30,31] and (3) isolation rates are higher in kennel populations than among household dogs.[30,31]

6. Cats

Kittens are more frequently culture-positive for *C. jejuni* than adult cats[30] and highest isolation rates are in kennel populations.[31]

7. Rodents

Healthy rodents frequently excrete campylobacters. Laboratory-raised hamsters, mice and rats, and wild rodents may excrete in their feces organisms resembling *C. jejuni, C. coli,* and *C. faecalis*.[32]

8. Other Animals

Unlike *Salmonella, C. jejuni* has not been isolated from reptiles and other poikylotherms. Salmonellas may multiply at ambient temperatures, but *C. jejuni* will not multiply below approximately 30°C. Several studies of primate colonies have shown that *Campylobacter* infection is common,[33,34] although infection rates among animals in the bush or at a zoo are lower.[1] Other zoo animals that have low *C. jejuni* isolation rates include felines (2%) and ungulates (6%). Pigeons (17%) and other birds at a zoo (10%) had higher rates.[1] The reported serotypes of *Campylobacter* isolates from zoo animals are similar to those from humans.[1]

In conclusion, the animals with which man is in most frequent contact, including those used for food production and those that are domestic pets, are frequently reservoirs for *Campylobacter*.

B. Humans as Reservoirs

In the developed countries, infected humans who are prolonged carriers constitute, at most, a minor reservoir for *C. jejuni*. In developing countries, prolonged human carriage could play a larger role in the transmission of infection, a topic discussed further in Section IV.

C. Inanimate Reservoirs

1. Water

The sources of *Campylobacter* organisms found in water are not known but may be due to fecal contamination by wild or domestic animals. *C. jejuni* is not adapted for free-living existence in water. When kept at several different temperatures in stream water from several locations, it would not multiply. In seeding experiments in water, organisms would survive for up to 4 weeks when kept at 4°C; however, survival at higher temperatures was reduced.[35]

In nature, *C. jejuni* has been isolated from stream and river water,[36] from the effluent of a turkey processing plant,[10] and from seawater.[37] In a study sampling river water from a single site in England over the course of a year, isolations peaked during June and July.[38]

In a study of isolations from riverine water in the Southampton area in England, campylobacters were isolated from 50.4% of 540 water samples. No isolations were made from the water samples unless *Escherichia coli* was also present. After heavy rains isolation rates diminished, which suggests that perhaps the water itself could have been the primary reservoir. Brackish water had lower isolation rates than fresh water, and in estuarial waters, isolation rates were also higher at low tide.[36]

2. Soil

Studies of mud from contaminated rivers in Southampton failed to show campylobacters. However, isolations from both mud and sewage sludge have been made in the U.S.[39] In seeding experiments, campylobacters were found to survive in soil for at least 10 days, and for 20 days when the ambient temperature decreased to 6°C.[40] Similarly, *C. jejuni* in feces from infected humans and dogs survived for 3 weeks when kept at 4°C. Freezing has been shown to kill *C. jejuni*.[41] Thus, under moderately cold ambient conditions, fecal contamination of soil could constitute a potential reservoir for infection.

III. MODES OF TRANSMISSION

Campylobacter may be transmitted from its animal reservoir to humans in numerous ways. Transmission may follow direct contact with contaminated animals or animal carcasses; more commonly, *C. jejuni* is transmitted indirectly through the ingestion of contaminated food or water. Person-to-person transmission can occur from humans with active infections who are excreting large numbers of *C. jejuni*. As the conditions and circumstances under which these modes of transmission have occurred are described, information about the infective dose, the spectrum of illness, and the excretion pattern will be discussed.

A. Transmission Following Direct Animal Contact

Campylobacter enteritis has been reported in children and young adults who before their illnesses had close contact with infected puppies.[42,43] In some instances, identical *Campylobacter* sero- or biotypes have been isolated from both animal and human.[44] *Campylobacter* may occasionally be isolated from the stools of adult dogs as well as puppies, but excretion in adult dogs has not been associated with human diarrhea.[12,30] Diarrheal stools from puppies appear to be most important in transmitting *C. jejuni* to humans.

Cats also may transmit *Campylobacter* to humans but appear to be less important than dogs in this respect.[45] In England, for example, 3 cases of *Campylobacter* enteritis associated with cats were reported in a series in which 97 cases were reported to be associated with dogs.[46] In the three cases, the infection occurred in young children who were in intimate contact with kittens; in each case the same serotype was isolated from both child and animals.[46,47] As with puppies, the major risk appears to be from newly acquired kittens, particularly if the animal develops diarrhea. A possible exception is a human case in which an asymptomatic adult cat was found to be excreting the same serotype that was isolated from its owner.[48] Although it could not be determined when the cat had become infected, frequent contact of the human with the cat's feces supports the causal relationship.

Direct contact with domestic farm animals and their feces may also be important in transmitting infection. There are times when excretion is high and transmission is more likely to occur. For example, calves that develop enteritis, and cattle during times of stress or during calving, may excrete higher numbers of *C. jejuni*. A case was reported in which a young man developed *Campylobacter* enteritis after beginning employment in a cattle feedlot.[49] In sheep, *C. jejuni* causes an asymptomatic enteric infection that is associated with transient bacteremia which, in a pregnant ewe, may lead to fetal death or abortion.[29] Thus, handling pregnant sheep and lambs could also increase the risk of transmission.[50]

Serologic surveys have suggested that persons in frequent contact with animals or carcasses have a higher rate of seropositivity to *Campylobacter* antigens. In England, complement-fixing antibodies to *Campylobacter* were detected in 2% of persons living in urban areas, in 5% of people living in rural areas, in 18% of veterinary assistants, and from 27 to 60% of persons working in poultry and meat processing plants.[51] It was not determined if any increased rate of illness occurred as a result of this exposure.

Laboratory animals may also act as a reservoir of *Campylobacter* and serve as a possible source of infection to their handlers. In a study initiated when an animal technician developed severe *Campylobacter* enteritis after handling newly imported primates, *Campylobacter* was found in 18% of healthy monkeys and 60% of monkeys with diarrhea.[33]

Laboratory technicians who handle fecal specimens and cultures can acquire the infection.[52,53] A 31-year-old hospital microbiology technician spilled a broth solution containing *C. jejuni* on her hands and work bench. Although she promptly disinfected the workbench and washed her hands, she developed severe *Campylobacter* enteritis 2 1/2 days after exposure.[52]

B. Ingestion of Contaminated Foods

In areas of the world where *Campylobacter* isolations from humans are systematically reported, the majority of *Campylobacter* infections occur sporadically with an undetermined mode of transmission. Given the similarities in the animal reservoirs for *Campylobacter* and *Salmonella*, one must consider that the vehicles may be similar to those found for *Salmonella* These include uncooked or poorly cooked meat and poultry products, unpasteurized dairy products, and uncooked foods which may be contaminated by meat and poultry products or with untreated sewage. For example, since poultry intended for human consumption is frequently contaminated with *C. jejuni* (Table 1). it has been suggested that it is an important vehicle for human infection. Other sources or vehicles may be inferred on the basis of a convincing clinical situation, such as when it is assumed that a butcher became infected as a result of handling contaminated meat. The third and most convincing approach, used most successfully in the investigation of outbreaks, is to epidemiologically identify a vehicle by its association with the ill population and to isolate *Campylobacter* from the implicated vehicle.

Epidemiologic studies have thus far been most successful in demonstrating an association between *Campylobacter* enteritis with the ingestion of raw (unpasteurized) milk. Although attempts to isolate the organism from the milk have not generally been successful, *Campylobacter* of the same serotype has been isolated from the cattle and milk filters at implicated dairies. In almost all other outbreaks in which food-borne transmission was believed to have occurred, the vehicle has been implicated without actually isolating *Campylobacter* from the suspected source.

C. Milk-Borne Transmission

In England and Wales, unpasteurized milk is the most frequently identified vehicle of transmission in human outbreaks of *C. jejuni* infection. Robinson et al. described 13 outbreaks involving an estimated 4500 persons in the period 1978 to 1980.[25,54,55] In the U.S., illness in persons who have ingested unpasteurized milk has also been commonly reported. The first outbreak associated with a *Campylobacter*-like organism was in 1938 at a prison in Illinois.[56] More recently, now that *C. jejuni* can be easily isolated from human fecal specimens, outbreaks have been reported in California in 1977,[57] Colorado in 1978,[58] New Mexico in 1979, and Oregon,[59] Kansas,[60] Arizona,[17] Georgia,[27] and Minnesota in 1981, with a total of over 500 persons affected.

In England, outbreaks have been associated with unpasteurized and with improperly pasteurized milk. In the U.S., outbreaks have been reported only affecting persons who prefer to drink raw milk. The U.S. outbreaks have been epidemiologically associated with raw milk ingestion and in some cases the serotype isolated from the humans could be isolated from the implicated herd, but campylobacters have not been isolated from milk samples or from milk filters obtained from the implicated dairy. In England, the implicated serotype has been isolated from the herd[61] and from milk filters.[54] To demonstrate the disease potential of *Campylobacter* in milk, two volunteers have drunk milk experimentally contaminated with *C. jejuni*. The first such volunteer drank 10^6 organisms in a glass of milk and subsequently experienced a mild illness 3 days later.[62] The second volunteer became mildly ill after drinking 180 mℓ of pasteurized milk that contained 500 organisms of a strain of *C. jejuni* isolated from humans in a milk-associated outbreak.[63] Milk experimentally inoculated with *Campylobacter* and kept at 4°C yielded viable organisms for up to 3 weeks.[35] The source of the organisms in nature is thought to be from fecal contamination of the milk by the cow, but bovine mastitis could play a role.[26]

D. Transmission by Other Foods

Raw or undercooked poultry has been one of the most frequently suggested vehicles of food-borne *Campylobacter* enteritis.[5,12] Raw or inadequately cooked chicken eaten by military recruits during a military training exercise was the vehicle in an outbreak in the Netherlands.[64] Ingestion of poorly cooked chicken was thought to be an important exposure in an outbreak occurring in England after a catered banquet,[65] and was determined to be a risk factor for sporadically occurring illness in the Netherlands.[66] In the U.S., processed turkey was shown to be epidemiologically associated with 11 cases of *C. jejuni* bacteremia in California.[67] In a recent outbreak in Colorado, ingestion of undercooked chicken at a barbecue was associated with illness.[117]

Raw hamburger was implicated in another outbreak in the Netherlands,[68] and beef tartare was suspected in an outbreak at a cooking school in New York. An outbreak described at a boy scout camp was associated with cake icing.[69] *C. jejuni* could not be isolated from any of the icing ingredients, but the icing could have been contaminated at the time of preparation by the cook. Raw clams, probably contaminated by sewage, were described as the vehicle in an outbreak in New Jersey.[67]

Campylobacter enteritis has been reported frequently in travelers returning from tropical countries. In Sweden and Finland, 73 and 50% of the cases of enteritis, respectively, in which *C. jejuni* was isolated occurred in persons who became ill while traveling outside of Scandanavia.[70,71] In the Swedish study, 8% of the travelers from whom *C. jejuni* was isolated also had one another enteric pathogen, most commonly *Salmonella*. In Panamanian visitors to Mexico, *C. jejuni* accounted for 11% of the episodes of diarrhea that were acquired while traveling.[72] In these studies, the cases occurred sporadically and in none was a vehicle determined. This suggests that multiple vehicles that are only intermittently contaminated may be involved. The concomitant isolation of *Salmonella* in 8% of travelers suggests that some vehicles may be similar for these two pathogens. Among persons who developed travelers' diarrhea in Mexico, Food-borne transmission was believed to be predominant.[73,74]

E. Ingestion of Contaminated Water

Water has not only been indirectly implicated as a source of contamination in an outbreak of *Campylobacter* infection associated with shellfish but has been directly implicated in several large outbreaks of *Campylobacter* enteritis traced to drinking water. In June, 1978 an outbreak of *Campylobacter* enteritis involved nearly 3000 (19%) of the residents of Bennington, Vt.[75] At the time of the outbreak, *C. jejuni* was isolated from five of nine rectal swabs that were obtained from ill persons. One week later *C. jejuni* was isolated from an additional 10 of 33 rectal swabs. Isolates serotyped were subsequently found to be either Penner serotype 36/23 or 13/16. Fourteen convalescent sera from both culture positive and negative ill persons were tested by the IFA assay. All of these had IgG antibody titers of 1:16 or greater to one of the two epidemic serotypes, but none of the sera of 20 healthy control subjects contained similar antibody titers. The epidemiologic investigation suggested that the outbreak was related to the 80-year-old municipal water system. Shortly before the outbreak, there had been heavy rains that increased runoff which may have led to contamination of a portion of the town water system by surface water.

In October 1980, another municipal outbreak occurred in Sweden involving an estimated 2000 persons, approximately 20% of the community at risk,[76]. During this outbreak, *C. jejuni* was isolated from rectal swabs from 221 (84%) of 263 persons cultured. The cases were geographically distributed along the water distribution system and had no other common exposures besides the same drinking water source. The water system utilized deep ground water, but for irrigation purposes the water main was cross-connected with river water which, during times of low pressure, could backflow into the drinking water. A time of low water pressure had occurred just before the outbreak.

These two waterborne outbreaks associated with municipal water supplies occurred in communities with less than 15,000 inhabitants. In these communities, from 10 to 20% of the population had a mild diarrheal illness that rarely required hospitalization or treatment with antibiotics. The sources of contamination appeared to be surface water that entered municipal water systems that already had insufficient chlorination. The Vermont and Swedish outbreaks may have differed with respect to secondary cases. There were no secondary cases in Vermont, while in Sweden 26 of 48 cases investigated after the peak date of illness occurred in houses where primary cases had occurred in the week before. Surprisingly, 10 of the late cases may have been acquired from an ill adult. Late cases imply either a continuing common source or prolonged incubation periods, or if spread is actually secondary from adults, extremely close contact or secondary food contamination.

In rural areas where surface water is used for drinking, communities and individuals are at risk for waterborne *Campylobacter* enteritis. An outbreak occurred in a small community in northern British Columbia in July 1980, where an estimated 700 persons had *Campylo-*

bacter enteritis. The outbreak correlated geographically with the unchlorinated town water supply, which was gravity fed from a reservoir or pumped directly from a creek.[77]

C. jejuni has been isolated from both salt- and freshwater sources in places where humans have acquired the infection.[37] Pure cultures of *C. jejuni* inoculated into unchlorinated water maintained at 4°C will remain viable for weeks.[35] Reports of illness among scouts in Wales[78] and backpackers in Wyoming have suggested that *Campylobacter* enteritis can be acquired from the ingestion of contaminated stream water in mountainous areas. In western Wyoming, where backpackers and tourists visiting in national parks make up a large percentage of the summer population, *C. jejuni* accounted for 25% of the diarrheal disease.[67] Illness occurred sporadically and was caused by diverse *Campylobacter* serotypes, but the ingestion of untreated surface water was a significant exposure, particularly for young adults. *Campylobacter* was isolated frequently from a number of birds and lowland water sources, but only an isolate from a high mountain stream, where backpackers had been, had a serotype identical to that of strains isolated from humans.

F. Person-to-Person Transmission

Campylobacter jejuni has been reported as a cause of neonatal gastroenteritis and, rarely, as a cause of septic abortion. When neonatal infections occur, there is often a history of diarrheal disease among members of the infected mother's household. This suggests that maternal infection can be acquired by the fecal-oral route and then be transmitted to the fetus through fecal contamination of the birth canal, or via the placenta after septicemia in the mother.

Sexual transmission of *Campylobacter* from man to woman has not been demonstrated to be a cause of vaginal infection or subsequent neonatal infection in a pregnant female. There have been two cases of *Campylobacter* proctitis reported in homosexual men.[79,80] One male had a 9-day history of bloody diarrhea, rectal pain, and rectal discharge; the other case had no recent history of diarrhea. Both men gave a history of passive anal intercourse and oral-anal contact. Homosexual men have a higher incidence of proctitis caused by sexually transmitted agents such as *Neisseria gonorrheae* and *Herpes simplex* virus and by enteric pathogens such as *Entamoeba histolytica* and *Shigella*. However, *C. jejuni* was not isolated from 167 homosexual men without gastrointestinal symptoms.[30] It is not known if the disease is sexually transmitted through oral-anal contact, or whether other modes of transmission are more important in this population.

G. Perinatal Transmission

Although most infections due to *C. jejuni* are associated with mild enteritis, bacteremia in pregnant women may be associated with a severe systemic infection in the fetus. A 24-year-old Canadian woman, who was 18 weeks pregnant, had a febrile illness that lasted 3 weeks but had no diarrhea or vaginal discharge. Her fetus died *in utero* during the time of her infection. *C. jejuni* was subsequently isolated from multiple maternal blood cultures, placenta, and fetal spleen.[81]

Neonatal meningitis was reported from Wales in a 12-day-old boy, and *C. jejuni* was isolated from the cerebrospinal fluid, but blood cultures were negative.[82] The child recovered after treatment with chloramphenicol and gentamicin. His mother had had a diarrheal illness 6 weeks before delivery. High antibody titers to the organism isolated from the child were detected in both the mother's and the child's serum. The IgG antibody detected in the child may have been maternal.

In England, a newborn child delivered by Cesarean section developed diarrhea on the third day after delivery. *C. jejuni* was isolated from the feces of the mother and baby. A blood culture taken from the baby shortly after delivery yielded *C. jejuni*, which suggests that the infection may have been acquired *in utero*.[83]

In most cases, the neonate acquires the infection during or shortly after delivery from fecal contamination of the vaginal passage. A 19-year-old mother was healthy at the time of delivery, but two other persons in the household had diarrhea.[84] One day after delivery the mother was febrile, and on the second day she had mild diarrhea that lasted for 2 days. On the third day her baby developed mucoid and bloody diarrhea from which *C. jejuni* was isolated. Both mother and child developed antibody responses to the same strain of *Campylobacter*.

Over a 2-year period at a hospital in Colorado, 8 cases of *Campylobacter* enteritis in neonates were observed.[85] All of the neonates showed irritability, and 7 had bloody diarrhea that began between 2 to 11 days of age. Only one was seriously ill and none had bacteremia. *C. jejuni* was isolated from fecal cultures obtained from four of the mothers and, in these families, from two of the fathers. One mother's vaginal culture was positive. None of the mothers had a recent history of diarrhea, although several remembered a diarrheal illness within the past few months. In a case of *Campylobacter* enteritis in a 3-day-old girl[86] the mother had diarrhea a month before delivery, but by 1 week after delivery her stool culture was negative. She had an elevated antibody titer to the strain of *C. jejuni* isolated from the baby. The father developed bloody diarrhea 12 days before delivery and was still culture-positive 5 days after delivery; he also had an elevated antibody titer against the same strain.

H. Transmission during Childhood

In a study at Montreal Children's Hospital, an illness compatible with *Campylobacter* infection was documented in 18 o 72 household contacts of infected children in 6 of 24 families. Ten were bacteriologically confirmed. Although no specific vehicles could be determined, the dates of onset suggest that both common-source exposure and person-to-person transmission had occurred.[87] In a large milk-borne outbreak in England involving 2500 children 2 to 7 years old, secondary transmission to siblings or adults occurred frequently, possibly accounting for 20% of the total infected population.[55]

An outbreak of *Campylobacter* enteritis occurred in a nursery in Japan. Over a 7-day period, 35 of 74 children 1 to 5 years old became ill with diarrhea and fever. *C. jejuni* was isolated from 13 of 33 stool cultures. Epidemiologic survey indicated a point-source exposure, but the vehicle of transmission could not be determined. In this outbreak only one episode of diarrheal illness, in a child's father, was thought to represent secondary spread.[88]

Another outbreak in school children from 6 to 12 years old occurred in Japan in May 1980; 800 of the 2500 children who attended the same school over a 2 to 3 day period became ill, and many asymptomatic infections were suspected. Each child ate the same lunch so food-specific attack rates could not be determined, but vinegared pork was thought to be the most likely vehicle. No secondary transmission among household contacts was noted.[89]

IV. INCIDENCE AND PREVALENCE

A. By Location

1. Developed Countries

C. jejuni is an important cause of diarrheal illness in the developed countries of the world. The magnitude of the problem is difficult to assess because fecal cultures are obtained from only a fraction of persons who have acute diarrheal illnesses. Nevertheless, by comparing campylobacter isolation rates with those of other well-known pathogens, a relative estimate of incidence can be made. *Campylobacter* infections have been reported to the Communicable Disease Surveillance Center in England since 1977,[90] and by now most laboratories in the U.K. that culture fecal specimens for *Salmonella* and *Shigella* do so for *Campylobacter*. In

1980, about 9500 isolations of campylobacters, 3800 isolations of shigellae, and 10,500 isolations of salmonellae were reported. Using 1980 population data, the reported annual isolation rate for *Campylobacter* was about 21/100,000. In 1981, about 12,500 *Campylobacter* isolates were reported, corresponding to an annual isolation rate of about 28/100,000. Among persons with acute enteritis who reported to a single general practice in England, the projected annual rate of *Campylobacter* infection was 1100/100,000.[91] These data further confirm the vast underreporting of this infection.

In the U.S., national reporting of *Campylobacter* infection has just begun, and no data are available. However, studies conducted in several different localities in the U.S. and Canada again show that *Campylobacter* infections are at least as common as *Salmonella* or *Shigella* infections among patients with diarrheal diseases (Table 3). In a study done in the Denver metropolitan area, the reported annual incidence was 17.2/100,000 population.[118] In Europe and Australia data from individual medical centers illustrate a relative parity between the numbers of isolates of *Campylobacter*, *Salmonella*, and *Shigella*.

Among persons without diarrheal illness in developed countries, *Campylobacter* infection is uncommon. Several survey of healthy persons have shown the isolation rate from fecal specimens to be less than 1% (Table 3). Data on duration of excretion of *Campylobacter* after infection in developed countries show a median of 2 to 3 weeks.[100] Combining these data with the data from the culture surveys of healthy persons suggests that in the developed countries the incidence of *Campylobacter* infections, whether symptomatic or not, is about 1 to 2%/year. Kendall and Tanner's observations in a general practice in England support such estimates.[91] A similar incidence of *Salmonella* infections in the U.S. has been calculated.[119]

2. Developing Countries

Much of the data on prevalence of infection in developing countries have been presented in Chapter 3; however, several points deserve reemphasis. *Campylobacter* infection appears common in children in developing areas regardless of climate; frequent infections have been noted among Alaskan natives and children in South India. The prevalence of infection is much greater than in the developed countries, although the infection-to-illness ratio is probably higher. The vast majority of infections occur in the first 5 years of life, especially the first 2 years. Since data on the relative virulence of various serotypes and variation in the frequency of various serotypes from different regions are not yet available, it is uncertain if this high infection-to-illness ratio is serotype-related. However, *C. jejuni* infections are among the most common of the various enteric pathogens reported among travelers returning with acute diarrheal illnesses from developing areas.[70]

B. By Age and Sex

Most data on age- and sex-specific incidence of *Campylobacter* infections in the developed countries are based on culture surveys of patients with diarrhea. These data are thus affected by age-specific differences in the infection-to-illness ratio, and by differences in the age-specific rates for which fecal cultures are obtained from ill persons.

In several surveys based on clinical microbiology laboratory isolations, the highest rates of positivity were from specimens submitted from persons 10 to 29 years old.[103] In studies of submitted stool specimens done in Denver, 12.2% from persons 10 to 29 years old, and 1.4% from children less than 1 year old were positive.[105] In one study reported by Butzler and Skirrow,[106] when the number of positive cultures for an age group were divided by the population for that age group, the highest age-specific incidence of documented disease was found to be in young children, especially those less than 1 year old. In this same study over 60% of the isolates were from adults. In another population-based study done in the Denver area, for children under 1 year old the annual incidence of culture-positive illness was 149/100,000 whereas for persons 20 to 29 years old it was 31.8/100,000.[118] Data from a single

Table 3
ISOLATION OF *CAMPYLOBACTER* FROM FECAL CULTURES FROM PATIENTS WITH DIARRHEA AND FROM HEALTHY CONTROLS IN THE DEVELOPED COUNTRIES, 1973 TO 1981

				Patients with diarrhea				Healthy controls	
Author	Ref.	Location	Population	Number studied	% with *C. jejuni*	% with *Salmonella*	% with *Shigella*	Number studied	% with *C. jejuni*
Butzler	92	Belgium	Children	800	5.1	—	—	1,000	1.3
Skirrow	43	England	All ages	803	7.1	—	—	194	0
Brunton	93	Scotland	All ages	196	8.7	2.5	6.7	50	0
Bruce	94	England	All ages	280	13.9	4.3	3.9	156	0.6
Severin	95	Netherlands	All ages	584	10.8	10.0	—	120	0
Pai	87	Canada	Children	1,004	4.3	5.1	1.4	176	0
Blaser	96	U.S.	All ages	2,670	4.6	3.4	2.9	157	0
Lopez-Brea	97	Spain	All ages	446	4.5	12.1	1.3	—	—
Delorme	98	France	All ages	100	9.0	0	1.0	330	0
Watson	99	New Zealand	All ages	122	4.9	0.8	0.8	—	—
Graf	100	Switzerland	All ages	665	5.7	12.6	0.9	800	0
McAlister	101	Australia	All ages	69	8.7	11.6	4.3	—	—
Young	102	U.S.	All ages	998	4.7	1.0	5.3	181	1.2
Wright[a]		England	All ages	695	4.6	—	—	1300	0.4
Blaser	103	U.S.	All ages	7,531	4.9	2.4	1.0	—	—

[a] Personal communication: E. P. Wright.

Table 4
EXTRAPOLATION OF MENTZING'S EPIDEMIOLOGIC DATA GENERATED DURING A WATERBORNE OUTBREAK OF *CAMPYLOBACTER* ENTERITIS IN SWEDEN, ILLUSTRATING AGE-SPECIFIC SURVEILLANCE ARTIFACTS[76]

Age group (years)	Estimated no. of cases	No. of reported cases	Reporting rate (per 100 estimated cases)
0—9	187	101	53.9
10+	1530	262	17.1
Total	1717	363	21.1

Table 5
ISOLATION OF *C. JEJUNI* FROM HEALTHY PERSONS IN SOUTH INDIA BY AGE — NOVEMBER 1980 TO FEBRUARY 1981[110]

Age group	No. cultured	Percent positive
0—4	54	37.0
5—11	71	19.7
12—17	37	5.4
18+	143	6.3
Total	305	14.8

general practice in England showed a bimodal distribution with peak incidences projected in children under 5 years old (5400/100,000) and in persons 15 to 24 years old (2000/100,000).[91]

However, review of data generated during a waterborne outbreak of *Campylobacter* infection[76] demonstrated some surveillance artifacts often present in data available from hospitals and clinics. Mentzing[76] showed that during a large common-source outbreak, the largest number of patients brought to medical attention were children under 10 years old. However, a cross-sectional survey of the affected population showed that attack rates were approximately equal in all age groups studied. Extrapolating from his data showed that a higher proportion of the cases in children (54%) than in adults (17%) were reported (Table 4).

In summary, although there may be surveillance artifacts, there appears to be a bimodal distribution of *Campylobacter* infection by age with the highest incidence in infants and in persons 20 to 29 years old. Presumably the relative importance of various modes and vehicles for infection may be different in the two age groups.

In areas of developing countries where hygiene is poor, the prevalence of infection appears to be highest in young children.[107-110] Studies of healthy persons in South India illustrate this trend and also demonstrate that infection is prevalent in all age groups (Table 5). In Gambia, infection of healthy children under 5 years old was less common (5.7%) and none of 104 healthy persons ≥5 years old were culture positive.[109] In contrast, in developing countries where hygienic conditions are improved, such as Indonesia[111] and Saudi Arabia,[112] isolation of *C. jejuni* from healthy persons is uncommon and rates resemble those seen in

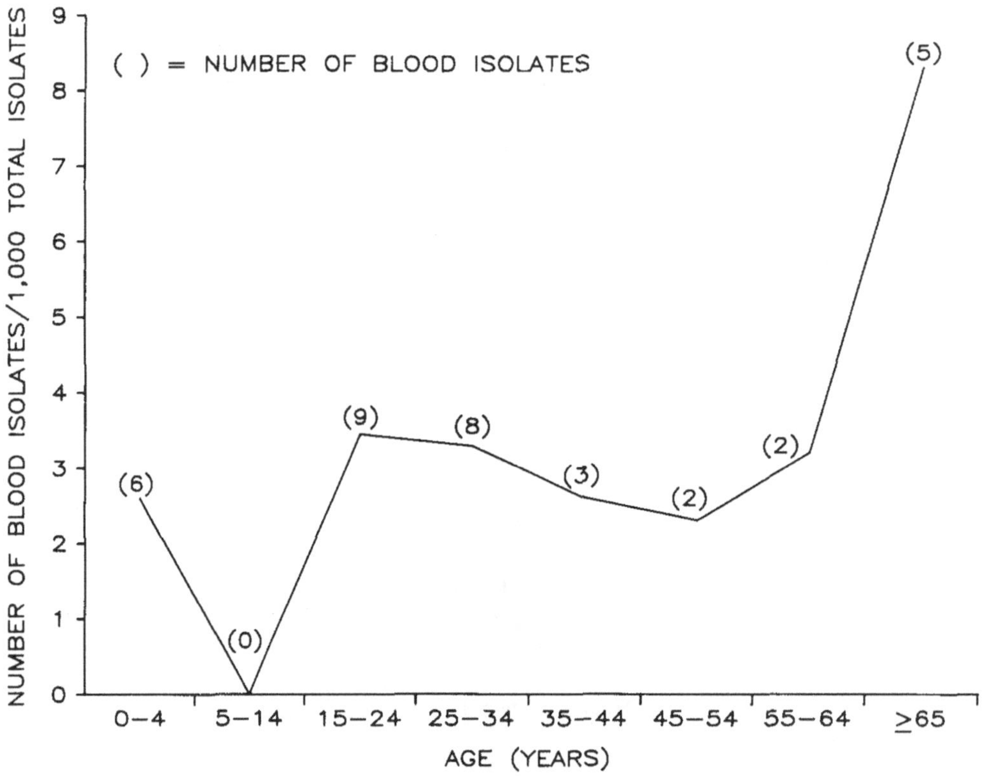

FIGURE 1. Ratio of *Campylobacter* blood isolates to total *Campylobacter* isolates reported to the Communicable Disease Surveillance Centre (CDSC) by patient's age, 1980.

the developed countries. In a study done in South Africa, 13.4% of black children with diarrhea under 2 years old had *C. jejuni* isolated from stools, whereas for a comparable group of Caucasian children the rate was 4.9%.[113]

Campylobacter enteritis is sometimes associated with bacteremia. Review of data from the Communicable Disease Surveillance Centre on numbers of blood and stool isolates reported in 1980[90] has enabled us to construct age-specific ratios of blood-to-total isolates (Figure 1). As shown in our analysis of their data, for most age groups the ratio of blood to total isolates was relatively constant, but there was an increase in persons over the age of 65 years. A similar phenomenon has been observed in surveillance data of *Salmonella* infections.

The older literature implied that the male sex was a risk factor for *Campylobacter (Vibrio fetus)* infection.[114] The data addressing this question for *C. jejuni* are conflicting. Of 3319 isolates from English patients in 1978, 52.8% were from males, and of patients below the age of 15 years, 57.7% were from males; whether males were cultured more frequently is not known. Data from an eight-hospital cooperative study in the U.S. showed that males were cultured more frequently, but the isolation rates were approximately the same for males and females.[103] However, in that study the seasonal distribution of infection by sex was appreciably different. For males during June through September, the isolation rate was 9.3% while for the other months the rate was 4.5%. For females during the same time period the rates were 7.0 and 5.7%, respectively.

CDSC data show that 61% of isolates from persons ≥65 years are from females. A similar female predominance in that age group has been noted for *Shigella* infections as well, but both these findings may reflect the fact that by age 65 more women than men are alive.

C. Seasonality

Studies in England, Belgium, the U.S., and South Africa have shown a summertime peak of *Campylobacter* infection.[103,106,113] These studies have shown that both the absolute number of isolates and the isolation rate from submitted stools are increased in the warm months. In Sweden, where most of the reported infections occur in travelers returning from abroad, late summer and fall peaks were also noted, corresponding to seasonal habits for traveling abroad.[70] From the CDSC data, a second smaller peak in isolations during the winter months has been reported.[90] In Zaire, where mean temperatures are constant throughout the year, isolation of *C. jejuni* from patients with diarrhea was much more frequent in the wet than in the dry season.[115]

D. Occupation

No relation of *Campylobacter* infection to occupation per se has been shown; however, the seroepidemiologic studies of Jones and Robinson, cited earlier, are suggestive.[51] Using a complement-fixation assay, only 2 to 5% of sera from healthy young women had significant titers to *C. jejuni*, but 27 to 68% of workers with poultry and cattle were similarly positive. These data suggest that previous infection was more common among the meat workers; however, there are no data at present to suggest that illness due to *C. jejuni* was more common in the latter group. Another survey of stools of 410 meat workers in Australia showed none to be culture positive.[62] Nevertheless, there have been anecdotal reports in which a strong temporal relationship was shown between *Campylobacter* infection and new occupational contact with cattle and sheep, or with beef and chicken carcasses.[33,49,50,96] Nosocomial transmission of *Campylobacter* infection appears to be rare, and there is no information at present to suggest that hospital personnel are at increased risk for infection.

V. CONCLUSIONS

Much that we have learned about the campylobacters is reminiscent of what is known about the salmonellae. As with salmonellae, the major identified reservoirs are animals and products obtained from animals. Although milk has been most frequently shown to be a vehicle for *Campylobacter*, one anticipates that future investigations will identify poultry, poultry products, and meats obtained from other animals as major reservoirs and vehicles. Although dogs, cats, and birds have been shown to harbor salmonellae, they are not significant reservoirs or vehicles of human infection. Similarly, although *Campylobacter* from puppies and kittens may infect some humans, it is unlikely that pets will be the major reservoir or vehicle for *Campylobacter*.

The seasonality of *Campylobacter* infections in humans may relate to the frequency of contamination of animal products used for human food, but may also relate to the fact that in warmer weather larger infectious doses occur, with higher disease-to-infection ratios. As has been found with the salmonellae, however, it is probable that for some *Campylobacter* serotypes there will be little seasonality.

Although serotypes of *Campylobacter* have not yet been identified which have specific association with vehicles or with magnitude of clinical illness, such associations have been found with salmonellae and may be expected with campylobacters. Campylobacters have been found frequently in the developing world, while salmonellae are infrequently found. This difference deserves exploration, since it suggests differences between transmission of *Campylobacter* and *Salmonella* not yet identified.

REFERENCES

1. **Luechtefeld, N. W., Cambre, R. C., and Wang, W. L. L.,** Isolation of *Campylobacter*-fetus subsp. *jejuni* from zoo animals, *J. Am. Vet. Med. Assoc.*, 179, 1119, 1980.
2. **Smith, M. V. and Muldoon, O. J.,** *Campylobacter fetus* subspecies *jejuni (Vibrio fetus)* from commercially processed poultry, *Appl. Microbiol.*, 27, 995, 1974.
3. **Bruce, D., Zochowski, W., and Ferguson, I. R.,** Campylobacter enteritis, *Br. Med. J.*, 2, 1219, 1977.
4. **Ribeiro, C.,** Campylobacter enteritis, *Lancet*, 2, 270, 1978.
5. **Simmons, N. A. and Gibbs, E. J.,** Campylobacter spp. in oven-ready poultry, *J. Infect.*, 1, 159, 1979.
6. **Grant, I. H., Richardson, N. J., and Bokkenheuser, V. D.,** Broiler chickens as potential source of *Campylobacter* infections in humans, *J. Clin. Microbiol.*, 11, 508, 1980.
7. **Goren, E. and de Jong, W. A.,** Campylobacter fetus subspecies jejuni in chickens, *Tijdschr. Diergeneesk*, 105, 724, 1980.
8. **Eiden, J. J. and Dalton, H. P.,** An animal reservoir for *Campylobacter fetus* ss. *jejuni*, presented at 20th Intersci. Conf. Antimicrob. Agents and Chemother., Am. Soc. Microbiol., New Orleans, September 1980 (Abstr. 694).
9. **Park, C. E., Stankiewicz, Z. K., Love, H. J., and Hunt, J.,** Incidence of *Campylobacter jejuni* in fresh eviscerated whole market chickens, *Can. J. Microbiol.*, 27, 841, 1981.
10. **Luechtefeld, N. W. and Wang, W.-L. L.,** *Campylobacter fetus* subsp. *jejuni* in a turkey processing plant, *J. Clin. Microbiol.*, 13, 266, 1981.
11. **Prescott, J. F., and Bruin-Mosch, C. W.,** Carriage of *Campylobacter jejuni* in healthy and diarrheic animals, *Am. J. Vet. Res.*, 42, 164, 1981.
12. **Svedhem, A. and Kaijser, B.,** Isolation of *Campylobacter jejuni* from domestic animals and pets: Probable origin of human infection, *J. Infect.*, 3, 37, 1981.
13. **Cruickshank, J. G., Egglestone, S. I., Gawler, A. H. L., and Lanning, D.,** Campylobacters and the broiler chicken, in *Campylobacter: Epidemiology, Pathogenesis, and Biochemistry*, Newell, D. G., Ed., MTP Press, Lancaster, U.K., 1982, 263.
14. **Mehle, J., Gubina, M., and Gliha, B.,** Contamination of chicken meat with *Campylobacter jejuni* during the process of industrial slaughter, in *Campylobacter: Epidemiology, Pathogenesis, and Biochemistry*, Newell, D. G., Ed., MTP Press, Lancaster, U.K., 1982, 267.
15. **Luechtefeld, N. W., Blaser, M. J., and Wang, W. L. L.,** Isolation of *Campylobacter fetus* subsp. *jejuni* from migratory waterfowl, *J. Clin. Microbiol.*, 12, 406, 1980.
16. **Robinson, D. A.,** Campylobacter infection in milking herds, in *Campylobacter: Epidemiology, Pathogenesis, and Biochemistry*, Newell, D. G., Ed., MTP Press, Lancaster, U.K., 1982, 274.
17. **Taylor, D. N., Porter, B. W., Williams, C. A., Miller, H. E., Bopp, C. A., and Blake, P. B.,** A large outbreak of *Campylobacter* enteritis traced to commercially produced raw milk, *West. J. Med.*, 137, 365, 1982.
18. **Bryner, J. H., O'Berry, P. A., Estes, P. C., and Foley, J. W.,** Studies of vibrios from gallbladder of market sheep and cattle, *Am. J. Vet. Res.*, 33, 1439, 1972.
19. **Luechtefeld, N. W. and Wang, W. L. L.,** Animal reservoirs of *Campylobacter jejuni*, in *Campylobacter: Epidemiology, Pathogenesis, and Biochemistry*, Newell, D. G., Ed., MTP Press, Lancaster, U.K., 1982, 249.
20. **Stern, N. J.,** Recovery rate of *Campylobacter fetus* spp. *jejuni* on eviscerated pork, lamb, and beef carcasses, *J. Food Sci.*, 46, 1291, 1981.
21. **Hudson, W. R. and Roberts, T. A.,** The occurrence of campylobacter on commercial red meat carcasses from one abbattoir, in *Campylobacter: Epidemiology, Pathogenesis, and Biochemistry*, Newell, D. G., Ed., MTP Press, Lancaster, U.K., 1982, 273.
22. **Turnbull, P. C. B. and Rose, P.,** *Campylobacter jejuni* in raw red meats. A Public Health Laboratory Service Survey, in *Campylobacter: Epidemiology, Pathogenesis, Biochemistry*, Newell, D. G., Ed., MTP Press, Lancaster, U.K., 1982, 271.
23. **Oosterom, J.,** The presence of *Campylobacter fetus* subspecies *jejuni* in normal slaughtered pigs, *Tijdschr. Diergeneesk*, 105, 49, 1980.
24. **Stichtgroh, V.,** Campylobacter in healthy slaughter pigs: A possible source of infection for man, *Vet. Rec.*, 110, 104, 1982.
25. **Robinson, D. A. and Jones, D. M.,** Milk-borne campylobacter infection, *Br. Med. J.*, 282, 1374, 1981.
26. **Lander, K. P. and Gill, K. P. W.,** Experimental infection of the bovine udder with *Campylobacter coli/jejuni*, *J. Hyg. (Cambridge)*, 84, 421, 1980.
27. **Potter, M. E., Blaser, M. J., Sikes, R. K., Kaufmann, A. F., and Wells, J. G.,** Human campylobacteriosis associated with certified raw milk, *Am. J. Epidemiol.*, 117, 475, 1983.
28. **Doyle, M. P. and Roman, D. J.,** Growth and survival of *Campylobacter fetus* subsp. *jejuni* as a function of temperature and pH, *J. Food Prot.*, 44, 596, 1981.

29. **Smibert, R. M.,** The genus *Campylobacter, Ann. Rev. Microbiol.*, 32, 674, 1978.
30. **Blaser, M. J., LaForce, F. M., Wilson, N. A., and Wang, W.-L. L.,** Reservoirs for human campylobacteriosis, *J. Infect. Dis.*, 141, 665, 1980.
31. **Bruce, D., Zochowski, W., and Fleming, G. A.,** Campylobacter infections in cats and dogs, *Vet. Rec.*, 107, 200, 1980.
32. **Fernie, D. S. and Park, R. W. A.,** The isolation and nature of Campylobacters (microaerophilic vibrios) from laboratory and wild rodents, *J. Gen. Microbiol.*, 10, 325, 1977.
33. **Tribe, G. W. MacKenzie, P. S., and Fleming, M. P.,** Incidence of thermophilic Campylobacter species in newly imported simian primates with enteritis, *Vet. Rec.*, 105, 333, 1979.
34. **Lauwers, S., Berge, E., Naessens, A., and Butzler, J. P.,** Monkeys as a reservoir for *Campylobacter jejuni*, abstr. C217 presented at the 81st Annu. Meet. Am. Soc. Microbiology, Dallas, Tex., 1981.
35. **Blaser, M. J., Hardesty, H. L., Powers, B., and Wang, W.-L. L.,** Survival of *Campylobacter fetus* subsp. *jejuni* in biological milieus, *J. Clin. Microbiol.*, 1980, 11: 309—13.
36. **Knill, M. J., Suckling, W. G., and Pearson, A. D.,** Campylobacters from water, in *Campylobacter: Epidemiology, Pathogenesis, and Biochemistry,* Newell, D. G., Ed., MTP Press, Lancaster, U.K., 1982, 281.
37. **Knill, M. J., Suckling, W. G., and Pearson, A. D.,** Environmental isolation of heat-tolerant Campylobacter in the Southampton area, *Lancet,* 2, 1002, 1978.
38. **Khan, M. S.,** An epidemiological study of a campylobacter enteritis outbreak involving dogs and man, in *Campylobacter: Epidemiology, Pathogenesis, an Biochemistry,* Newell, D. G., Ed., MTP Press, Lancaster, U.K., 1982, 256.
39. **Ottolenghi, A. C. and Hamparian, V. V.,** Bacteriology of sewage sludge: examination for the presence of *Salmonella* and *Campylobacter*, abstr. presented at the 82nd Annu. Meet. Am. Soc. Microbiology, Atlanta, Ga., March 7 to 12, 1982.
40. **Lindenstruth, R. W. and Ward, B. O.,** Viability of *Vibrio fetus* in hay, soil and manure, *J. Am. Vet. Med. Assoc.*, 113, 163, 1948.
41. **Luechtefeld, N. W. and Wang, W.-L. L.,** Presentation of *Campylobacter jejuni* in fecal specimens frozen at −20 and −70°C, abstr. presented at the 82nd Annu. Meet. Am. Soc. Microbiology, Atlanta, Ga., March 7 to 12, 1982.
42. **Blaser, M., Powers, B. W., Cravens, J., and Wang, S. L.,** Campylobacter enteritis associated with canine infection, *Lancet*, 2, 979, 1978.
43. **Skirrow, M. B.,** Campylobacter enteritis: A ''new'' disease, *Br. Med. J.*, 2, 9, 1977.
44. **Blaser, M. J., Penner, J. L., and Wells, J. G.,** Diversity of serotypes involved in outbreaks of *Campylobacter enteritis, J. Infect. Dis.*, 146, 826, 1982.
45. **Gruffydd-Jones, T. J., Marston, M., and White, E.,** *Campylobacter jejuni* enteritis from cats, *Lancet,* 2, 366, 1980.
46. **Skirrow, M. B., Turnbull, G. L., Walker, R. E., and Young, S. E.,** *Campylobacter jejuni* enteritis transmitted from cat to man, *Lancet,* 2, 1188, 1980.
47. **Svedhem, A. and Norkrams, G.,** *Campylobacter jejuni* enteritis transmitted from cat to man, *Lancet,* 1, 713, 1980.
48. **Blaser, M. J., Weiss, S. H., and Barrett, T. J.,** *Campylobacter* enteritis associated with a healthy cat, *J. Am. Med. Assoc.*, 247, 816, 1982.
49. **Blaser, M. J., Parsons, R. B., and Wang, W. L. L.,** Acute colitis caused by *Campylobacter fetus* ss. *jejuni, Gastroenterology*, 78, 448, 1980.
50. **Duffell, S. J. and Skirrow, M. B.,** Shepherd's scours and bovine campylobacter abortion — a ''new'' zoonosis?, *Vet. Rec.*, 103, 144, 1978.
51. **Jones, D. M. and Robinson, D. A.,** Occupational exposure to *Campylobacter jejuni* infection, *Lancet,* 1, 440, 1981.
52. **Oates, J. D. and Hodgin, U. G., Jr.,** Laboratory-acquired *Campylobacter* enteritis, *South. Med. J.*, 74, 83, 1981.
53. **Prescott, J. F. and Karmali, M. A.,** Attempts to transmit *Campylobacter* enteritis to dogs and cat, *Can. Med. Assoc. J.*, 4, 1001, 1978.
54. **Porter, I. A. and Reid, T. M. S.,** A milk-borne outbreak of *Campylobacter* infection, *J. Hyg. (Cambridge)*, 84, 415, 1980.
55. **Jones, P. H., Willis, A. T., Robinson, D. A., Skirrow, M. B., and Josephs, D. S.,** Campylobacter enteritis associated with the consumption of free school milk, *J. Hyg. (Cambridge)*, 87, 155, 1981.
56. **Levy, A. J.,** A gastro-enteritis outbreak probably due to a bovine strain of *Vibrio, Yale J. Biol. Med.*, 18, 243, 1946.
57. **Taylor, P. R., Weinstein, W. M. and Bryner, J. H.,** *Campylobacter fetus* infection in human subjects: Association with raw milk, *Am. J. Med.*, 68, 779, 1979.
58. **Blaser, M. J., Cravens, J., Powers, B. W., LaForce, F. M., and Wang, W.-L. L.,** Campylobacter enteritis associated with unpasteurized milk, *Am. J. Med.*, 67, 715, 1979.

59. Centers for Disease Control, Raw-milk-associated illness — Oregon, *MMWR*, 30, 90, 1981.
60. Centers for Disease Control, Outbreak of *Campylobacter* enteritis associated with raw milk — Kansas, *MMWR*, 30, 218, 1981.
61. **Robinson, D. A., Edgar, W. M., Gibson, G. L., Matchett, A. A., and Robertson, L.,** Campylobacter enteritis associated with the consumption of unpasteurized milk, *Br. Med. J.*, 2, 1171, 1979.
62. **Steele, T. W. and McDermott, S.,** Campylobacter enteritis in South Australia, *Med. J. Austr.*, 2, 404, 1978.
63. **Robinson, D. A.,** Infective dose of *Campylobacter jejuni* in milk, *Br. Med. J.*, 1, 1584, 1981.
64. **Brouwer, R., Mertens, M. J. A., Siem, T. H., and Katchaki, J.,** An explosive outbreak of *Campylobacter- enteritis in soldiers, Antonie Van Leeuwenhoek J.*, 45, 517, 1979.
65. **Skirrow, M. B., Fidoe, R. B., and Jones, D. M.,** An outbreak of presumptive food-borne campylobacter enteritis, *J. Infect.*, 3, 234, 1981.
66. **Severin, W. P. J.,** Epidemiology of campylobacter infection, in *Campylobacter: Epidemiology, Pathogenesis, and Biochemistry*, Newell, D. G., Ed., MTP Press, Lancaster, U.K., 1982, 285.
67. **Blaser, M. J., Feldman, R. A., and Wells, J. G.,** Epidemiology of endemic and epidemic campylobacter infections in the U.S., in *Campylobacter: Epidemiology, Pathogenesis, and Biochemistry*, MTP Press, Lancaster, U.K., 1982, 3.
68. **Oosterom, J., Beckers, H. J., van Noorle Jansen, L. M., and van Schothorst, M.,** Een explosie van Campylobacter-infectie in een Kazerne, waarschijnlijk veroorzaakt door rauwe tartaar, *Ned. Tijdschr. Geneeskd.*, 27, 1631, 1980.
69. **Blaser, M. J., Checko, P., Bopp, C., Bruce, A., and Hughes, J. M.,** *Campylobacter* enteritis associated with foodborne transmission, *Am. J. Epidemiol.*, 116, 886, 1982.
70. **Svedhem, A. and Kaijser, B.,** *Campylobacter fetus* subspecies *jejuni:* A common cause of diarrhea in Sweden, *J. Infect. Dis.*, 142, 353, 1980.
71. **Pitkanen, T., Pettersson, T., Ponka, A., and Kosunen, T. U.,** Clinical and serological studies in patients with *Campylobacter fetus* ssp. *jejuni* infection. I. Clinical findings, *Infection*, 9, 274, 1981.
72. **Ryder, R. W., Oquist, C. A., Greenberg, H., Taylor, D. N., Orskov, F., Orskov, I., Kapikian, A. Z., and Sack, R. B.,** Travelers' diarrhea in Panamanian tourists in Mexico, *J. Infect. Dis.*, 144, 442, 1981.
73. **Merson, M. H., Morris, G. K., Sack, D. A., Wells, J. G., Feeley, J. C., Sack, R. B., Creech, W. B., Kapikian, A. Z., and Gangarosa, E. J.,** Travelers' diarrhea in Mexico, *N. Engl. J. Med.*, 294, 1299, 1976.
74. **Tjoa, W. S., DuPont, H. L., Sullivan, P., Pickering, L. K., Holguin, A. H., Olarte, J., Evans, D. G., and Evans, D. J.,** Location of food consumption and travelers' diarrhea, *Am. J. Epidemiol.*, 106, 61, 1977.
75. **Vogt, R. L., Sours, H. E., Barrett, T., Feldman, R. A., Dickinson, R. J., and Witherell, L.,** Campylobacter enteritis associated with contaminated water, *Ann. Intern. Med.*, 96, 292, 1982.
76. **Mentzing, L. O.,** Waterborne outbreaks of Campylobacter enteritis in Central Sweden, *Lancet*, 2, 352, 1981.
77. Health and Welfare, Canada, possible waterborne Campylobacter outbreak — British Columbia, *Can. Dis. Wkly. Rep.*, 7(45), 223, 1981.
78. Rhyl Public Health Laboratory, Campylobacter enteritis amongst scouts, CDR 80/53, Communicable Disease Surveillance Center, London, January 9, 1981, 1.
79. **Quinn, T. C., Corey, L., Chaffee, R. G., Schuffler, M. D., and Holmes, K. K.,** Campylobacter proctitis in a homosexual man, *Ann. Intern. Med.*, 93, 458, 1980.
80. **Carey, P. B. and Wright, E. P.,** *Campylobacter jejuni* in a male homosexual, *Br. J. Vener. Dis.*, 55, 381, 1979.
81. **Gribble, M. J., Salit, I. E., Isaac-Renton, J., and Chow, A. W.,** Campylobacter infections in pregnancy, *Am. J. Obstet. Gynecol.*, 140, 423, 1981.
82. **Thomas, K., Chan, K. N., and Ribeiro, C. D.,** Campylobacter jejuni/coli meningitis in a neonate, *Br. Med. J.*, 2, 1301, 1980.
83. Campylobacter in a mother and baby, CDR 79/17, Communicable Disease Surveillance Center, London, 1981, 4.
84. **Vesikari, T., Huttunen, L., and Maki, R.,** Perinatal *Campylobacter fetus* s.s. *jejuni* enteritis, *Acta Paediatr. Scand.*, 70, 261. 1981.
85. **Anders, B. J., Lauer, B. A., and Paisley, J. W.,** Campylobacter gastroenteritis in neonates, *Am. J. Dis. Child.*, 135, 900, 1981.
86. **Karmali, M. A. and Tan, T. Y.,** Neonatal campylobacter enteritis, *Can. Med. Assoc. J.*, 122, 192, 1980.
87. **Pai, C. H., Sorger, S., Lackman, L., Sinai, R. E., and Marks, M. I.,** Campylobacter gastroenteritis in children, *J. Pediatr.*, 94, 589, 1979.

88. **Itoh, T., Saito, K., Maruyama, T., Sakai, S., Ohashi, M., and Oka, A.**, An outbreak of acute enteritis due to *Campylobacter fetus* subspecies *jejuni* at a nursery school in Tokyo, *Microbiol. Immunol.*, 24, 371, 1980.
89. **Yanagisawa, S.**, Large outbreak of campylobacter enteritis among schoolchildren, *Lancet*, 2, 153, 1980.
90. Review of Campylobacter reports to CDSC 1977—80 CDR 81/12, Communicable Disease Surveillance Centre, London, 1981, 3.
91. **Kendall, E. J. C. and Tanner, E. I.**, *Campylobacter* enteritis in general practice, *J. Hyg.*, 88, 155, 1982.
92. **Butzler, J. P., Dekeyser, P., Detrain, M., and Dehaen, F.**, Related vibrio in stools, *J. Pediatr.*, 82, 493, 1973.
93. **Brunton, W. A. T. and Heggie, D.**, Campylobacter-associated diarrhoea in Edinburgh, *Br. Med. J.*, 2, 956, 1977.
94. **Bruce, D., Zochowski, W., and Ferguson, I. R.**, Campylobacter enteritis, *Br. Med. J.*, 2, 1219, 1977.
95. **Severin, W. P. J.**, Campylobacter enteritis, *Ned. Tijdschr. Geneeskd.*, 122, 499, 1978.
96. **Blaser, M. J., Berkowitz, I. D., LaForce, F. M., Craens, J., Reller, L. B., and Wang, W. L. L.**, Campylobacter enteritis: clinical and epidemiologic features, *Ann. Intern. Med.*, 91, 179, 1979.
97. **Lopez-Brea, M., Molina, D., and Baquero, M.**, Campylobacter enteritis in Spain, *Trans. R. Soc. Trop. Med. Hyg.*, 37, 474, 1970.
98. **Delorme, I., Lambert, T., Branger, C., and Acar, J. F.**, Enteritis due to *Campylobacter jejuni* in the Paris area, *Med. Mal. Infect.*, 9, 675, 1979.
99. **Watson, L. A., Brooks, H. J. L., and Scrimgeour, G.**, Campylobacter enteritis and Yersinia enterocolitics infection in New Zealand, *N.Z. Med. J.*, 90, 240, 1979.
100. **Graf, J., Schar, G., and Heinzer, I.**, *Campylobacter jejuni* enteritis in der Schweiz, *Schweiz Med. Wochenschr.*, 110, 590, 1980.
101. **McAlister, T. V.**, Brief note. Campylobacter jejuni gastroenteritis in Brisbane, *Pathology*, 11, 299, 1979.
102. **Young, J. R., Callahan, P., Drew, W. L., and Hadley, W. K.**, Diarrheal disease associated with isolation of *Campylobacter fetus* subsp. *jejuni* in adults, in *Current Chemotherapy and Infectious Disease*, American Society for Microbiology, Washington, D.C., 1980, 939.
103. **Blaser, M. J., Wells, J. G., Feldman, R. A., and Pollard, R. A.**, *Campylobacter* enteritis in the United States. A multicenter study, *Ann.* Int. Med., 98, 360, 1983.
104. **Wright, E. P.**, Duration of excretion period of *Campylobacter jejuni* in human subjects, in *Campylobacter: Epidemiology, Pathogenesis, and Biochemistry*, Newell, D. G., Ed., MTP Press, Lancaster, U.K., 1982, 294.
105. **Blaser, M. J., Reller, L. B., Luechtefeld, N. W., and Wang, W. L.**, Campylobacter enteritis in Denver, *West, J. Med.*, 136, 287, 1982.
106. **Butzler, J. P. and Skirrow, M. B.**, Campylobacter enteritis, *Clin. Gastroenterol.*, 8, 737, 1979.
107. **Bokkenheuser, V. D., Richardson, N. J., Bryner, J. H., Roux, D. J., Schutte, A. B., Koornhof, H. J., Freiman, I., and Hartman, E.**, Detection of enteric campylobacteriosis in children, *J. Clin. Microbiol.*, 9, 227, 1979.
108. **Blaser, M. J., Glass, R. I., Huq, M. I., Stoll, B., Kibriya, G. M., and Alim, A. R. M. A.**, Isolation of *Campylobacter fetus* spp. *jejuni* from Bangladeshi children, *J. Clin. Microbiol.*, 12, 744, 1980.
109. **Billingham, J. D.**, Campylobacter enteritis in the Gambia, *Trans. R. Soc. Trop. Med. Hyg.*, 75, 641, 1981.
110. **Rajan, D. P. and Mathan, V. I.**, Prevalence of *Campylobacter fetus* subsp. *jejuni* in healthy populations in southern India, *J. Clin. Microbiol.*, 15, 749, 1982.
111. **Ringertz, S., Rockhill, R. C., Ringertz, O., and Sutomo, A.**, *Campylobacter fetus* subsp. *jejuni as a cause of gastroenteritis in Jakarta, Indonesia, J. Clin. Microbiol.*, 12, 539, 1980.
112. **Chowdhury, M. N. H. and Mahgoub, E. S.**, Gastroenteritis due to *Campylobacter jejuni* in Riyadh, Saudi Arabia, *Trans. R. Soc. Trop. Med. Hyg.*, 75, 359, 1981.
113. **Mauff, A. C. and Chapman, S. R.**, Campylobacteritis in Johannesburg, S. Afr. Med. J., *59, 217, 1981.*
114. **Communicable Disease Surveillance Center**, Campylobacter infections, 1977-80, *Br. Med. J.*, 282, 1484, 1981.
115. **Bokkenheuser, V.**, *Vibrio fetus* infection in man. I. Ten new cases and some epidemiologic observations, *Am. J. Epidemiol.*, 91, 400, 1970.
116. **De Mol, P., Brasseur, D., Lauwers, S., Zissis, G., and Butzler, J. P.**, *Campylobacter — An Important Enteropathogen in a Tropical Area*, American Society for Microbiology, Washington, D.C., 1980.
117. **Istre, G. and Blaser, M. J.**, unpublished data.
118. **Olmstead, R.**, personal communication.
119. **Blaser, M. J.**, unpublished data.

Chapter 14

CAMPYLOBACTER IN FOODS

Michael P. Doyle

TABLE OF CONTENTS

I. Introduction 164

II. Occurrence in Foods 164
- A. Poultry 164
- B. Red Meats 164
- C. Raw Milk 164

III. Factors Affecting Survival and Growth 165
- A. Temperature 165
  - 1. Growth 165
  - 2. Refrigeration 167
  - 3. Freezing 167
  - 4. Thermal Inactivation 168
- B. pH 169
- C. Sodium Chloride 170
- D. Dehydration 174
- E. Oxygen and Carbon Dioxide 174
- F. Disinfectants 174
- G. Raw Foods 176

IV. Isolation Procedures 177

References 179

I. INTRODUCTION

The recent recognition of *Campylobacter jejuni* being a foodborne pathogen, coupled with statistics that *Campylobacter* enteritis is among the leading causes of bacterial gastroenteritis in humans, puts the food processor on the defensive and in need of information regarding the organism's association with foods. Information that is needed includes knowledge of: (1) the types of foods with which campylobacters are associated, (2) how well the organism survives and grows under different conditions, (3) what treatments eliminate campylobacters from foods, and (4) what methods may be used to detect *C. jejuni* in foods. This chapter will review our current knowledge of each of these topics.

II. OCCURRENCE IN FOODS

A. Poultry

Poultry parts and carcasses have been found to be major sources of *C. jejuni* (Table 1). The organism has been isolated from as many as 92% of poultry carcasses and 85 and 89% of livers and gizzards surveyed, respectively. Approximately 50% of the contaminated livers and gizzards contained greater than 1100 *C. jejuni* per gram.[7] Many of these surveys were done on poultry tissue obtained at the retail level; hence the homemaker and food service personnel should be exceptionally careful in handling poultry meat, taking precautions to avoid cross contamination when preparing meals.

B. Red Meats

Surveys of red meats have found *C. jejuni* most commonly associated with porcine carcasses (Table 2). The organism was isolated from 38 to 59% of carcasses of freshly slaughtered swine; however, refrigerated storage appears to reduce the number of campylobacter-positive carcasses. Rosef[9] reported a decrease from 56% campylobacter-positive carcasses sampled soon after slaughter to 32% campylobacter-positive carcasses after 4 days at 5°C. Hudson and Roberts[10] reported greater recovery of campylobacters from moist than dry areas of refrigerated porcine carcasses. They observed a reduction from 59% campylobacter-positive carcasses sampled soon after slaughter to 26 and 2% campylobacter-positive carcasses sampled from moist and dry carcasses, respectively, after 24 hr at 0°C. Additionally, they reported that when porcine carcasses were contaminated, the number of *C. jejuni* was generally very low, i.e., less than 1 CFU/cm^2.

In contrast to swine, *C. jejuni* was rarely isolated from carcasses of freshly slaughtered beef (Table 2) suggesting that beef is less likely to be a source of *Campylobacter* than either pork or poultry meat.

Although studies are limited, the incidence of *C. jejuni* on retail red meats appears to considerably less than on retail poultry meat. As part of a large survey conducted by 31 public health laboratories, Turnbull and Rose[11] reported the isolation of *C. jejuni* from only 49 of 4933 (1.0%) retail red meat samples. Interestingly, the organism was more often isolated from minced beef than minced pork (Table 2). This is somewhat surprising since *Campylobacter* is more commonly isolated from porcine than bovine carcasses.

C. Raw Milk

Although foodborne outbreaks of *Campylobacter* enteritis have been associated most frequently with consumption of unpasteurized milk, relatively few studies have been done to assess the incidence of *C. jejuni* in raw milk. Doyle and Roman,[12] using an enrichment procedure capable of detecting 0.1 to 1.0 *C. jejuni* per milliliter of raw milk, isolated the organism from 1 of 108 (0.9%) samples of unpasteurized milk obtained from bulk tanks of Grade A dairy farms. Using a less sensitive recovery procedure, Christopher et al.[7] were

Table 1
INCIDENCE OF *CAMPYLOBACTER JEJUNI* ON RETAIL OR RETAIL-READY PROCESSED POULTRY

Animal	Tissue sampled	No. sampled	% Positive	Ref.
Chicken	Neck	121	2	1
Chicken	Liver	25	0	1
Chicken	Carcass	19	5	1
Chicken	Carcass	35	80	2
Turkey	Carcass	12	92	2
Chicken	Carcass	750	0	3
Turkey	Heart, livers, and gizzards	24	33	4
Turkey	Carcass	83	34	4
Chicken	Carcass (U.S.)	50	54	5
Chicken	Carcass (Canada)	50	62	5
Chicken	Carcass (frozen)	82	22	6
Chicken	Liver	60	85	7
Chicken	Gizzard	64	89	7
Turkey	Liver	86	0	7
Turkey	Gizzard	86	1	7

Table 2
INCIDENCE OF *CAMPYLOBACTER JEJUNI* ON RED MEATS

Animal	Tissue sampled	No. sampled	% Positive	Ref.
Beef	Carcass (freshly slaughtered)	58	2	8
Sheep	Carcass (freshly slaughtered)	59	24	8
Swine	Carcass (freshly slaughtered)	58	38	8
Swine	Carcass (freshly slaughtered)	100	56	9
Swine	Carcass (stored at 5°C for 4 days)	100	32	9
Swine	Liver (freshly slaughtered)	100	43	9
Beef	Carcass (freshly slaughtered)	100	0	10
Lamb	Carcass (freshly slaughtered)	100	0	10
Swine	Carcass (freshly slaughtered)	100	59	10
Swine	Carcass (stored at 0°C for 24 hr, moist)	50	26	10
Swine	Carcass (stored at 0°C for 24 hr, dry)	50	2	10
Beef	Minced meat (retail)	2015	1	11
Swine	Minced meat (retail)	342	0.3	11
Beef and/or swine	Sausage/sausage meat (retail)	1448	0.1	11

unable to recover campylobacters from any of 100 bulk tank milk samples. Results from these studies suggest that the incidence of *C. jejuni* in raw milk is low.

III. FACTORS AFFECTING SURVIVAL AND GROWTH

A. Temperature

1. Growth

Growth of *C. jejuni* in laboratory culture media is restricted to temperatures greater than 30°C and less than 47°C (Figure 1).[13] At temperatures beyond these limits campylobacters die during extended incubation, suggesting that the organism is rather fragile and does not survive well at normal room temperature. Of three strains studied, the optimum temperature for growth was 42 to 45°C (Table 3).[13] Growth at 42°C was approximately twice as rapid

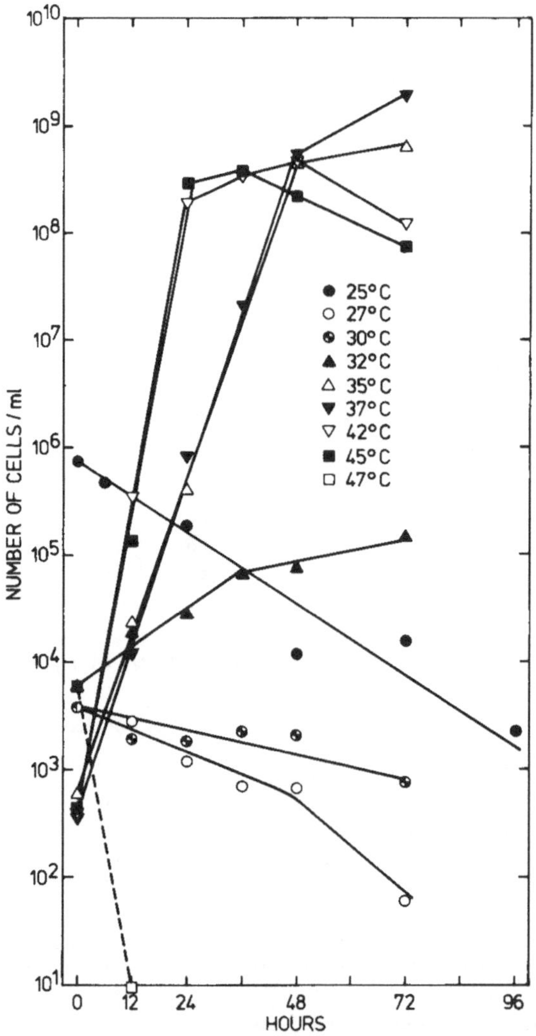

FIGURE 1. Growth or inactivation of *Campylobacter jejuni* at different temperatures in brucella broth plus 0.1% agar. (From Doyle, M. P. and Roman, D. J., *J. Food Protect.*, 44, 596, 1981. With permission.)

as at 35 to 37°C; an important factor that should be considered when attempting to isolate and culture *C. jejuni*. This is supported by the observations of Helstad and Janssen[14] who isolated 48 strains of *C. jejuni* from 580 stool specimens cultured at 42°C and recovered only 31 strains (65%) at 35°C.

Conflicting reports have been published regarding the ability of *C. jejuni* to grow on foods at temperatures which allow growth in laboratory media. Christopher et al.[15] reported death of the organism in both sterile skim milk and beef held at 40°C, with survival in beef being superior to that in skim milk. An approximate 3-log decrease occurred after 72 hr in beef, whereas a 5- to 6-log decrease to an undetectable level occurred by 24 hr in skim milk, both at 40°C. Svedhem et al.[16] reported an increase (~2 logs after 8 hr) in numbers of *C. jejuni* on chicken held at 42°C for 1 day, followed by a rapid decrease to undetectable levels after 2 days. No growth occurred at 42°C in minced meat of pork and beef, and *C. jejuni* could not be recovered (>6-log reduction) from such meat 2 days after inoculation. Blan-

Table 3
RATES OF GROWTH OF *CAMPYLOBACTER JEJUNI* AT DIFFERENT TEMPERATURES[a]

Temperature (°C)	Doubling time (hr) FRI-CF3	FRI-CF6	FRI-CF8
30	D[b]	D	D
32	6.54	5.88	11.1
35	3.32	2.84	2.96
37	2.79	2.69	2.16
42	1.30	1.32	1.24
45	1.34	1.31	1.28
47	D	D	D

[a] Three human isolates were each grown in brucella broth plus 0.1% agar and an atmosphere of 5% O_2:10% CO_2:85% N_2.
[b] D = number of organisms declined.

From Doyle, M. P. and Roman, D. J., *J. Food Protect.*, 44, 596, 1981. With permission.

kenship and Craven[17] studied the response of *C. jejuni* in sterile, autoclaved chicken breast meat at 37 and 43°C. Growth was only observed at 37°C, and growth in meat in an atmosphere of 5% O_2 occurred sooner and increased by approximately 1 log CFU/g more than occurred in meat stored in air.

2. Refrigeration

Campylobacter jejuni survives better in foods at refrigeration temperature than at room temperature. Blaser et al.[18] reported that the organism may remain viable in sterile milk at 4°C for up to 22 days, whereas at 25°C no viable organisms could be recovered after 3 days. Studying a wider range of temperatures, Christopher et al.[15] reported an approximate 6-log death of *C. jejuni* per milliliter in sterile skim milk within 48 hr at 30°C, within 60 hr at 20°C, and within 14 days at 10°C; only a 4 log/mℓ reduction occurred after 14 days at 1°C. At these same temperatures, *C. jejuni* survived better in beef than in milk. A decrease of only 2 to 3 logs of viable organisms per gram occurred in beef after 48 to 72 hr at 30°C, after 72 hr at 20°C, and after 14 days at 1 or 10°C.[15] Svedhem et al.[16] studied the survival response of *C. jejuni* on poultry and minced meat at 4 and 20°C. Their results were similar for both foods. Approximately the same number of organisms as was originally inoculated was recovered from both meats after 7 days at 4°C, whereas at 20°C organisms were rapidly inactivated and were no longer recoverable (>5 log/4 cm^2 reduction) after 3 days. Similar trends were observed by Blankenship and Craven[17] for survival of *C. jejuni* in sterile ground chicken meat. At 4°C, <2 log/g decrease occurred after 18 days, whereas a 5-log/g reduction occurred by 18 days at 23°C.

3. Freezing

Campylobacter jejuni is sensitive to freezing at temperatures commonly used for frozen foods.[15,19-21] For example, Luechtefeld and Wang[19] reported that storing fecal specimens containing *C. jejuni* (7×10^3 to 6×10^7/g) at −20°C eliminated all viable campylobacters by 37 days. Several factors influence the survival of *Campylobacter* during freezing and frozen storage. These include: (1) the composition of the freezing medium, (2) the environmental conditions, (3) the initial number of *Campylobacter*, and (4) the strain of *C. jejuni*. Mills et al.[20] compared the effects of different freezing procedures on survival of *Campylobacter*. They found that campylobacters survived better when frozen at −65°C than at −5°C, and in the presence of 10% glycerol than in brucella broth. Recovery rates of approximately 65% were obtained when cells were stored for 30 days at −65°C in either 10% glycerol or brucella broth; however, only 37% of the cells frozen at −5°C in 10 % glycerol and none of the cells frozen at −5°C in brucella broth were recovered. *C. jejuni* appears to be more sensitive to freezing in liquid than solid foods, as shown by the studies of Christopher et al.[15] They reported no survivors (>6 log/mℓ decrease of viable cells) in sterile skim milk after 8 days at −20°C, whereas only a 4- to 5-log/g decrease occurred in beef after 30 days at the same temperature. As with frozen beef, *C. jejuni* may survive for several weeks in frozen ground beef liver and on frozen chicken carcasses. Hänninen[21] observed relatively slow rates of death, e.g., a 2- to 3-log decrease of *Campylobacter* per gram of ground beef liver and a <0.5- to 2-log decrease per gram of chicken skin after 12 weeks at −20°C. Interestingly, human isolates of *C. jejuni* survived better on chicken carcasses than did animal isolates. At the retail level, Svedhem et al.[16] reported isolating *C. jejuni* from 6 of 7 store-bought chickens that had been stored frozen for 3 months, and Simmons and Gibbs[2] recovered the organism from 6 of 14 previously identified campylobacter-positive chickens after 3 weeks of frozen storage. These studies further indicate that, although *C. jejuni* is sensitive to frozen storage, the organism may survive on frozen meats for several weeks.

4. Thermal Inactivation

Knowledge of the time-temperature treatments needed to inactivate *C. jejuni* in foods is essential to assure food processors and the public that viable campylobacters are eliminated from foodstuffs before consumption. Foods of particular concern include raw milk, beef, and poultry as all three foods have been implicated as vehicles of human *Campylobacter* infections. Doyle and Roman[13] reported $D_{55°C}$ values, or times at 55°C required to inactivate 90% of the population, of 1.0 min or less for inactivation of several strains of *C. jejuni* in skim milk. Christopher et al.[15] found that 10^6 to 10^7 cells of *C. jejuni* per milliliter do not survive heating in skim milk at 60°C for 1 min. Hence pasteurization by the vat process of holding milk at 62.8°C for 30 min should be sufficient treatment to free milk of unusually large numbers of *C. jejuni*. Results reported by Gill et al.[22] confirm that the high temperature-short time (HTST) process (71.7°C, 15 sec), which is a treatment commonly used to pasteurize milk in the U.S., is sufficient to inactivate 2.3×10^6 campylobacters per milliliter of milk. It is unlikely that contaminated milk would contain more than 10^6 to 10^7 campylobacters per milliliter as mastitic milk derived from *Campylobacter*-infected udders contained $\leq 10^5$ campylobacters per milliliter[23] and fecal material entering milk at the farm would likely be diluted sufficiently to reduce the number of campylobacters to $<10^7$ cells per milliliter. Hence, properly pasteurized milk should be free of viable cells of *C. jejuni*.

Christopher et al.[15] studied the heat resistance of *C. jejuni* in beef roasts. An inoculum of 6.6×10^6 campylobacters per gram was injected into 6.4-cm diameter roasts at depths of 3.2 and 2.5 cm. The roasts were heated in an oven at 177°C (350°F) and cooked until their internal temperature reached 50 or 55°C. No survivors were recovered when the final internal temperature of the roasts was 57°C. However, survivors were recovered at the 10-CFU/g level for one of two strains in roasts having a final internal temperature of 53°C. These authors concluded that few if any survivors of *C. jejuni* can be expected in beef that has been heated to and kept at 60°C for several minutes.

To assess the thermal sensitivity of *C. jejuni* in poultry, Blankenship and Craven[17] determined the thermal death times of several strains in autoclaved ground chicken breast meat. D-values of a composite of five strains at 55 and 57°C were 2.25 and 0.98 min, respectively.

Table 4
EFFECT OF pH ON RATES OF GROWTH OF THREE STRAINS OF *CAMPYLOBACTER JEJUNI*[a]

		Doubling time (hr)		
Initial pH	pH after 12-hr incubation	FRI-CF3	FRI-CF6	FRI-CF8
4.7	4.7	NG[b]	NG	NG
4.9	4.9	NG	NG	7.41
5.3	5.3	1.71	1.96	2.28
5.5	5.5	1.45	1.43	1.49
6.0	6.0	1.35	1.34	1.37
6.5	6.5	1.31	1.31	1.24
6.9	6.9	1.31	1.30	1.21
7.5	7.3	1.33	1.28	1.13
8.0	7.7	1.33	1.33	1.30
8.5	8.0	1.75	1.67	1.44
9.0	8.4	R[c]	5.56	1.73
9.5	8.6	R	R	4.37

[a] Cultures were inoculated into brucella broth plus 0.1% agar adjusted to pH 4.7 to 9.5 with HCl or NaOH and held at 42°C in an atmosphere of 5% O_2:10% CO_2:85% N_2.
[b] NG = no growth when incubated for up to 72 hr.
[c] R = reduction in number of organisms during the first 12 hr of incubation; the number of organisms increased after 12 hr.

From Doyle, M. P. and Roman, D. J., *J. Food Protect.*, 44, 596, 1981. With permission.

Hence, poultry meat heated to and held at 60°C for 10 min should be sufficient treatment to free the meat of unusually large numbers of viable *C. jejuni*.

B. pH

The response of *C. jejuni* to pH varies with the strain and is influenced by the temperature and type of acid used to adjust the pH. Doyle and Roman[13] studied the effect of pH on growth of three human isolates at 42°C in semisolid brucella agar adjusted from pH 4.7 to 9.5 with HCl or NaOH. The best growth of all three strains occurred in media initially adjusted to pH 6.0 to 8.0 (Table 4). One strain grew at pH 4.9. The pH of media adjusted to pH 7.5 and above decreased during incubation, probably due to the formation of carbonic acid by CO_2 in the atmosphere. Hence, the maximum pH at which *C. jejuni* can grow could not be determined. However, results indicate that growth of *C. jejuni* in media at either extreme in pH is strain dependent.

Survival of *C. jejuni* in media adjusted to pH values below the pH limit of growth is temperature dependent.[13] Doyle and Roman[13] determined rates of inactivation in culture media adjusted to pH 3.0 to 5.0, using HCl, and stored at 4, 25, and 42°C. At the same pH, rates of death were greatest at 42°C, intermediate at 25°C, and least at 4°C. The time required to inactivate 3 logs of *C. jejuni* per milliliter (initial population of 10^6 CFU/mℓ) in a medium adjusted to pH 4.5 was 8 hr at 42°C, greater than 24 hr but less than 48 hr at 25°C, and 4 days at 4°C. If the medium was pH 5.0 and held at 4°C, a 3-log reduction resulted after 14 days suggesting that, given a large population, *C. jejuni* may remain viable for weeks when refrigerated in a low-acid environment.

Christopher et al.[15] observed that, of two strains of *C. jejuni* evaluated, none of 10^5 to 10^6 CFU/mℓ survived 24 hr at 37°C in brucella broth adjusted to pH 5.0 with lactic acid.

Table 5
RESPONSE OF *CAMPYLOBACTER* TO SODIUM CHLORIDE AT 42°C[a]

NaCl concentration (%)	Doubling time or D-value (hr)			
	C. jejuni CF6	*C. jejuni* CF8	*C. jejuni* CF74C	NARTC CF31P
0.0	14.5[b,c]	7.20[b,c]	2.94[b,c]	NG[d]
0.5	1.20[b]	1.18[b]	0.97[b]	1.00[b]
1.0	1.49[b]	1.32[b]	1.45[b]	1.10[b]
1.5	10.5[b]	1.74[b]	3.94[b]	1.28[b]
2.0	4.95[e]	4.91[e]	5.72[e]	2.86[b,f]
2.5	3.14[e]	3.28[e]	2.35[e]	NG[d]
4.5	1.91[e]	1.78[e]	1.54[e]	6.66[e]

[a] Cultures were inoculated into brucella broth plus 0.1% agar and held in an atmosphere of 5% O_2:10% CO_2:85% N_2. Sodium chloride-free brucella medium was specially formulated for the 0.0% NaCl treatment.
[b] Doubling time.
[c] Growth after 12-hr lag.
[d] NG = no growth or death up to 30 hr.
[e] D-value = time required to inactivate 90% of the population.
[f] Growth after 6-hr lag.

From Doyle, M. P. and Roman, D. J., *Appl. Environ. Microbiol.*, 43, 561, 1982. With permission.

Table 6
RESPONSE OF *CAMPYLOBACTER* TO SODIUM CHLORIDE AT 25°C[a]

NaCl concentration (%)	D-value (hr)			
	C. jejuni CF6	*C. jejuni* CF8	*C. jejuni* CF74C	NARTC CF31P
0.0	16.5	17.9	14.1	16.3
0.5	26.4	23.1	19.4	16.7
1.0	19.1	15.8	18.0	19.9
1.5	14.9	11.7	16.4	24.2
2.0	10.1	8.7	15.3	29.7
2.5	10.3	5.6	11.2	19.6
4.5	11.4	9.7	12.9	16.8

[a] Cultures were inoculated into brucella broth plus 0.1% agar and held in an atmosphere of 5% O_2:10% CO_2:85% N_2. Sodium chloride-free brucella medium was specially formulated for the 0.0% NaCl treatment.

From Doyle, M. P. and Roman, D. J., *Appl. Environ. Microbiol.*, 43, 561, 1982. With permission.

In contrast, Doyle and Roman[13] observed growth, albeit slow, of one strain at 42°C in brucella medium adjusted to pH 5.0 with HCl. Although direct comparisons cannot be made, these results suggest that *Campylobacter* may be more sensitive to an acid environment when the pH is adjusted with lactic acid than HCl.

C. Sodium Chloride

Sodium chloride is one of the most important food adjuncts in food preservation, hence

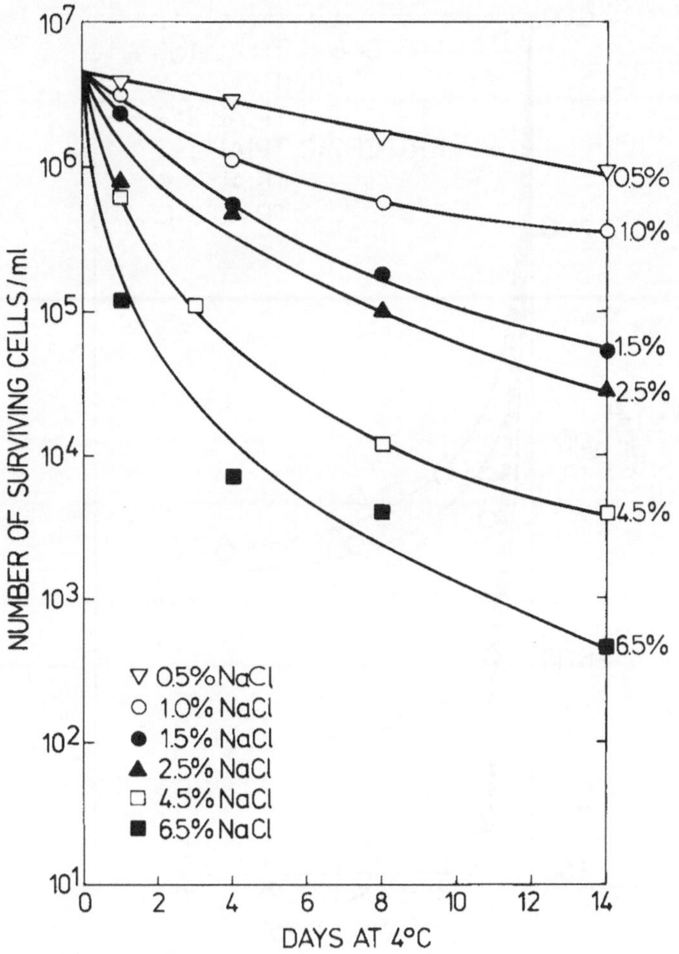

FIGURE 2. Inactivation at 4°C of *C. jejuni* CF8 in brucella broth containing 0.1% agar and different amounts of NaCl. (From Doyle, M. P. and Roman, D. J., *Appl. Environ. Microbiol.*, 43, 561, 1982. With permission.)

it is useful to know how *C. jejuni* will respond in its presence. To fully assess this, Doyle and Roman[24] determined the response of three strains of *C. jejuni* and one strain of nalidixic acid-resistant thermophilic *Campylobacter* (NARTC) in the presence of different sodium chloride concentrations at 4, 25, and 42°C. All three strains of *C. jejuni* could grow at 42°C in the presence of 1.5% NaCl, but not 2.0% NaCl (Table 5). Similar observations were made by Hänninen[25] who studied the NaCl tolerance of two strains of *C. jejuni/coli* at 35°C. One strain grew in the presence of 1.75% NaCl whereas the other grew, but only slightly, in 1.5% NaCl. The NARTC strain could grow in 2.0% NaCl at 42°C and was substantially more tolerant to 2.5 and 4.5% NaCl than was *C. jejuni*.[24] All four strains grew poorly in the absence of added NaCl and grew best in the presence of 0.5% NaCl (Table 5). Hence the presence of 0.5% NaCl in media used to culture *C. jejuni* is recommended.

At 25°C, NaCl concentrations of 1.0 to 2.5% were protective to NARTC, but the same concentrations of salt generally enhanced the rate of death of *C. jejuni* (Table 6). At 4°C, both *C. jejuni* and NARTC were sensitive to 1.0% or more NaCl (Figure 2); however, the rates of death at this temperature were substantially less than those that occurred at 25°C. In 4.5% NaCl, a 3-log decrease of cells occurred after 1.2 to 2.1 days at 25°C, whereas a

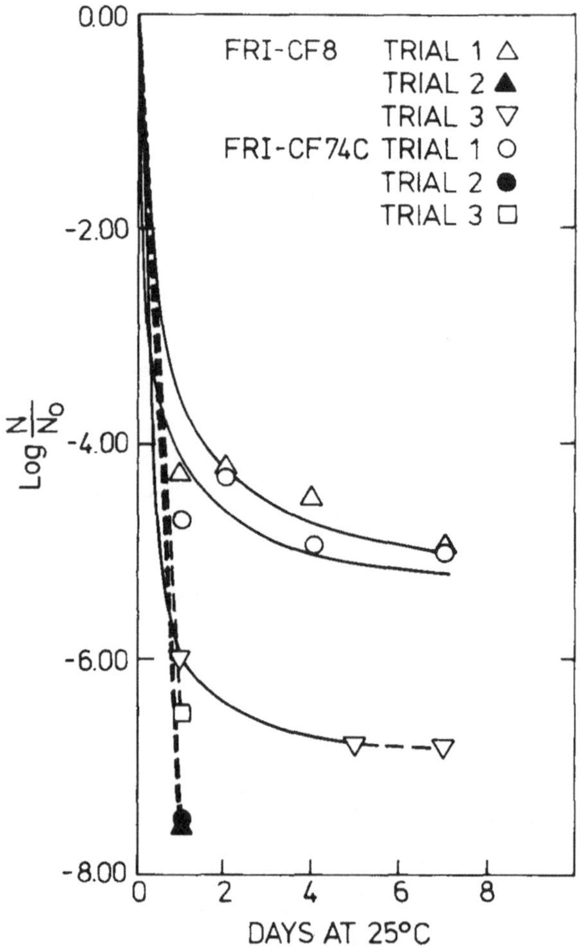

FIGURE 3. Survival curves of *C. jejuni* CF8 and CF74C when maintained in an anhydrous environment (<1% RH) in the presence of skim milk at 25°C. (From Doyle, M. P. and Roman, D. J., *J. Food Protect.*, 45, 507, 1982. With permission.)

similar reduction of cells took approximately 2 weeks at 4°C. The data indicate that as the environmental temperature decreases from the organism's optimum temperature for growth to refrigeration temperature, *C. jejuni* becomes increasingly tolerant to bactericidal concentrations of sodium chloride.

Reports by Sticht-Groh[26,27] exemplify some of the practical applications of these data. In Germany, bowels from pigs are often used for making sausages. In preparing the bowels for stuffing, butchers soak the tissues in salt using a concentration that can only be guessed at. After an overnight soaking, the bowels are washed and ready for sausage making. Sticht-Groh assayed several such pig bowels for campylobacters following an overnight soaking in salt and, based on the results of the Doyle and Roman study,[24] it is not surprising that 30% or more of the samples were found to contain campylobacters. Although *C. jejuni* is sensitive to NaCl, unless the temperature is unusually warm, a simple overnight soaking in salt would not be sufficient to eliminate large numbers of the organism from pig bowels. Sticht-Groh[26] speculated that in countries where much pork sausage and related products are consumed, this may be an additional cause of digestive tract infection in man.

D. Dehydration

Several factors have been shown to influence the rate of inactivation of *C. jejuni* during drying and storage in an anhydrous environment.[28] These include the strain, the temperature and humidity of the environment, and the medium used to suspend the organism. Doyle and Roman[28] studied the tolerance of several strains of *Campylobacter* to drying on a glass surface under different conditions. Of four strains tested, three isolates of *C. jejuni* survived drying better than did a strain of NARTC. Rates of inactivation were temperature dependent, with death occurring much more rapidly when cells were dried at 25°C than at 4°C. Only two of the four strains (initial population of 10^7) survived longer than 24 hr when held in an anhydrous environment at 25°C. Those strains that survived longer did so inconsistently, as is shown in Figure 3. When held in an anhydrous environment at 4°C, all four strains were recoverable (3- to 6-log reduction) after 29 days (Figure 4). Rates of inactivation were also dependent on the medium in which campylobacters were suspended during drying. Doyle and Roman[28] reported greater survival when *C. jejuni* was dried in the presence of brucella broth than in skim milk. After 29 days at 4°C, there were approximately 0.5 to 1.0 log more cells of each strain inactivated when dried in skim milk than in brucella broth. The humidity of the environment in which drying occurred had a profound effect on survival (Figure 5). When *C. jejuni* was held at 4°C in environments ranging from <1 to 75% relative humidity (RH), survival was greatest in the presence of 14% RH and least in 59% RH. Luechtefeld et al.[29] have also found *C. jejuni* to be quite sensitive to drying at room temperature. They reported that the organism did not survive more than 2 hr when dried in the presence of turkey fecal specimens on strips of filter paper. Results suggest that *C. jejuni* is generally quite sensitive to drying and storage at room temperature, but at refrigeration temperature and the appropriate humidity large numbers may survive dehydration and remain viable for several weeks.

E. Oxygen and Carbon Dioxide

Campylobacter jejuni is a microaerophile that requires both oxygen and carbon dioxide for growth. The organism's requirement for oxygen is a concentration considerably less than that present in air, and normal atmospheric levels of oxygen are inhibitory to growth.[30,31] Fletcher and Plastridge[32] studied the gaseous requirements of several different microaerophilic campylobacters and concluded that, for most strains, growth was optimal in the presence of 5% oxygen and 10% carbon dioxide. Similar findings have been reported by others.[33] Hence an atmosphere comprised of 5% O_2, 10% CO_2, and 85% N_2 is commonly used to culture *C. jejuni*. Addition of low levels of norepinephrine, high levels of iron salts, or a combination of ferrous sulfate, sodium bisulfite, and sodium pyruvate to growth medium greatly enhances the aerotolerance of *C. jejuni*, allowing many strains to grow in the presence of 21% O_2.[30,31] Oxygen tolerance is also greatly enhanced by addition of bovine superoxide dismutase, and to a lesser extent catalase, to growth broth or agar.[34] All of these agents share the ability to quench either superoxide anions or hydrogen peroxide. Hoffman et al.[34] suggest that *C. jejuni* is more sensitive to exogenous superoxide anions and hydrogen peroxide than are aerotolerant bacteria, even though the organism possesses superoxide dismutase and catalase activities.[35] They further suggest that compounds which enhance oxygen tolerance in *C. jejuni* act by quenching superoxide anions and hydrogen peroxide which occur spontaneously in culture media.

F. Disinfectants

Several waterborne outbreaks of *Campylobacter* enteritis have been reported. Since drinking water is often chlorinated to eliminate waterborne bacterial pathogens, Wang et al.[36] studied the effect of chlorine on *C. jejuni* in water. They observed that 10^3 to 10^4 *C. jejuni* per milliliter were inactivated by 1.25 and 0.625 ppm of chlorine after 1 and 30 min,

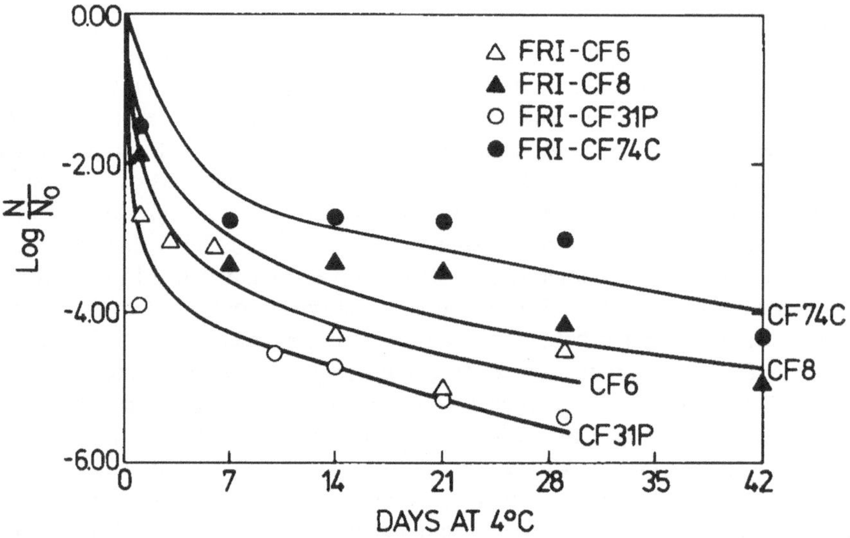

FIGURE 4. Survival curves of *C. jejuni* CF6, CF8, and CF74C and NARTC CF31P when maintained in an anhydrous environment (<1% RH) in the presence of skim milk at 4°C. (From Doyle, M. P. and Roman, D. J., *J. Food Protect.*, 45, 507, 1982. With permission.)

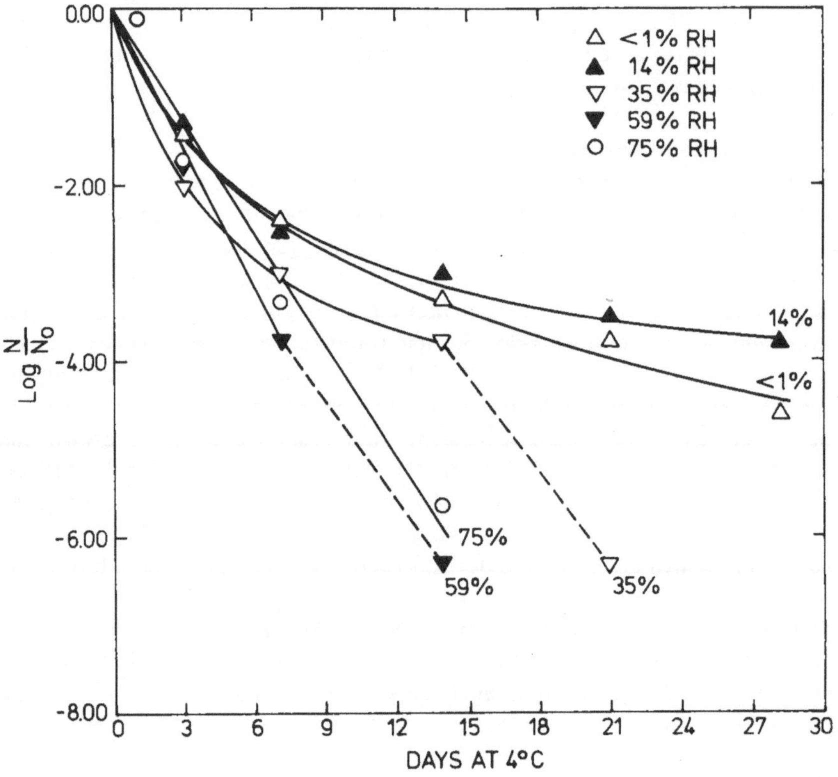

FIGURE 5. Survival curves of *C. jejuni* CF8 when maintained in different environmental humidities at 4°C in the presence of brucella broth. (From Doyle, M. P. and Roman, D. J., *J. Food Protect.*, 45, 507, 1982. With permission.)

respectively. However, using a higher level of inoculum, 10^6 to 10^7 CFU/mℓ, 5.0 ppm of chlorine were required to totally inactivate 2 of 3 strains in 1 min. None of 10^6 to 10^7 *C. jejuni* per milliliter were recovered after 30 min in 2.5 ppm chlorine. The authors suggest that properly chlorinated drinking water should be free of *C. jejuni*.

Studies on the use of chlorinated water to inactivate *C. jejuni* on poultry carcasses and giblets have yielded conflicting results. Luechtefeld and Wang[4] determined the incidence of *C. jejuni* on turkey carcasses before and after chilling with chlorinated ice water (20 to 50 ppm chlorine). They found that before chilling 94% of 33 carcasses were positive for *Campylobacter* and after an overnight soaking in chlorinated ice water 34% of 83 carcasses were still positive. Increasing the chlorine concentration from 50 to 340 ppm did not decrease the proportion of positive carcasses, as 9 of 20 (45%) turkeys were still positive after soaking in ice water with a chlorine content of 340 ppm. Likewise, Simmons and Gibbs[2] found that *C. jejuni* could survive on poultry carcasses treated with chlorinated water; however, details of the chlorination treatment were not reported. Christopher et al.[7] determined the incidence of *C. jejuni* on poultry giblets before and after treatment with chlorinated water (50 to 100 ppm). They reported that *Campylobacter* was present on 85% of 60 chicken livers and 89% of 64 chicken gizzards immediately after evisceration, but on none of 25 chicken giblets washed with chlorinated water. These studies suggest that, when present on poultry carcasses and edible viscera, *C. jejuni* is sensitive to chlorine ($\geq$50 ppm) treatment; however, treatment with chlorinated water may only partially reduce the *Campylobacter* population on poultry carcasses.

Several other disinfectants have also been found to be effective against *C. jejuni*. The organism (10^6 to 10^7 CFU/mℓ) is killed within 1 min by 70% ethyl alcohol, 10 ppm iodophor, and 0.125% glutaraldehyde, and within 15 min by 2.5% formalin.[36] The recommended use concentrations of these commonly used laboratory disinfectants should be sufficient to eliminate *C. jejuni* from the environment.

G. Raw Foods

Since *C. jejuni* is primarily associated with raw foods of animal origin, it is useful to know how well campylobacters can survive and compete with the indigenous flora of such foods. Four types of food that have been implicated as vehicles of outbreaks of campylobacteriosis have been studied. These include unpasteurized milk, fresh ground beef, raw chicken, and uncooked ground beef liver.

Barrell[37] reported the survival profiles of different strains of *C. jejuni* in unpasteurized milk held at 4 or 21°C. He observed that, from milk having initial counts of 10^5 to 10^7 *C. jejuni* per milliliter, four of seven strains were recoverable after 48 hr at refrigeration temperature and two of seven strains were recoverable after 48 hr at 21°C. One strain was still detected after 5 days at refrigeration temperature. Studies by Doyle and Roman[12] showed that there were large differences in the rates of inactivation of *C. jejuni* in unpasteurized milk at 4°C (Figure 6). They found that one strain was still recoverable at the 250 CFU/mℓ level after 21 days in the presence of milk which developed an aerobic plate count (APC) of 8.1×10^8 CFU/mℓ and pH of 5.9, whereas another strain was inactivated to a nondetectable level (>6 log/mℓ reduction) after 7 days in milk which developed an APC of 5.5×10^7 CFU/mℓ and pH of 6.3. In general, a 5- to 6-log/mℓ decrease of recoverable campylobacters occurred within 14 days for most isolates held in unpasteurized milk at 4°C. However, one strain, which was originally isolated from the feces of a dairy cow, survived exceptionally well in refrigerated raw milk. Less than a 2-log/mℓ reduction occurred after 14 days in milk which developed an APC of 5.8×10^8 CFU/mℓ and a pH of 6.0. In all instances, as the population of campylobacters decreased in unpasteurized milk, the APC increased and the pH decreased. However, no absolute correlation could be made between the rate of change in APC or pH of the milk and the rate of inactivation of *C. jejuni*. Inactivation of campylobacters in unpasteurized milk parallelled but was greater than that

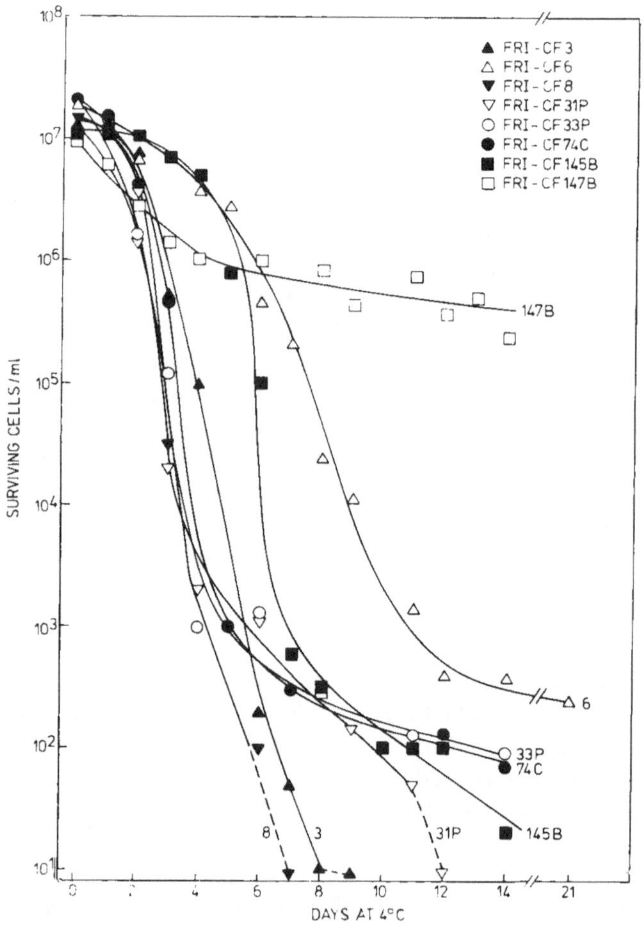

FIGURE 6. Survival of *C. jejuni* and NARTC (FRI-CF31P) in unpasteurized milk at 4°C. (From Doyle, M. P. and Roman, D. J., *Appl. Environ. Microbiol.*, 44, 1154, 1982. With permission.)

which occurred in sterile milk, suggesting that the psychrotrophic microflora of raw milk produced metabolites toxic to *C. jejuni*.[12] The change of pH may contribute to, but was not thought to be the sole factor influencing, the organism's rate of inactivation.

Campylobacter jejuni survived substantially better at 4°C in fresh ground beef than in unpasteurized milk (Figure 7).[38] The initial APC of the ground beef ranged from 1.4×10^4 to 2.0×10^7 CFU/g, and after 14 days at 4°C increased to 2.1×10^8 to 2.1×10^9 CFU/g. The initial pH of the meat ranged from 5.7 to 6.2 and after 14 days increased to pH 6.5 to 6.9. In contrast to the wide differences that were observed in the ability of *Campylobacter* to survive in raw milk, all but one of the strains survived equally well in ground beef. After 14 days, less than a 1 log/g decrease occurred for seven of eight strains (Figure 7). One strain, which was studied for as long as 35 days, decreased by only 1.5 log/g after 5 weeks.

Similarly, *C. jejuni* survived well on uncooked chicken[17] and in ground beef liver[21] stored at 4°C. Blankenship and Craven[17] reported an approximate 2 log decrease of *C. jejuni* per cm^2 of raw chicken skin after 14 to 21 days at 4°C. Survival was even greater when the meat was packaged in an atmosphere of CO_2 and stored at the same temperature. Hänninen[21] found that storage of ground beef liver at 4°C for 6 days had little effect on the survival of any of five strains of *C. jejuni/coli*. Less than a 0.5 log decrease of *Campylobacter* cells occurred by day 6, whereas the APC and lactobacilli count increased to 1.8×10^7 and 1.0

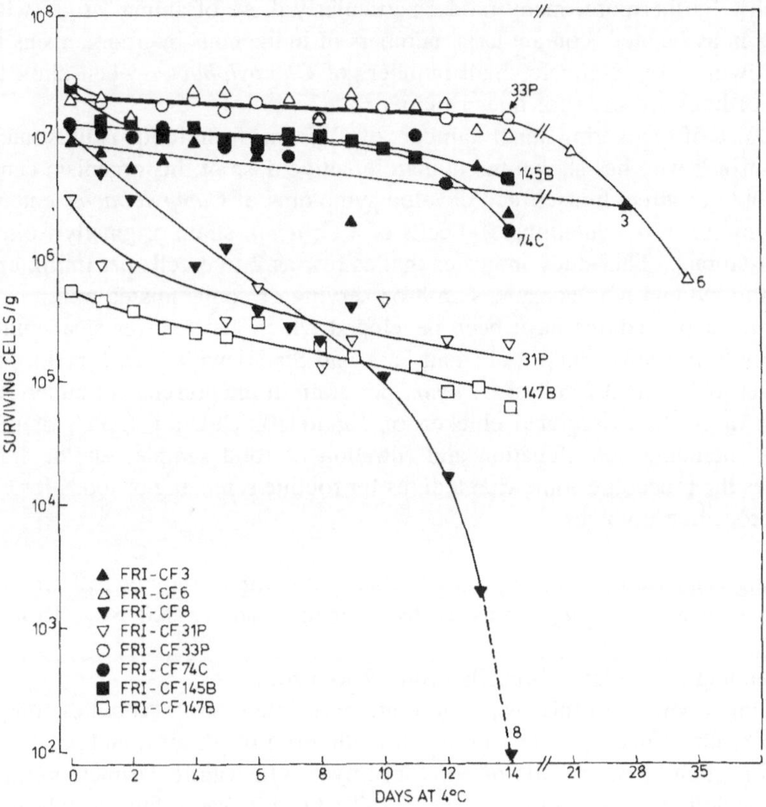

FIGURE 7. Survival of *C. jejuni* and NARTC (FRI-CF31P) in fresh ground beef at 4°C. (From Doyle, M. P., *J. Food Protect.*, 1983. With permission.)

$\times$ 10^7 CFU/g, respectively. The pH had decreased from an initial pH of 6.3 to 5.75 by day 6. By day 6, the liver had deteriorated to such a state that it was no longer suitable for human consumption.

These studies indicate that, depending on the strain of *C. jejuni*, the initial number of cells, and the environmental conditions, particularly the temperature of storage, campylobacters may survive in raw foods for long periods of time. Thus, individuals consuming foods of animal origin that have not been heated sufficiently to inactivate *Campylobacter* run the risk of *Campylobacter* infection.

IV. ISOLATION PROCEDURES

Many of the media and procedures used to isolate *C. jejuni* from foods were originally developed to isolate the organism from stool specimens. These methods generally involve passing the supernatant fluid of a centrifuged extract of stool specimen through a series of filters and plating the filtrate onto blood or chocolate agar plates,[39,40] or directly plating a few drops (~0.1 mℓ) of a stool specimen onto plates of blood agar containing antibiotics.[41-43] Grant et al.[40] reported that approximately 99% of *C. jejuni* cells are retained by the 0.65-μm filter used during the filtration technique. Direct plating of specimens onto antibiotic-containing blood agar plates is limited in that only a small amount of sample can be surface plated and solid specimens must be suspended in liquid and homogenized, thereby diluting any campylobacters that may be present. Hence neither method is conducive to recovering small numbers of *Campylobacter*. Unlike stool specimens of infected individuals which generally contain large numbers of *C. jejuni*, foods may contain relatively few *Cam-*

pylobacter cells. Furthermore, many foods, particularly those of animal origin with which *C. jejuni* is often associated, contain large numbers of indigenous microorganisms that may restrict the growth or out-compete small numbers of *Campylobacter* when clinical microbiological procedures are used for their isolation.

The importance of recovering small numbers of *C. jejuni* from foods is illustrated by the study of Robinson[44] who has shown the human infective dose of this organism can be very low. He was able to infect himself and develop symptoms of *Campylobacter* enteritis after ingesting 180 mℓ of milk containing 500 cells of a *C. jejuni* strain originally isolated from a milk-borne outbreak. This study indicates that as few as 2 to 3 cells per milliliter of milk may infect an individual who consumes an 8-oz serving of contaminated milk.

Two enrichment procedures have been developed which can recover small numbers of campylobacters from foods. Park et al.[5] and Park and Stankiewicz[45] described a procedure that can recover as few as 0.2 cell of *C. jejuni* per gram in the presence of either 10^4 to 10^6 CFU/g (APC) of fresh, eviscerated chicken or 10^2 to 10^4 CFU/mℓ (APC) of raw milk. Several steps, including centrifugation and filtration of food samples and/or enrichment cultures, makes the procedure somewhat tedious for routine screening of foods for *C. jejuni*. The general procedure involves:

1. Removing campylobacters from a food sample (200 mℓ of milk or 250 mℓ of nutrient broth used to wash poultry carcasses) by centrifugation (16,000 × g, 15 to 30 min, 4°C)
2. Resuspending the pellet in brucella broth (2 to 5 mℓ)
3. Inoculating a portion of this suspension into enrichment broth (100 mℓ) comprised of brucella broth, vancomycin (8 μg/mℓ), trimethoprim (4 μg/mℓ), and polymyxin B (8 U/mℓ) for poultry; or brucella broth, vancomycin (15 μg/mℓ), trimethoprim (10 μg/mℓ), cephalothin (10 μg/mℓ), 0.2% $FeSO_4 \cdot 7H_2O$, 0.025% sodium metabisulfite, and 0.05% sodium pyruvate for milk
4. Incubating (37°C, 3 days for poultry or 42°C, 2 to 3 days for milk) the broth under a constant flow (5 to 10 mℓ/min) of a gas mixture (5% O_2:10% CO_2:85% N_2)
5. Filtering the enrichment culture (~5 mℓ) through a 0.65-μm membrane filter
6. Plating several dilutions of the filtrate onto Skirrow's selective agar[42] (42°C, 48 hr, 5% O_2:10% CO_2:85% N_2) for isolation and subsequent identification of developing colonies

A second procedure, developed by Doyle and Roman,[46] can recover as few as 0.1 cell of *C. jejuni* per gram of fresh ground beef, raw milk, or chicken skin having indigenous microflora of 10^5 to 10^9 CFU/g (APC). Inoculation studies found the procedure to be effective in recovering 50 different isolates of *Campylobacter*, including strains of *C. jejuni*, *C. coli*, and NARTC. The technique involves:

1. Introducing a food sample (10 or 25 g) into enrichment medium (90 or 100 mℓ, respectively) comprised of brucella broth, 7% lysed horse blood, 0.3% sodium succinate, 0.01% cysteine hydrochloride, vancomycin (15 μg/mℓ), trimethoprim (5 μg/mℓ), polymyxin B (20 IU/mℓ), and cycloheximide (50 μg/mℓ)
2. Macerating solid samples in a Stomacher (15 sec)
3. Transferring the medium to a 250-mℓ sidearm Erlenmeyer flask
4. Evacuating the flask three times to 20 in. of Hg and replacing the atmosphere with 5% O_2:10% CO_2:85% N_2
5. Incubating the flasks (42°C, 16 to 18 hr) with agitation (100 gyrations/min)
6. Plating 0.1 mℓ and 0.1 mℓ each of two serial (1:10) dilutions onto Campy-BAP selective agar[41] (42°C, 48 hr, 5% O_2:10% CO_2:85% N_2) for isolation and subsequent identification of *C. jejuni*[46]

This procedure was most effective for recovering campylobacters from raw milk and ground beef as all of the 50 isolates evaluated were recovered from these foods at the 1 to 4 cells per gram level and 41 and 40 isolates were recovered from the ground beef and raw milk, respectively, at the 0.1 to 0.4 cells per gram level. The enrichment was least effective for recovering campylobacters from chicken skin as 43 and 24 of 50 isolates were recovered at the 1 to 4 and 0.1 to 0.4 cell per gram levels, respectively.

REFERENCES

1. **Smith, M. V., II, and Muldoon, P. J.,** *Campylobacter fetus* subspecies *jejuni (Vibrio fetus)* from commercially processed poultry, *Appl. Microbiol.*, 27, 995, 1974.
2. **Simmons, N. A. and Gibbs, F. J.,** *Campylobacter* spp. in oven-ready poultry, *J. Infect.*, 1, 159, 1979.
3. **Goren, E. and de Jong, W. A.,** *Campylobacter fetus* subspecies *jejuni* bij pluimvee, *Tijdschr. Diergeneesk.*, 105, 724, 1980.
4. **Luechtefeld, N. W. and Wang, W.-L. L.,** *Campylobacter fetus* subsp. *jejuni* in a turkey processing plant, *J. Clin. Microbiol.*, 13, 266, 1981.
5. **Park, C. E., Stankiewicz, Z. K., Lovett, J., and Hunt, J.,** Incidence of *Campylobacter jejuni* in fresh eviscerated whole market chickens, *Can. J. Microbiol.*, 27, 841, 1981.
6. **Norberg, P.,** Enteropathogenic bacteria in frozen chicken, *Appl. Environ. Microbiol.*, 42, 32, 1981.
7. **Christopher, F. M., Smith, G. C., and Vanderzant, C.,** Examination of poultry giblets, raw milk and meat for *Campylobacter fetus* subsp. *jejuni*, *J. Food Protect.*, 45, 260, 1982.
8. **Stern, N. J.,** Recovery rate of *Campylobacter fetus* ssp. *jejuni* on eviscerated pork, lamb, and beef carcasses, *J. Food Sci.*, 46, 1291, 1981.
9. **Rosef, O.,** *Campylobacter fetus* subsp. *jejuni* as a surface contaminant of fresh and chilled pig carcasses, *Nord. Vet. Med.*, 33, 538, 1981.
10. **Hudson, W. R. and Roberts, T. A.,** The occurrence of Campylobacter on commercial red meat carcasses from one abattoir, in *Proc. Int. Workshop on Campylobacter Infections,* Public Health Laboratory System, London, 1981, 34.
11. **Turnbull, P. C. B. and Rose, P.,** *Campylobacter jejuni* and *Salmonella* in raw red meats, *J. Hyg.*, 88, 29, 1982.
12. **Doyle, M. P., and Roman, D. J.,** Prevalence and survival of *Campylobacter jejuni* in unpasteurized milk, *Appl. Environ. Microbiol.*, 44, 1154, 1982.
13. **Doyle, M. P. and Roman, D. J.,** Growth and survival of *Campylobacter fetus* subsp. *jejuni* as a function of temperature and pH, *J. Food Protect.*, 44, 596, 1981.
14. **Helstad, A. G. and Janssen, D. M.,** Isolation of *Campylobacter fetus* ssp. *jejuni* from human fecal specimens by incubation at 35°C and 42°C, *J. Clin. Microbiol.*, 16, 398, 1982.
15. **Christopher, F. M., Smith, G. C., and Vanderzant, C.,** Effect of temperature and pH on the survival of *Campylobacter fetus*, *J. Food Protect.*, 45, 253, 1982.
16. **Svedhem, Å., Kaijser, B., and Sjögren, E.,** The occurrence of *Campylobacter jejuni* in fresh food and survival under different conditions, *J. Hyg.*, 87, 421, 1981.
17. **Blankenship, L. C. and Craven, S. E.,** *Campylobacter jejuni* survival in chicken meat as a function of temperature, *Appl. Environ. Microbiol.*, 44, 88, 1982.
18. **Blaser, M. J., Hardesty, H. L., Powers, B., and Wang, W.-L. L.,** Survival of *Campylobacter fetus* subsp. *jejuni* in biological milieus, *J. Clin. Microbiol.*, 11, 309, 1980.
19. **Luechtefeld, N. W. and Wang, W.-L. L.,** Preservation of *Campylobacter jejuni* in fecal specimens frozen at −20 and −70°C, *Abstr. Annu. Meet. Am. Soc. Microbiol.*, C-159, 298, 1982.
20. **Mills, C., Gall, P., and Gherna, R. L.,** Comparison of freezing and freeze-drying techniques in the preservation of the genus *Campylobacter*, *Abstr. Annu. Meet. Am. Soc. Microbiol.*, A-22, 213, 1982.
21. **Hänninen, M. L.,** Survival of *Campylobacter jejuni/coli* in ground refrigerated and in ground frozen beef liver and in frozen broiler carcasses, *Acta Vet. Scand.*, 22, 566, 1981.
22. **Gill, K. P. W., Bates, P. G., and Landers, K. P.,** The effect of pasteurization on the survival of *Campylobacter* species in milk, *Br. Vet. J.*, 137, 578, 1981.
23. **Lander, K. P. and Gill, K. P. W.,** Experimental infection of the bovine udder with *Campylobacter coli/jejuni*, *J. Hyg.*, 84, 421, 1980.
24. **Doyle, M. P. and Roman, D. J.,** Response of *Campylobacter jejuni* to sodium chloride, *Appl. Environ. Microbiol.*, 43, 561, 1982.

25. **Hänninen, M. L.,** The effect of NaCl on *Campylobacter jejuni/coli, Acta Vet. Scand.*, 22, 578, 1981.
26. **Sticht-Groh, V.,** Campylobacters in pig faeces, *Vet. Rec.*, 108, 42, 1981.
27. **Sticht-Groh, V.,** Campylobacter in healthy slaughter pigs: a possible source of infection for man, *Vet. Rec.*, 110, 104, 1982.
28. **Doyle, M. P. and Roman, D. J.,** Sensitivity of *Campylobacter jejuni* to drying, *J. Food Protect.*, 45, 507, 1982.
29. **Luechtefeld, N. W., Wang, W.-L. L., Blaser, M. J., and Reller, L. B.,** Evaluation of transport and storage techniques for isolation of *Campylobacter fetus* subsp. *jejuni* from turkey cecal specimens, *J. Clin. Microbiol.*, 13, 438, 1981.
30. **George, H. A., Hoffman, P. S., Smibert, R. M., and Krieg, N. R.,** Improved media for growth and aerotolerance of *Campylobacter fetus, J. Clin. Microbiol.*, 8, 36, 1978.
31. **Bowdre, J. H., Krieg, N. R., Hoffman, P. S., and Smibert, R. M.,** Stimulatory effect of dihydroxy phenyl compounds on the aerotolerance of *Spirillum volutans* and *Campylobacter fetus* subspecies *jejuni, Appl. Environ. Microbiol.*, 31, 127, 1976.
32. **Fletcher, R. D. and Plastridge, W. N.,** Gaseous environment and growth of microaerophilic vibrios, *J. Bacteriol.*, 87, 352, 1964.
33. **Smibert, R. M.,** The genus *Campylobacter, Annu. Rev. Microbiol.*, 32, 673, 1978.
34. **Hoffman, P. S., George, H. A., Krieg, N. R., and Smibert, R. M.,** Studies of the microaerophilic nature of *Campylobacter fetus* subsp. *jejuni*. II. Role of exogenous superoxide anions and hydrogen peroxide, *Can. J. Microbiol.*, 25, 8, 1979.
35. **Hoffman, P. S., Krieg, N. R., and Smibert, R. M.,** Studies of the microaerophilic nature of *Campylobacter fetus* subsp. *jejuni*. I. Physiological aspects of enhanced aerotolerance, *Can. J. Microbiol.*, 25, 1, 1979.
36. **Wang, W.-L. L., Powers, B. W., Blaser, M. J., and Luechtefeld, N. W.,** Laboratory studies of disinfectants against *Campylobacter jejuni, Abstr. Annu. Meet. Am. Soc. Microbiol.*, Q-140, 233, 1982.
37. **Barrell, R. A. E.,** The survival of *Campylobacter coli/jejuni* in unpasteurized milk, *J. Infect.*, 3, 348, 1981.
38. **Doyle, M. P.,** Survival of *Campylobacter jejuni* in fresh and heated red meat, *J. Food Protect.*, in press.
39. **Dekeyser, P. M., Gossuin-Detrain, M., Butzler, J. P., and Sternon, J.,** Acute enteritis due to related vibrio: first positive stool cultures, *J. Infect. Dis.*, 125, 390, 1972.
40. **Grant, I. H., Richardson, N. J., and Bokkenheuser, V. D.,** Broiler chickens as potential source of *Campylobacter* infections in humans, *J. Clin. Microbiol.*, 11, 508, 1980.
41. **Blaser, M. J., Berkowitz, I. D., LaForce, F. M., Cravens, J., Reller, L. B., and Wang, W-L. L.,** *Campylobacter* enteritis: clinical and epidemiological features, *Ann. Intern. Med.*, 91, 179, 1979.
42. **Skirrow, M. B.,** *Campylobacter* enteritis: a "new" disease, *Br. Med. J.*, 2, 9, 1977.
43. **Lauwers, S., DeBoeck, M., and Butzler, J. P.,** Campylobacter enteritis in Brussels, *Lancet*, 1, 604, 1978.
44. **Robinson, D. A.,** Infective dose of *Campylobacter jejuni* in milk, *Br. Med. J.*, 282, 1584, 1981.
45. **Park, C. E. and Stankiewicz, Z. K.,** A sensitive procedure for the isolation of thermophilic *Campylobacter* from milk, *Abstr. Annu. Meet. Assoc. Off. Anal. Chem.*, 212, 73, 1981.
46. **Doyle, M. P. and Roman, D. J.,** Recovery of *Campylobacter jejuni* and *Campylobacter coli* from inoculated foods by selective enrichment, *Appl. Environ. Microbiol.*, 43, 1343, 1982.

Chapter 15

BOVINE GENITAL CAMPYLOBACTERIOSIS

Joseph Dekeyser

TABLE OF CONTENTS

I. Historical 182

II. Pathogenesis and Clinical Signs 182
A. Enzootic Infertility (Venereal Bovine Campylobacteriosis) 182
B. Sporadic Abortion 183

III. Causal Agents 183
A. Morphology and Colonial Characteristics 185
B. Culture 185
C. DNA Base Composition; Cellular Fatty Acids; Biochemical Properties 185
D. Antigenic Studies 186
E. Sensitivity to Antibiotics 186

IV. Transmission and Dissemination 186

V. Diagnosis 187
A. Direct Culture 187
B. Immunofluorescence 187
C. Vaginal Mucus Agglutination Test 187

VI. Treatment 188

References 189

I. HISTORICAL

The first observations on abortion in cattle and sheep due to *Campylobacter fetus* were reported in 1913 by McFadyean and Stockman.[33] Some years later Smith and Taylor[53] described the isolation of spiral organisms from the stomach and gut contents of aborted fetuses coming from a farm which, although it was free of brucellosis, did have numerous cases of infectious abortion. They called these bacteria "vibrio". Plastridge et al.[42] were the first to suggest that *Vibrio fetus* caused a form of infectious sterility.

It was, however, the Dutch investigators Stegenga and Terpstra[54] who proved that "enzootic sterility" was indeed caused by *Vibrio fetus* and that the infection was venereally transmitted. In 1952 a report on these infections in England was published by Lawson and MacKinnon.[28a] Some years later Akkermans et al.[3] stated that the strains of *Vibrio fetus* which were isolated from farms with enzootic infertility produced no H_2S, while those isolated from farms on which only sporadic abortion occurred were good producers of H_2S.

Florent[25] drew attention to the fact that venereally transmitted enzootic infertility was caused by a variety of *Vibrio fetus* which he called *venerealis* and that sporadic abortions were caused by a variety of intestinal origin which he called *Vibrio fetus intestinalis*. He also described the biochemical tests by which the two varieties could be differentiated. *V. fetus venerealis* did not produce H_2S and would not grow in a semisolid medium containing 1% glycine, whereas *V. fetus intestinalis* was a good producer of H_2S and was resistant to glycine, as had earlier been described by Lecce.[30] In 1962 however, Bryner et al.,[10] Mohanty et al.,[35] and Park et al.[38] came to the conclusion that, although enzootic infertility on the great majority of farms was caused by *V. fetus venerealis*, some strains were good producers of H_2S and that it also could sometimes be caused by *V. fetus intestinalis* serotype O_2. Marsh and Firehammer,[31] Mitscherlich and Liess,[34] and Morgan[36] developed a classification based on antigenic differences and in 1971 both the biochemical and antigenic characteristics of *V. fetus venerealis* and *V. fetus intestinalis* were defined by Berg et al.[4] Dedié et al.[19] pointed out that there were some strains that were definitely *V. fetus venerealis* but were relatively resistant to glycine and could only be differentiated from *V. fetus intestinalis* serotype O_1 by the selenite reduction test.

Sebald and Veron[49] drew attention to the differences that existed in the proportions of the bases in vibrio DNAs. They proposed that the genus *Vibrio* should be limited to those bacteria in which the guanine + cytosine (G + C) content was about 47%. For the vibrios important in veterinary pathology with a G + C content of between 30 and 34% they proposed a new genus *Campylobacter* with *Campylobacter fetus* as the type species.

In 1973 the results of a taxonomic study of the genus *Campylobacter* were published by Veron and Chatelain.[58] They proposed that *V. fetus intestinalis* should be renamed *C. fetus* subspecies *fetus* and that *V. fetus venerealis* should be renamed *C. fetus* subspecies *venerealis*. This nomenclature was accepted by Skerman et al.[51]

Smibert,[52] however, in the 8th edition of *Bergey's Manual*, had proposed a quite contrary renaming, with *V. fetus venerealis* becoming *C. fetus* subsp. *fetus* and *V. fetus intestinalis* becoming *C. fetus* subsp. *intestinalis*.

The officially recognized nomenclature for campylobacters published in the "Approved List of Bacterial Names"[51] is in accordance with that to be used in the 9th edition of *Bergey's Manual of Determinative Bacteriology*.

II. PATHOGENESIS AND CLINICAL SIGNS

A. Enzootic Infertility (Venereal Bovine Campylobacteriosis)

Bulls most commonly become infected by serving infected female animals but contact infection from infected bedding is also possible.[48] This contact infection is the basic reason why up to 50% or more of the bulls in an artificial insemination center may be affected. Bulls 4 years old or more are certainly more susceptible than younger ones. A possible explanation for this in the view of Samuelson and Winter[47] is that the crypts in the epithelium of the penis become deeper and more numerous with age and thus provide a more favorable milieu for *Campylobacter*. This infection of the prepuce always remains strictly localized and produces no local or general symptoms. Spontaneous recovery from a natural infection very rarely occurs. Bier et al.[5] explain this phenomenon as due to the very slight antigenic variation which the organisms undergo so that only a local and minimal immunological response occurs.

In a susceptible female animal served by an infected bull, *Campylobacter* multiplies rapidly in the vagina. The organisms penetrate, after a week, through the cervix into the uterus and there continue to spread over a period of 6 to 8 weeks. Vandeplassche et al.[57] also found that in 25% of naturally infected heifers the organisms reached the oviducts. In the cervix, and even more so in the vagina, they can survive for months and sometimes for the whole period of the pregnancy, during which time the animal seems to find it very difficult to get rid of the infection.

Histological examination shows a progressive endometritis with inflammatory damage reaching a peak about the 8th to 13th week after infection. Estes et al.[22] describe it as a subacute diffuse mucopurulent endometritis, characterized by the accumulation of exudate in the lumen of the uterine glands and by heavy periglandular infiltration with lymphocytes.

Because of these inflammatory phenomena the embryo finds itself in a very unsatisfactory environment. According to Ware,[59] the oxygen supply to the embryo may become so restricted that it dies, and as a result a large percentage of the animals have a prolonged cycle of 24 to 40 days or more. The symptoms are most pronounced in heifers, more than 75% of whom come into heat again, usually with a prolonged cycle. Sometimes a 2- to 3-month-old fetus may be dropped while still in its membranes, and at a still later stage repeated abortions occur usually about the 4th to 6th month.[26]

The whole process of recovery lasts 4 to 5 months. Thereafter the animals are relatively immune. Nevertheless, fertility never returns to its normal level and in individual animals bilateral salpingitis leads to permanent sterility. Exceptionally, a female animal remains a permanent vaginal carrier of *Campylobacter*.

B. Sporadic Abortion

This form of genital infection is much less spectacular as it arises from an accidental dissemination of *Campylobacter* present in the gut or gallbladder. When such an occasional bacteriemia occurs, the already well-formed placenta of the pregnant uterus is a selective target for any organisms that may be circulating in the blood, for example, *Corynebacterium pyogenes, Salmonella, Listeria,* and also *Campylobacter fetus* subsp. *fetus*. The placentitis caused by these organisms leads to death of the fetus through anoxia and this is inevitably followed by a late abortion. This is the only symptom of the infection, of which the female genital system, moreover, rapidly rids itself. In contrast to its pathogenic role in sheep, *C. jejuni* does not interfere in any way with reproduction in cattle.

III. CAUSAL AGENTS

The cause of enzootic infertility is *C. fetus* subspecies *venerealis* in more than 95% of cases. With them must also be included the biochemical variants which produce H_2S in cysteine-enriched media. Some of these variants are, moreover, relatively tolerant of glycine but none of them reduce selenite. Antigenically, they all belong to serotype A.

C. fetus subspecies *fetus* is responsible for less than 5% of cases. They all belong to serotype B (Table 1). During extensive investigations carried out on farms where the

Table 1
CAMPYLOBACTER FETUS

Subspecies	Serotype	Biotype	Sensitive H_2S test	Growth in glycine 1% 0.6%	1%	1.3%	1.9%	Selenite 0.1%	Diseases caused	Commensal in
venerealis	A	1	−	+	−	−	−	−	Enzootic infertility	—
venerealis	A	Sub 1	+	+	−	−	−	−	Enzootic infertility	—
venerealis	A	Dedié	−	+	+	−	−	−	Enzootic infertility	—
fetus	A	2	+	+	+	+	−	+	Sporadic abortion	Prepuce, gut, gallbladder
fetus	B	—	+	+	+	+	+	+	Enzootic infertility; sporadic abortion	Gut, gallbladder

fertility must be regarded as normal, the only vibrios that could be isolated either from bulls or heifers were catalase negative. This is in direct contrast to the pathogenic vibrios which are all catalase positive. These saprophytic vibrios were called *V. bubulus* by Thouvenot and Florent.[55]

A. Morphology and Colonial Characteristics

C. fetus subsp. *venerealis* and *C. fetus* subsp. *fetus* cannot be differentiated with certainty by microscopical examination. They are Gram-negative, slender, short, comma- or S-shaped rods which may form a helix. They do not form spores. The proportion of short and S-shaped forms to the longer spirals is very much influenced by the constitution of the medium on which they are grown. These microorganisms are best examined under phase-contrast or dark-ground illumination where not only is their shape particularly well seen but so also is the rapid to and fro movement characteristic of the shorter forms.

When examined by electron microscopy these organisms show extremities which gradually taper to a point from which always one, and sometimes two flagella arise. The cell wall shows the three-layered structure typical of all Gram-negative organisms.[46]

The colonies on blood agar have a diameter of 1 to 2 mm, are convex, and are clearly raised above the surface of the medium. They are colored grayish white to light brown and are nonhemolytic. Bryner et al.[10] describe "smooth", "cut glass", and "rough" colonies which on Albimi brucella agar may even appear on primary isolation. The colonies of *C. fetus* subsp. *fetus* are already well developed by 48 hr, but *C. fetus* subsp. *venerealis* needs at least 72 hr. For primary isolation it is best to use a rich base for the blood agar, such as thioglycollate or brain-heart infusion agar.

B. Culture

C. fetus is a microaerophilic organism which will only grow on solid media in an atmosphere with reduced oxygen and increased CO_2; this is particularly important for primary isolation. The best results are obtained when two thirds of the normal air is replaced by a mixture of 95% nitrogen and 5% carbon dioxide. When these organisms are grown in a rich broth the resulting turbidity is only moderate and cannot be seen except in good lighting. They are, however, often grown in semisolid media where a ring of growth appears just below the surface. In contrast to the ring of growth seen with *C. sputorum* subspecies *bubulus*, a nonpathogen frequently present in the genital system, it does not extend into the depths of the medium. In a strictly anaerobic atmosphere these organisms grow very poorly or not at all.

The optimum temperature for development is about 37°C. Growth always takes place also at 25°C, but at 42°C the growth of most strains is completely inhibited.[23,29]

C. DNA Base Composition; Cellular Fatty Acids; Biochemical Properties

The G+C content of the DNA is 33 to 36 mol %.[14,49]

By gas-liquid chromatography analysis of cellular fatty acids, Blaser et al.[6] found that *C. fetus* subsp. *venerealis* and *C. fetus* subsp. *fetus* both lacked a 19-carbon cyclopropane acid which was present in *C. jejuni*. Both subspecies reduce nitrates, do not attack carbohydrates, and are nonproteolytic. They are catalase and oxidase positive.

The vast majority of the strains of *C. fetus* subsp. *venerealis* responsible for enzootic infertility produce no H_2S in semisolid thioglycollate broth enriched with 0.1% cysteine and do not grow in 1% glycine. They belong to serotype A, biotype 1. A minority are good producers of H_2S in the sensitive test although glycine sensitive (serotype A, biotype 1).[4] There are also strains that produce no H_2S but are relatively resistant to glycine.[19] However, not a single one of the above-mentioned strains will reduce sodium selenite.

Isolated cases of enzootic infertility are, nevertheless, due to *C. fetus* subsp. *fetus* serotype

B. These strains are good producers of H_2S in cysteine-enriched media, they grow in 1% glycine, and they reduce 1% sodium selenite.

The strains of *C. fetus* that are isolated from cases of sporadic abortion where no question of enzootic infertility arises belong to the subspecies *intestinalis*. They always produce H_2S in the sensitive test, always grow in the presence of glycine, and they reduce sodium selenite. Serologically they may belong either to group A or group B.

D. Antigenic Studies

Marsh and Firehammer[31] divided their *C. fetus* strains into five serological groups, using "whole cells antigen". Sometime later Mitscherlich and Liess[34] and then Morgan[36] described their division into two groups, based on heat-stable O antigens. The former used complement fixation and the latter an agglutination test.

The classification brought out by Berg et al.[4] was based on a combination of the biochemical and the serological characters of *C. fetus* strains. This was later extended by Dedié et al.[19] to include a few strains which special characteristics (H_2S negative and tolerant of 1% glycine).

All the strains that belong to the serotype A of Morgan[36] or the serotype I of Mitscherlich and Liess[34] can be venereally transmitted and cause enzootic infertility, with the exception of those strains that reduce 0.1% sodium selenite and are tolerant of 1.3% glycine. Dedié et al.[19] found the latter type of *C. fetus* subsp. *fetus* serotype A in the prepuce of 15 to 20% of the bulls they examined. With none of these strains could they set up an experimental vaginal infection in heifers, and the fertility of the bulls from which they were isolated was normal.

All the strains of serotype B can be found as commensals in the gastrointestinal tract and belong to *C. fetus* subsp. *fetus*. They can cause sporadic abortion, and in occasional rare cases of enzootic infertility are met with either in the prepuce of the bull or in the vaginal mucus of the animal served by the infected bull. An experimental vaginal infection with serotype B can also be set up, but usually disappears in a few weeks. The fact that there is little sharing of antigens between serotype A and serotype B means that immunization with a vaccine prepared from a strain of serotype A can usually neither prevent nor cure an infection in the bull due to a strain of serotype B.

E. Sensitivity to Antibiotics

In the diffusion test *C. fetus* subsp. *venerealis* is sensitive to gentamycin, neomycin, chloramphenicol, kanamycin, erythromycin, tetracycline, penicillin, and streptomycin.[13,44,48] In rare cases, streptomycin-resistant strains were isolated from preputial washings. *C. fetus* subsp. *venerealis* is, however, resistant to bacitricin and novobiocin and most strains are also resistant to polymyxin B. The MICs of bacitricin, novobiocin, and polymyxin B were greater than 50 μg/mℓ.

These last three are, in fact, used in selective media designed for the isolation of *C. fetus* subsp. *venerealis* from heavily contaminated material. *C. fetus* subsp. *fetus* shows the same pattern of antibiotic sensitivities except that it is not sensitive to penicillin.

IV. TRANSMISSION AND DISSEMINATION

From the many publications on enzootic infertility, whether dealing with cases in practice or experimental investigations, it stands out clearly that transmission of the infection occurs principally along venereal paths. For female animals it is indeed the only way in which they become infected. Bulls may, however, be indirectly infected by contact with infected material or equipment used in the A.I. centers for taking semen as well as by contact with infected bedding. Schutte[48] reported that three of the six bulls put on bedding where infected bulls had previously been, themselves became positive in 39, 42, and 55 days, respectively.

This observation provided a logical explanation for the fact that in numerous A.I. centers a number of bulls were found to be infected before they had ever served a single cow or heifer.[1] The infection can, of course, be even more easily spread if artificial insemination is carried out with infected semen. The additions of penicillin and streptomycin to semen, whether used fresh or stored deep-frozen, proved to be very effective in preventing spread of the disease by A.I.[1,32]

In the sporadic cases of abortion due to *C. fetus* subsp. *fetus*, dissemination of the disease at the time of the abortion itself, and for a week afterwards, is very great, but one rarely sees other cattle in the herd aborting as a result.

V. DIAGNOSIS

A very thorough investigation of a farm affected by enzootic infertility often leads to a clinical diagnosis, but it must be confirmed by laboratory investigation.

A. Direct Culture

Campylobacter, whether in the prepuce or the vagina, is in a milieu containing large numbers of other organisms. Media were, therefore, looked for at an early stage which would suppress the accompanying flora as thoroughly as possible. Florent and Vandeplassche[24] used brilliant green and Plastridge et al.[43] used bacitricin and novobiocin, both in solid media. Plumer et al.[45] used a 0.65-μm Millipore® filter to get rid of interfering organisms with success, and Shepler et al.[50] combined Millipore® filtrations with a medium comprised of brain-heart infusion agar containing 10% ox blood, to which was added bacitricin 15 IU/mℓ, novobiocin 5 μg/mℓ, and polymyxin 1IU/mℓ. The presence of polymyxin also prevents the development of any *V. bubulus* that penetrate the filter with the pathogens. It is also useful to add actidione (50 μg/mℓ).

B. Immunofluorescence

That is a convenient, quick, and accurate method for detection of carrier bulls. Dufty,[21] Winter et al.,[60] Philpott,[41] and Schutte[48] found that a combination of culture media and a fluorescent antibody test (FAT) revealed 98% of infected bulls. The test will, of course, give a false negative result if only a serum against serotype A is used and the infection is due to *C. fetus* subsp. *fetus* serotype B. Similarly it will give a false positive result if the bull's prepuce has been parasitized with *C. fetus* subsp. *fetus* serotype A.

C. Vaginal Mucus Agglutination Test

Following natural infection, female animals produce local antibodies and they are present in the vaginal mucus from 6 weeks up to 7 months afterward, on the average. These antibodies will agglutinate a suitable antigen in a tube test. This technique also allows an antibovine globulin test (ABGT) to be carried out, whereby the possible presence of "incomplete" antibody may be demonstrated and the sensitivity of the test increased.

In the individual animal a negative result does not exclude infection. Furthermore, the sample of mucus must be free of blood and transudate as the commonly used antigens usually give a fairly significant titer if the test is carried out on a serum instead of vaginal mucus. Nevertheless when the mucus agglutination test, together with the ABGT, is carried out on a large number of samples from all the animals in a herd where there has been a serious and recent infertility problem, a decisive answer will be obtained on the presence or not of enzootic infertility.

Vaginal infection has no influence on the serum antibody titers. It does give rise, however, to the formation of local antibodies belonging to immunoglobulin classes A, G, and M.[18,39,57]

The diagnostic techniques described above have led to the virgin heifer test mating and the vaginal inoculation of suspect material in guinea pigs being definitely abandoned.

VI. TREATMENT

When a diagnosis of enzootic infertility has been made on a farm, there is no point trying to cure the infected female animals unless they were served less than 3 weeks earlier. In the curative treatment of bulls with antibiotics, Schutte[48] had the best results from the application of an ointment containing 10 g neomycin and 4 g erythromycin in 200 g Carbowax® to the mucosa on the exteriorized penis and to the prepuce, after both had previously been thoroughly washed with 8 ℓ of warm water containing 1% hydrogen peroxide. These bulls, which were known to have been treated earlier with streptomycin or tetracycline without success, were cured after this combined therapy. The presence of multiresistant *Escherichia coli* in the prepuce was considered responsible for the production of enzymes which inactivated the streptomycin and tetracycline, as has been demonstrated in vitro by Okamoto and Suzuki.[37]

The great problem with bulls serving cattle naturally is, of course, reinfection. In countries where extensive cattle breeding is carried on, preventive and curative vaccination has been given to female animals since 1961, but not to bulls. Clark et al.,[15] however, successfully treated four infected bulls with two injections of their vaccine. Bouters et al.[8] were able to cure 41 infected bulls with an experimental vaccine, 30 after one dose and 11 after two doses 6 weeks apart. These bulls continued to serve regularly in an infected area and remained free of infection for more than a year. In the same infected area, 288 bulls negative for *V. fetus* were given one dose and similarly remained free of infection for more than a year. Clark et al.[16] confirmed these good results. Van Aert et al.[56] showed that after vaccination the preputial secretion contains antibodies of classes IgG, IgM, and IgA. The objection to vaccinating bulls was that they might, nevertheless, be able subsequently to transfer the infection from the vagina of an infected cow to that of an uninfected one during coitus. This objection was removed by the work of Clark et al.[17] Hoerlein et al.[28] reported that of 39 heifers vaccinated 3 months before being served by an infected bull, 74% became pregnant after three services, in contrast to 5% of the unvaccinated heifers.

Frank et al.[27] confirmed that vaccination of heifers produced a significant improvement of their breeding efficiency although protection against infection was not obtained. This observation was confirmed by Bouters et al.[9] They reported an almost normal fertility in their vaccinated test heifers which had been served by a naturally infected bull, in spite of the fact that nine of the ten animals acquired a vaginal infection.

Inoculated pregnant animals also can apparently only rid themselves of a vaginal infection with great difficulty. That they do, in fact, become pregnant can only be explained by the uterus being absolutely resistant, mainly due to the more ample supply of antibodies from the blood.

Border and Firehammer[7] demonstrated that only the K antigens stimulated the production of protective antibodies; heifers that were vaccinated with O and H antigens or with O antigens alone were not protected against infection.

A study of the comparative effectiveness of immunizing agents derived from *C. fetus* was set up in pregnant guinea pigs by Bryner et al.[11] They subsequently reported the results, having compared ten different vaccines.[12] It is evident that only high-quality immunizing agents, given in the proper way, will give good results.

REFERENCES

1. **Adler, H. C. and Lindegaard, L. E.,** Bovine genital vibriosis. Eradication from Danish A. I. Centres, *Nord. Vet. Med.*, 17, 237, 1965.
2. **Adler, H. C.,** Bovine genital vibriosis. The transmission through deepfrozen semen and its prevention by streptomycin treatment, *Nord. Vet. Med.*, 18, 30, 1966.
3. **Akkermans, J. P. W. M., Terpstra, J. I., and Van Waveren, H. G.,** Over de betekenis van verschillende vibrionen voor de steriliteit van het rund, *Tijdsch. Dierg.*, 81, 430, 1956.
4. **Berg, R. L., Jutila, J. W., and Firehammer, B. D.,** A revised classification of *Vibrio fetus*, *Am. J. Vet. Res.*, 32, 11, 1971.
5. **Bier, P. J., Hall, C. E., Duncan, J. R., and Winter, A. J.,** Experimental infections with *Campylobacter fetus* in bulls of different age, *Vet. Microbiol.*, 2, 13, 1977.
6. **Blaser, M. J., Moss, C. W., and Weaver, R. E.,** Cellular fatty acid composition of *Campylobacter fetus*, *J. Clin. Microbiol.*, 11, 448, 1980.
7. **Border, M. M. and Firehammer, B. D.,** Antigens of *Campylobacter fetus* subsp. *fetus* eliciting vaccinal immunity in heifers, *Am. J. Vet. Res.*, 31, 746, 1980.
8. **Bouters, R., Dekeyser, J., Vandeplassche, M., Van Aert, A., Brone, E., and Bonte, P.,** *Vibrio fetus* infection in bulls: curative and preventive vaccination, *Br. Vet. J.*, 129, 52, 1973.
9. **Bouters, R., Dekeyser, P., Vandeplassche, M., and Van Aert, A.,** Weerstand tagen infektie en vruchtbaarheid van vaarzen na vaccinatie tegen *Vibrio fetus*, *Vlaams Diergen. Tijdschr.*, 45, 207, 1976.
10. **Bryner, J. H., Frank, A. H., and O'Berry, P. A.,** Dissociation studies of vibriosis from the bovine genital tract, *Am. J. Vet. Res.*, 23, 32, 1962.
11. **Bryner, J. H., Foley, J. W., Hubbert, W. T., and Mattheus, P. J.,** Pregnant Guinea pig model for testing efficacy of *Campylobacter fetus* vaccines, *Am. J. Vet. Res.*, 39, 119, 1978.
12. **Bryner, J. H., Foley, J. W., and Thompson, K.,** Comparative efficacy of ten commercial *Campylobacter fetus* vaccines in the pregnant guinea-pig: challenge with *Campylobacter fetus* serotype A, *Am. J. Vet. Res.*, 40, 433, 1979.
13. **Butzler J. P., Dekeyser, P., and Lafontaine, T.,** Susceptibility of related Vibrios and *Vibrio fetus* to twelve antibiotics, *Antimicrob. Agents Chemother.*, 5, 86, 1973.
14. **Charlier, G., Dekeyser, P., Florent, A., Strobbe, R., and De Ley, J.,** DNA base composition and biochemical characters of *Campylobacter* strains, *J. Antonie van Leeuwenhoek*, 40, 145, 1974.
15. **Clark, B. L., Dufty, J. H., and Monsbourgh, M. J.,** Vaccination of bulls against bovine vibriosis, *Aust. Vet. J.*, 44, 530, 1968.
16. **Clark, B. L., Dufty, J. H., Monsbourgh, M. J., and Parsonson, I. M.,** Immunisation against bovine vibriosis. Vaccination of bulls against infection with *Campylobacter fetus* subsp. *venerealis*, *Aust. Vet. J.*, 50, 407, 1974.
17. **Clark, B. L., Dufty, J. H., Monsbourgh, M. J., and Parsonson, I. M.,** Studies on venereal transmission of *Campylobacter fetus* by immunised bulls, *Aust. Vet. J.*, 51, 531, 1975.
18. **Corbeil, L. B., Duncan, J. R., Schurig, G. G. D., Hall, C. E., and Winter, A. J.,** Bovine venereal vibriosis: variations in immunoglobulin class of antibodies in genital secretions and serum, *Infect. Immunity*, 10, 1084, 1974.
19. **Dedié, K., Pohl, P., and Reisshauer, K.,** Vorkommen und Pathogenität glycin-positiver Stämme des *Campylobacter (Vibrio) fetus* Serotyp 01, *Zbl. Vet. Med., Reihe B*, 24, 767, 1977.
20. **Dufty, J. H.,** Diagnosis of vibriosis in the bull, *Aust. Vet. J.*, 43, 433, 1967.
21. **Dufty, J. H., Clark, B. L., and Monsbourgh, M. J.,** The influence of age on the susceptibility of bulls to *Campylobacter fetus* subsp. *venerealis*, *Aust. Vet. J.*, 51, 294, 1975.
22. **Estes, P. C., Bryner, J. H., and O'Berry, P. A.,** Histopathology of bovine vibriosis and the effects of *Vibrio fetus* extracts on the female genital tract, *Cornell Vet.*, 55, 610, 1966.
23. **Firehammer, B. D. and Berg, R. L.,** The use of temperature tolerance in the identification of *Vibrio fetus*, *Am. J. Vet. Res.*, 26, 995, 1965.
24. **Florent, A. and Vandeplassche, M.,** Valeur comparée de la génisse d'épreuve et de la culture sur milieu sélectif du liquide préputial pour la confirmation de l'infection à *Vibrio fetus* du taureau, Proc. 3rd Int. Cong. Animal Reproduction, Cambridge, June 25 to 30, 1956.
25. **Florent, A.,** Les deux vibrioses génitales: La vibriose due à *V. fetus venerialis* et la vibriose d'origine intestinale due à *V. fetus intestinalis Med. Veeartsenijschool Rijksuniversiteit Gent*, 3, 60, 1959.
26. **Frank, A. H., Bryner, J. H., and O'Berry, P. A.,** Reproductive patterns of female cattle bred for successive gestations to *Vibrio fetus* infected bulls, *Am. J. Vet. Res.*, 25, 988, 1964.
27. **Frank, A. H., Bryner, J. H., and O'Berry, P. A.,** The effect of *Vibrio fetus* vaccination on the breeding efficiency of cows bred to *Vibrio fetus* infected bulls, *Am. J. Vet. Res.*, 28, 1237, 1967.
28. **Hoerlein, A. B., Carroll, E. J., Kramer, T., and Beckenhauer, W. H.,** Bovine vibriosis immunization, *J. Am. Vet. Med. Assoc.*, 146, 828, 1965.

28a. **Lawson, J. R. and MacKinnon, D. J.,** *Vibrio fetus* infection in cattle, *Vet. Rec.*, 64, 763, 1952.

29. **Leaper, S. and Owen, R. J.,** Identification of catalase-producing *Campylobacter* species based on biochemical characteristics and on cellular fatty acid composition, *Curr. Microbiol.*, 6, 31, 1981.
30. **Lecce, J. C.,** Some biochemical characteristics of *Vibrio fetus* and other related vibrios isolated from animals, *J. Bacteriol.*, 76, 312, 1958.
31. **Marsh, H. and Firehammer, B. D.,** Serological relationship of twenty-three ovine and three bovine strains of *Vibrio fetus*, *Am. J. Vet. Res.*, 14, 396, 1953.
32. **McEntee, K., Gilman, H. L., Highes, D. E., Wagner, W. C., and Dunn, H. O.,** 1959, Insemination of heifers with penicillin and dihydrostreptomycin treated frozen semen from *Vibrio fetus* carrier bulls, *Cornell Vet.*, 49, 175, 1959.
33. **McFadyean, J. and Stockman, S.,** Report of the Departmental Committee appointed by the Board of Agriculture and Fisheries to inquire into epizootic abortion. Appendix to Part III, Abortion in Sheep, Her Majesty's Stationery Office, London, 1913.
34. **Mitscherlich, E. and Liess, B.,** Die serologische differenzierung von *Vibrio fetus*-Stämmen, *Dtsche. Tierärtzl. Wochenschr.*, 65, 2, 1958.
35. **Mohanty, S. B., Plumer, G. J., and Faber, J. E.,** Biochemical and colonial characteristics of some bovine vibriosis, *Am. J. Vet. Res.*, 23, 554, 1962.
36. **Morgan, W. J. B.,** Studies on the antigenic structure of *Vibrio fetus*, *J. Comp. Pathol. Ther.*, 69, 125, 1959.
37. **Okamoto, S. and Suzuki, Y.,** Chloramphenicol-, dihydrostreptomycin-, and kanamycin-inactivating enzymes from multiple drug resistant *Escherichia coli* carrying episome "R", *Nature (London)*, 208, 1301, 1965.
38. **Park, R. W. A., Munro, I. B., Melrose, D. R., and Stewart, D. L.,** Observations on the ability of two biochemical types of *Vibrio fetus* to proliferate in the genital tract of cattle and their importance with respect to infertility, *Br. Vet. J.*, 118, 411, 1962.
39. **Pedersen, K. B., Aalund, O., Nansen, P., and Adler, H. C.,** Immunofluorescent Immunoglobulin differentiation of *Vibrio fetus* antibodies from bovine cervico-vaginal secretions, *Acta Vet. Scand.*, 12, 303, 1971.
40. **Philpott, M.,** Diagnosis of *Vibrio fetus* infection in the bull. I. A modification of Mellick's fluorescence antibody test, *Vet. Rec.*,82, 424, 1968.
41. **Philpott, M.,** Diagnosis of *Vibrio fetus* infection in the bull. II. An epidemiological survey using a fluorescent antibody test and comparing this with a cultural method, *Vet. Rec.*, 82, 458, 1968.
42. **Plastridge, W. N., Williams, L. F., and Petrie, Q.,** Vibrionic abortion in cattle, *Am. J. Vet. Res.*, 8, 178, 1947.
43. **Plastridge, W. N., Koths, M. E., and Williams, L. F.,** Antibiotic Mediums for the isolation of vibrios from bull semen, *Am. J. Vet. Res.*, 22, 867, 1961.
44. **Plastridge, W. N. and Trowbridge, D. G.,** Antibiotic sensitivity of physiologic groups of microaerophilic vibrios, *Am. J. Vet. Res.*, 25, 1295, 1964.
45. **Plumer, G. J., Duvall, W. C., and Shepler, V. M.,** A preliminary report on a new technic for isolation of *Vibrio fetus* from carrier bulls, *Cornell Vet.*, 52, 110, 1962.
46. **Ritchie, A. E., Keeler, R. F., and Bryner, J. H.,** Anatomical features of *Vibrio fetus:* electron microscopic survey, *J. Gen. Microbiol.*, 43, 427, 1966.
47. **Samuelson, J. D. and Winter, A. J.,** Bovine vibriosis: the nature of the carrier state in the bull, *J. Infect. Dis.*, 116, 581, 1966.
48. **Schutte, A. P.,** Some aspects of *Vibrio fetus* infection in bulls, *Med. Veeartsenijschool Rijksuniversiteit Gent*, 13, 88, 1969.
49. **Sebald, M. and Veron, M.,** Teneur en bases de l'ADN et classification des vibrions, *Ann. Inst. Pasteur*, 105, 897, 1963.
50. **Shepler, V. M., Plumer, G. J., and Faber, J. E.,** Isolation of *Vibrio fetus* from bovine preputial fluid using millipore filters and an antibiotic medium, *Am. J. Vet. Res.*, 24, 749, 1963.
51. **Skerman, V. D. B., McGowan, V., and Sneath, P. H. A.,** Approved lists of bacterial names, *Int. J. Syst. Bacteriol.*, 30, 270, 1980.
52. **Smibert, R. M.,** *Bergey's Manual of Determinative Bacteriology*, 8th ed., Buchanan, R. and Gibbons, F., Eds., Williams – Wilkins, Baltimore, 1974, 207.
53. **Smith, T. and Taylor, M. S.,** Some morphological and biological characteristics of the spirilla associated with disease of fetal membranes in cattle, *J. Exp. Med.*, 30, 299, 1919.
54. **Stegenga, Th. and Terpstra, J. I.,** Over *Vibrio fetus* infecties bij het rund en "enzoötische" steriliteit, *Tijdschr. Dierg.*, 74, 293, 1949.
55. **Thouvenot, H. and Florent, A.,** Étude d'un anaerobie du sperme du taureau et du vagin de la vache, *Vibrio bubulus* Florent 1953. *Ann. Inst. Pasteur*, 86, 237, 1954.
56. **Van Aert, A., Dekeyser, P., Brone, E., Bouters, R., and Vandeplassche, M.,** Nature of antibodies to *Campylobacter fetus* in preputial secretions from a vaccinated bull, *Br. Vet. J.*, 132, 615, 1976.

57. **Van Aert, A., Dekeyser, P., Florent, A. F., Bouters, R., Vandeplassche, M., and Brone, E.,** Nature of *Campylobacter fetus* agglutinins in vaginal mucus from experimentally infected heifers, *Br. Vet. J.*, 133, 88, 1977.

57a. **Vandeplassche, M., Florent, A., Bouters, R., Huysman, A., Brone, E., and Dekeyser, P.,** The pathogenesis, epidemiology, and treatment of *Vibrio fetus* infection in cattle, *I.W.O.N.L., Verslagen van Navorsingen*, 29, 18, 1963.

58. **Veron, M. and Chatelain, R.,** Taxonomic study of the genus *Campylobacter* Sebald and Veron and designation of the neotype strain for the type species, *Campylobacter fetus* (Smith and Taylor) Sebald and Veron, *Int. J. Syst. Bacteriol.*, 23, 122, 1973.

59. **Ware, D. A.,** Pathogenicity of *Campylobacter fetus* subspecies *venerealis* in causing infertility in cattle, *Br. Vet. J.*, 136, 301, 1980.

60. **Winter, A. J., Samuelson, J. D., and Elkana, M.,** A comparison of immunofluorescence and cultural techniques for demonstration of *Vibrio fetus, J. Am. Vet. Med. Assoc.*, 150, 499, 1967.

Chapter 16

ENTERIC INFECTIONS WITH CATALASE-POSITIVE CAMPYLOBACTERS IN CATTLE, SHEEP, AND PIGS

D. J. Taylor and R. R. Al-Mashat

TABLE OF CONTENTS

I. Introduction ... 194

II. Catalase-Positive Campylobacter Infections in the Enteric Tracts of Cattle ... 194
- A. *C. jejuni* Infections ... 195
- B. *C. fetus* subsp. *fetus* Infections ... 199
- C. *C. fecalis* Infections ... 199

III. Catalase-Positive Campylobacter Infections in the Enteric Tracts of Sheep ... 200
- A. *C. fetus* subsp. *fetus* Infections ... 200

IV. Catalase-Positive Campylobacter Infections in the Enteric Tracts of Pigs ... 201
- A. *C. jejuni* Infections ... 201
- B. *C. hyointestinalis* Infections ... 201
- C. *C. coli* Infections ... 201

V. Conclusions ... 204

References ... 205

I. INTRODUCTION

In contrast to the recent prominence given to catalase-positive campylobacters in man, these organisms have been associated unequivocally with enteric infections in animals for over 50 years. Prior to the work in the 1930s of Jones and co-workers[1-4] with "*Vibrio*" *jejuni* in cattle which provided firm evidence for the causal role of these organisms in enteric disease, there had been a number of reports of the presence of Gram-negative curved or spiral rods in the feces of animals with diarrhea and in enteric lesions. Few of these accounts were accompanied by isolation of the organisms seen and in many cases it is difficult to distinguish between spirillae, spirochaetes, and vibrios in the earlier papers on enteric disease in these species.

The two syndromes classically associated with infection by catalase-positive campylobacters in cattle, sheep, and pigs are winter diarrhea, a profuse watery diarrhea or dysentery of sudden onset and short duration in cattle seen particularly in winter, and swine dysentery, a muco-hemorrhagic colitis with a 7- to 14-day incubation period occurring in recently weaned pigs. In neither case has the catalase-positive campylobacter incriminated been shown to reproduce the syndrome, although in cattle some elements of the winter diarrhea syndrome result from experimental infections with *C. jejuni*. In sheep, no specific clinicopathological entity was classically associated with enteric campylobacter infections, and campylobacters isolated from sheep were more commonly associated with reproductive disturbances than with those of the enteric tract.

The poor correlation between catalase-positive campylobacter infections and clinical enteric disease in these species appears to result from three main causes which will be discussed in more detail below, but which are outlined here for the benefit of those who are unfamiliar with farm animal husbandry and enteric disease. The first and possibly the most important reason is that infections with catalase-positive campylobacters appear to produce few clinical signs, little depression of productivity, and little or no mortality in these three species. The second is that the tremendous variation in the conditions of animal husbandry, from the extensive range conditions of much sheep and some cattle rearing and little antimicrobial administration, to the highly intensive pig rearing enterprise with varying inputs of antimicrobials, do not favor the close observation of individual animals. The third reason is that the state of hygiene in animal enterprises favors the spread of enteric disease of all types, so that primary campylobacter infections comprise a minor part of the whole spectrum of enteric disease and occur very early in life against a background of widespread active or passive immunity to those commonly encountered enteropathogens.

The recent medical interest in catalase-positive campylobacters and the improvement in our ability to isolate these organisms has increased veterinary awareness of them and led to a reevaluation of studies carried out in the 1930s in cattle[1-4] and in the 1940s to 1960s[5-11] in pigs. Their role in enteric disease in animals as opposed to their public health significance is reviewed below under the individual species.

II. CATALASE-POSITIVE CAMPYLOBACTER INFECTIONS IN THE ENTERIC TRACTS OF CATTLE

Three species of catalase-positive campylobacters have been isolated from the gastrointestinal tracts of cattle, as have a number of isolates which do not fall into the established species groups but which include the aerotolerant campylobacter described by Neill et al.[12] The species which can be identified are *C. jejuni*, *C. fetus* subsp. *fetus* and *C. fecalis*.[14] All have been isolated from inflammatory lesions in the gastrointestinal tract of cattle and can also be isolated from the feces of animals with diarrhea initiated by a wide variety of agents. The organisms are present in the feces of both calves and adult cattle. All three species may

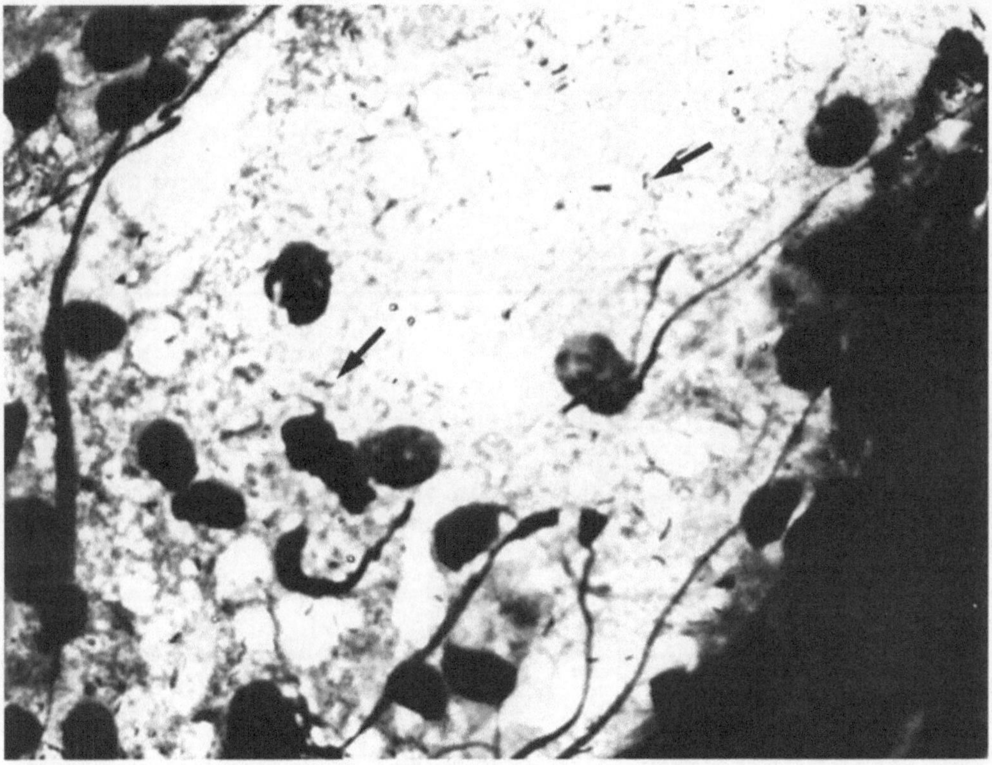

FIGURE 1. Smear of ileal mucosa of a clinical case of enteritis in a calf at a site from which *C. jejuni* was isolated. Note the presence of curved rods (arrow). (Gram stain; magnification × 1200.)

be present in the same gastroenteric tract and at least two species may be recovered from the same site. *C. fecalis* may be distinguished by its colonial morphology from the other two species, but *C. jejuni* and *C. fetus* subsp. *fetus* can only be distinguished subsequent to isolation. The presence of both species in a single site or sample then depends upon the number of colonies sampled. The clinical signs and pathological changes associated with all three species are broadly similar.[14-17] Since *C. jejuni* is the organism which has been studied in most detail, it will be considered first and at greatest length.

A. *C. jejuni* Infections

C. jejuni has been isolated from cases of winter dysentery or winter diarrhea[1-4] and from calves and older animals with diarrhea.[14] Mucus strands containing fresh blood are commonly present in the feces of animals from which profuse cultures of *C. jejuni* are isolated. These animals are frequently presented as possible cases of salmonellosis. Most cases of diarrhea in cattle from which *C. jejuni* is isolated in large numbers also yield other agents such as cryptosporidia, coccidia, rotaviruses, coronaviruses, enteropathogenic *Escherichia coli*, salmonellae, or nematodes. It is unusual to find *C. jejuni* as the only potential pathogen in an outbreak of diarrhea in cattle. In one study of diarrhea in calves, Firehammer and Myers[18] isolated *C. jejuni* from diarrheic calves in only 56% of herds studied although these results may be affected by the technique of recovery used (filtration).

Robinson[19] isolated *C. jejuni* from 10% of apparently normal cows in two milking herds. More than one serotype was present in the herd at any one time but infection with single serotypes continued for some months. Fecal shedding of *C. jejuni* was intermittent and few organisms were shed. It appears, therefore, that *C. jejuni* may be present in the feces of both diarrheic and healthy cattle, although usually in low numbers in the feces of the latter.

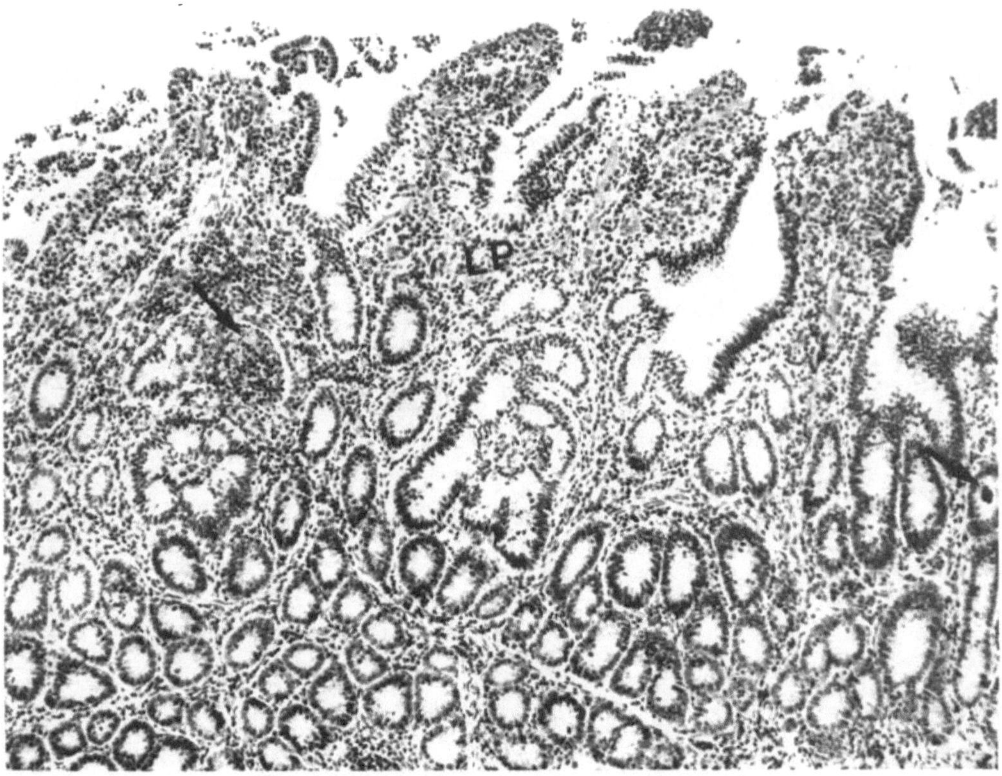

FIGURE 2. Ileum of clinically affected calf. Note the presence of material in the lumens of crypts (arrow). Note the hypercellularity of the lamina propria (L.P.). (H.E.; magnification × 120.)

When Al-Mashat and Taylor[14] surveyed the intestinal mucosa of 47 cattle of varying ages presented for postmortem examination, 7 of the 17 isolates of campylobacter obtained were of *C. jejuni*. The organisms were present in the mucosa of all levels of the gastrointestinal tract from the abomasum to the colon. The organism was isolated from grossly normal and from inflamed mucosa, although in all cases microscopic lesions were present. Other agents were present, in most cases, at some point in the gastrointestinal tract. Vibrios could be seen in smears made from the mucosal surface (Figure 1) at sites from which *C. jejuni* had been isolated. The changes seen included thickening of the terminal ileum and pallor of its wall, the presence of fluid mucoid ileal contents, mild inflammation of the mucosa, and enlargement of the mesenteric lymph nodes. Histological changes included shortening of the villi, the presence of inflammatory cells in the crypts, and lymphoid hyperplasia (Figure 2).

Pure cultures of *C. jejuni* have been used to reproduce enteric disease in both milk-fed[15] and ruminating calves.[1-3,15] The syndrome produced in animals of any age has an incubation period of 1 to 3 days after inoculation and began with slight fever (to 41°C), and the passage of diarrheic feces of a dark color, uniform consistency containing mucus and some blood. Altered feces were passed for 8 to 16 days and *C. jejuni* could be recovered from them daily in the earlier stages of the disease. Later, recovery of the organism became less regular. A rise in agglutinating antibody titer to 1:640 commonly occurred. The pathological changes seen at postmortem examination resembled those described above for chronic natural cases (Figure 3). *C. jejuni* was isolated from the mucosa of the ileum, cecum, and colon in all cases and less frequently from the jejunum, abomasum, and gall bladder.

When animals were killed at 24-hr intervals after oral inoculation, *C. jejuni* was recovered

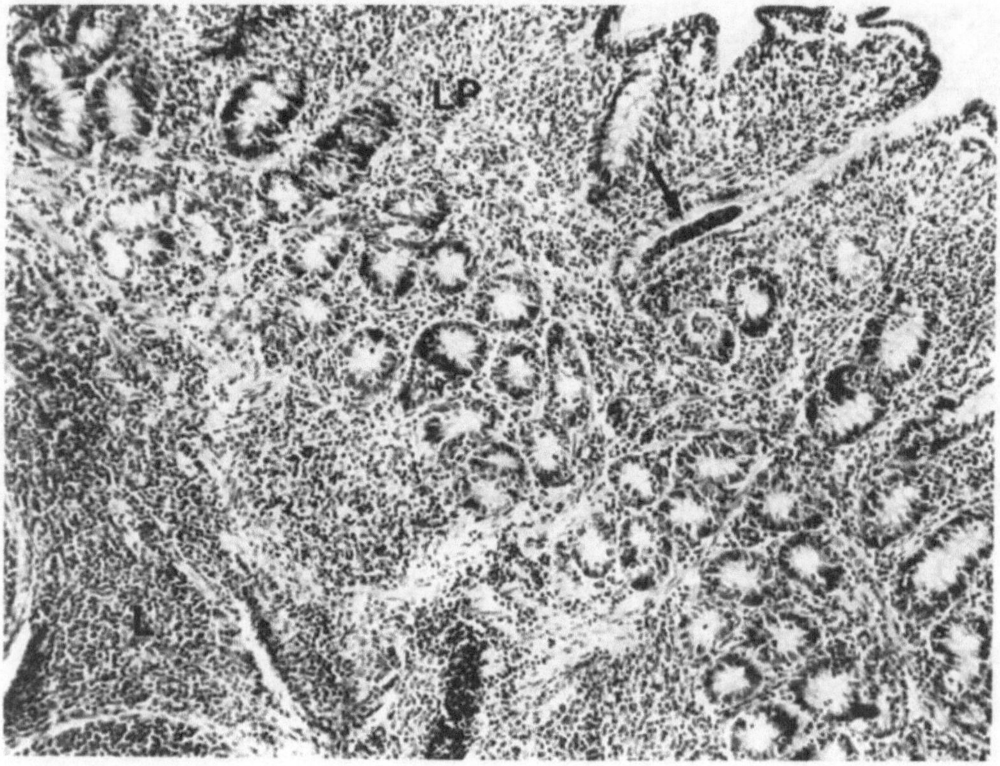

FIGURE 3. Ileal mucosa from a calf killed 20 days postinoculation with a pure culture of *C. jejuni*. Note the lymphoid hyperplasia (L), the hypercellularity of the lamina propria (L.P.) and the presence of cellular debris in the crypt (arrow). (H.E.; magnification × 120.)

from all levels of the gastrointestinal tract and from the mesenteric lymph nodes, gall bladder, lung, and spleen at 24 hr after inoculation, but only from the gastrointestinal tract, mesenteric lymph node, and spleen at 48 hr postinoculation. By 96 hr postinoculation *C. jejuni* could only be isolated from the ileum, cecum, and colon.

There were few gross changes other than the presence of mucoid ileal contents after 24 hr and the microscopical changes were slight. There was mild capillary dilatation, particularly in the abomasal mucosa, and bacteria were seen plugging the glands. They could not be identified as campylobacters by electron microscopy although *C. jejuni* had been isolated from adjacent tissue. Some villous atrophy and neutrophil infiltration of the lamina propria was seen and silver-stained bacteria, some of them apparently spiral, were seen in crypts, particularly in the ileum.

At 48 hr after inoculation the jejunal contents were watery and mucoid, as were those of the ileum. Contents of normal consistency were present in the colon. The mesenteric lymph nodes were enlarged and the ileal mucosa thickened. Microscopic changes included more obvious inflammation of the abomasal mucosa and the presence of inflammatory cells, mainly neutrophils, in the crypts of the jejunum and ileum. In these regions there was some villous atrophy and the lumenal epithelial cells were cuboidal. They were found to have shortened microvilli and damaged mitochondria by electron microscopy, but no bacteria could be seen within them or closely adjacent to their lumenal surfaces. Bacteria with the dimensions of campylobacters were, however, present in some of the neutrophils in the crypts. Similar organisms were also seen in the crypts of the cecal mucosa, but not within cells. There was marked degeneration of the cells lining the large intestinal crypts, and goblet-cell discharge was prominent in contrast to the appearance of similar cells from control calves.

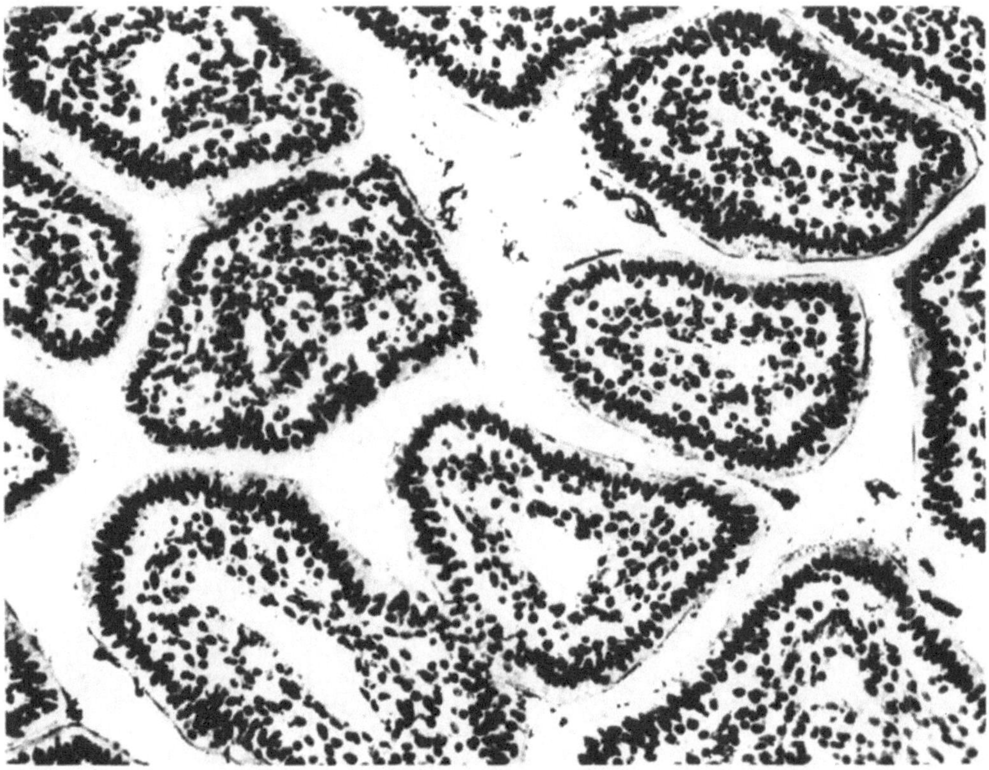

FIGURE 4. Ileal mucosa from a calf infected with *C. jejuni* 3 days postinfection. Note the presence of silver-stained bacteria between the villi. (Youngs stain; magnification × 400.)

At 72 hr postinoculation, the jejunum and ileum were flaccid and contained thin, cloudy mucus and fluid contents. The mucosa was thickened and inflamed. The cecal contents were mucoid and watery. Microscopic and ultrastructural changes resembling those seen above were found, but the inflammatory cells in the crypts were found by electron microscopy to be largely lymphocytes and macrophages and to contain few neutrophils and erythrocytes. Once again, no bacteria could be found within the epithelial cells (Figure 4). Agglutinating antibody to the inocular strain was present in the serum of this animal at a low titer.

Similar changes were seen in the animal killed at 5 days although the level of serum antibody was higher. The animal killed on the 6th day had a chronic lesion similar to those described by Al-Mashat and Taylor[14,15] and antibody titers of 1:640. It was heavily infected with coccidia, and in silver-stained sections there appeared to be invasion of damaged small intestinal mucosal epithelium by silver-stained curved or spiral bacteria. This has not yet been confirmed by electron microscopy.

It seems from this pathogenicity study that there may be a brief bacteremia immediately following oral infection with large inocula of *C. jejuni*, but that the organism can no longer be demonstrated outside the gastrointestinal tract 72 hr after infection. This may be due to the appearance of circulating antibody. The presence of the organism in the lung (actually in a consolidated pneumonic lesion) is also of interest.

The relationship of *C. jejuni* to the gut seemed to alter in time with its disappearance from the abomasum and jejunum to persist in the ileum, cecum, and colon. It appeared to be present adjacent to, but not closely adherent to the epithelial cells and was never seen within cells early in uncomplicated infections. A loose association with isolated brush borders was noted in in vitro studies but the adhesions were much looser than that with *E. coli* in control preparations. This appeared to confirm the results of the ultrastructural studies.

The host response to the presence of *C. jejuni* appeared to be mild inflammation including an initial outpouring of neutrophils followed by the presence of lymphocytes and macrophages in the crypts. This stimulation of the lymphoid system is a major feature of the gross and microscopic changes, especially in the ileum and its draining lymph nodes.

B. *C. fetus* Subsp. *fetus* Infections

C. fetus subsp. *fetus* was isolated from bovine feces by El Azhary[20] but was not considered to be an enteric pathogen. No enteric changes were recorded by Florent[21] following experimental infections, although the organism was isolated from the cecal contents and feces. An association with lesions was, however, indicated by the isolation of what appears to be this organism from a diarrheic calf[22] and the reproduction of loss of condition and fecal changes in experimental animals by Allsup et al.[23] Al-Mashat and Taylor[14] also found *C. fetus* subsp. *fetus* in gastrointestinal lesions similar to those from which *C. jejuni* was isolated and concluded that the organism might be capable of causing a syndrome similar to that caused by *C. jejuni*.

This did, in fact, prove to be the case when they fed pure cultures of *C. fetus* subsp. *fetus* to both milk-fed and ruminating calves.[17] The syndrome produced was slightly less obvious with less fever and less diarrhea but strands of bloody mucus were seen in the feces. The organism was reisolated, circulating antibody developed, and lesions similar to those produced by *C. jejuni* were also produced by *C. fetus* subsp. *fetus*. The organism was isolated from the lesions in profuse culture and also from the feces of affected animals.

C. *C. fecalis* Infections

C. fecalis was originally isolated from the feces of clinically normal sheep by Firehammer[24] but was found by Al-Mashat and Taylor[14] in inflammatory lesions of the alimentary tract of cattle. It is present in abomasal lesions much more frequently than the other campylobacters and can be isolated from lesions which are indistinguishable from those associated with the other two campylobacters described above. It may be present in lesions with other campylobacters.

Pure cultures of *C. fecalis* were fed by Al-Mashat and Taylor[16] to milk-fed calves and were found to cause few clinical signs, although the organism could be recovered from their feces after infection. Pathological changes similar to those caused by *C. jejuni* were seen in the abomasum and ileum and *C. fecalis* was isolated from them. Agglutinating antibodies to *C. fecalis* developed in the sera of the infected calves. Similar but more severe changes were seen in ruminating animals fed pure cultures of the organism, and thick strands of blood-stained mucus were demonstrated in the soft feces of the infected animals for some days after infection. The organism was isolated from the abomasum in all these animals.

C. fecalis can, therefore, initiate lesions with features of those of *C. jejuni* and cause some of the elements of *C. jejuni* disease, in particular the blood-stained mucus in loose feces. It appears not to be a harmless commensal as originally thought.

It appears from the experimental work described above that the campylobacters present in the bovine enteric tract are associated with lesions. Upon first infection they can initiate a syndrome consisting of low fever, passage of loose or diarrheic feces containing blood and mucus, but causing little apparent distress or loss of bodily condition. The lesions produced are most prominent in the ileum although there may be some effects on the colon which result in inability to reabsorb fluid. During the acute phase following initial infection the organism involved can be isolated readily from the feces, but later can be recovered only on occasions and in low numbers, much as described in naturally infected cattle by Robinson.[19] Such animals appear to harbor large numbers of organisms in the ileal, cecal, and colonic mucosa and should be regarded as recovered carriers rather than as "normal" animals. The duration of this mucosal carriage is unknown, but it is likely to be considerable.

C. jejuni, C. fetus subsp. *fetus* and *C. fecalis* have all been shown to be associated with such changes, and cause syndromes differing only in degree.[15-17] This finding is of profound significance when considering the emphasis presently placed on *C. jejuni* in man as it does not seem to be entirely justified in terms of disease in cattle. It is probable that, in cattle, pathogenicity testing of other biochemical variants would show that they could cause changes similar to those shown by *C. jejuni, C. fetus* subsp. *fetus* and *C. fecalis*.

Finally, it seems that in natural disease campylobacters can act as initiating agents in bovine enteric disease although in practice they are usually accompanied by other agents. They may also multiply in lesions caused by a wide variety of agents. In cattle, therefore, the picture of campylobacter infection is much more complex than it appears to be in man. This complexity means that enteric campylobacter infections are rarely diagnosed in cattle and thus are rarely treated specifically. No treatment trials have been carried out, but in vitro studies suggest that all bovine campylobacter isolates are sensitive to neomycin, a compound frequently used in the control of diarrhea in calves.

III. CATALASE-POSITIVE CAMPYLOBACTER INFECTIONS IN THE ENTERIC TRACTS OF SHEEP

Three species of catalase-positive campylobacters have been recorded from the feces or enteric tracts of sheep. They are *C. fetus* subsp. *fetus, C. jejuni* and *C. fecalis*. Other campylobacters with differing biochemical and temperature characteristics may also be isolated but *C. jejuni* and *C. fetus* subsp. *fetus* have been most commonly reported. *C. fecalis* was isolated from the feces of normal sheep and was considered both by Firehammer[24] and by Smibert[25,26] to be a normal inhabitant of the ovine gastrointestinal tract. There are no reports of experimental infections of sheep with this organism.

A. *C. fetus* Subsp. *fetus* Infections

C. fetus subsp. *fetus* has gained most attention as a cause of abortion in sheep, but has also been isolated from the intestine, feces, and gall bladder by a number of authors.[26] In none of these cases was there any description of pathological enteric lesions or enteric clinical signs. In studies of gastrointestinal lesions in sheep *C. fetus* subsp. *fetus* can occasionally be recovered from inflammatory lesions similar to those in which it was found in cattle by Al-Mashat and Taylor.[14]

The few experimental oral infections of sheep which have been described were intended to demonstrate carriage of the organism and the effects of oral infection upon the reproductive tract. For this reason the published accounts of oral infection merely indicate that *C. fetus* subsp. *fetus* could be recovered from the feces and that there was no diarrhea. There are no descriptions of enteric changes, and only the fact that the organism could be recovered from the gall bladder-infected animals indicates that the organism may infect sheep in the same way as it was shown by Al-Mashat and Taylor to affect cattle.[17]

C. jejuni may be recovered from the feces of clinically normal sheep and from lambs with diarrhea. In the latter it was present in the feces[18] and it may also be isolated from intestinal lesions. The lesions are generally mild in nature and resemble those described in cattle, although it can be isolated from hemorrhagic lesions at times. It can be isolated from any level of the alimentary tract from the abomasum to the colon but is most commonly found in the ileal mucosa. Changes similar to those seen in cattle may be seen in the small intestinal mucosa of animals with *C. jejuni* infection.

Experimental oral infections with *C. jejuni* have been carried out in lambs by Firehammer and Myers,[18] who recorded that blood and mucus could be demonstrated in the feces of infected animals but that diarrhea only developed in the presence of other enteropathogenic agents. They did not describe the pathological or serological findings. Other workers appear to have found that little clinical change results in the absence of other enteropathogens.

C. jejuni has been given to pregnant ewes by Shaw and Ansfield[27] but no account of any enteric consequences was described.

It appears that campylobacter infections in sheep are not uncommon, that the organisms can be isolated from normal feces, diarrheic feces, and intestinal lesions, and that *C. jejuni* is capable of causing the presence of blood and mucus in the feces following the experimental infection of lambs. The enteric consequences of oral infections of nonimmune adults with *C. jejuni* and of any class of sheep with *C. fetus* subsp. *fetus* and *C. fecalis* have not yet been adequately documented.

IV. CATALASE-POSITIVE CAMPYLOBACTER INFECTIONS IN THE ENTERIC TRACT OF PIGS

Three named species of catalse-positive campylobacter have been recovered from pigs. They are *C. coli*, *C. jejuni*, and recently, *C. hyointestinalis*. A number of other biochemical variants have been reported, but the organism most commonly encountered in pigs is *C. coli* and the majority of the information about campylobacter infections in pigs relates to this species or to organisms called *C. coli*.

A. *C. jejuni* Infections

C. jejuni has been reported from normal pig feces by Prescott and Bruin-Mosch[28] and from the intestines of slaughter pigs by Sticht-Groh,[29] but has only been associated with disease in pig herds in a few reports in Britain. The disease syndrome appears to be one in which blood and mucus were seen in the feces but in which deaths were uncommon and the causal agent of swine dysentery, *Treponema hypodysenteriae*, could not be demonstrated. *C. jejuni* was, however, isolated.

The failure of most investigators to evaluate their catalase campylobacter isolates in terms of the 1980 approved names[30] means that some confusion between *C. jejuni* and *C. coli* must have occurred. Some reports of *C. coli* prior to 1974 may have been of *C. jejuni*, and some between the publication of *Bergey's Manual*[25] and the redefinition of the criteria for *C. coli* by Skirrow and Benjamin[31] were called *C. jejuni* but were, in fact, *C. coli*.

Little experimental work has been carried out with *C. jejuni* in pigs. Prescott and Bruin-Mosch[28] found that in gnotobiotic pigs it could establish, but caused few clinical signs and only mild inflammatory changes which were most prominent in the large intestine.

B. *C. hyointestinalis* Infections

C. hyointestinalis has recently been described by Gebhart and Ward[32] as a catalase-positive campylobacter forming a yellow, circular convex colony 1 to 2 mm in diameter. The bacteria were long, arranged in loose spirals 0.35 to 0.55 μm in diameter, with a single polar flagellum.

C. hyointestinalis differs from *C. coli* in the production of H_2S on TSI, failure to grow in TTC, and resistance to nalidixic acid and cephalothin.

It was found in the lesions of proliferative intestinal adenopathy[33] but there is as yet no genuine evidence for its involvement as the cause of this condition. No reports of the reproduction of the condition with this organism have yet appeared and, in view of the difference between the type of lesion produced by catalase-positive campylobacters generally and the lesions of proliferative intestinal adenopathy, it seems unlikely that it should.

C. *C. coli* Infections

C. coli is commonly present in the feces of clinically normal pigs in small numbers and

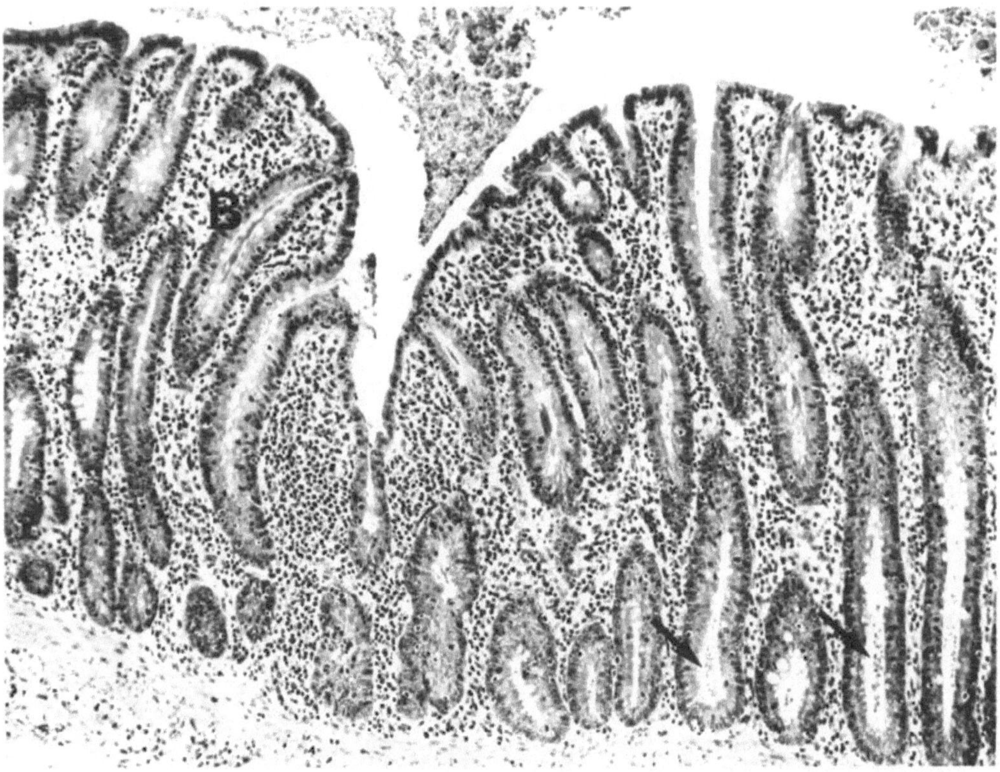

FIGURE 5. Colonic mucosa in a natural case of mucoid diarrhea in a pig from which *C. coli* and *Trichomonas* spp. were demonstrated. Note the dilated crypts containing bacteria (L) and protozoa (arrow). (H.E.; magnification × 120.)

appears in the diarrheic feces of pigs with certain types of enteritis, such as swine dysentery, in much larger numbers. It is present in virtually every pig herd and most animals are either actively immune to it or, if sucking, are receiving milk which probably contains specific IgA antibody. The sera of weaned pigs may contain low levels of antibody to *C. coli* when tested by the agglutination test.

In a normal immune pig herd, *C. coli* may be isolated in small numbers from the feces of piglets with normal feces from 7 to 10 days of age onwards. In animals with little immunity, i.e., those which have not taken colostrum or those with thymic atrophy, the organism may occur in larger numbers and earlier in life, and also when diarrhea of whatever cause occurs in younger pigs.

It can be isolated most commonly and consistently in younger pigs from 10 days to about 10 weeks of age and may be most easily isolated from the feces of recently weaned and mixed batches of pigs. It is present in the mucosa of the ileum, cecum, and colon in these carrier pigs.

C. coli has a long association with disease. Spiral microorganisms (both spirochetes and vibrios) were identified in the crypts of the affected colonic mucosa of pigs with swine dysentery by Whiting et al.[34] and were finally isolated by Doyle in 1944[5] and used by him to reproduce a syndrome which he claimed resembled swine dysentery in experimental pigs. In 1948 he described and named the organism *'Vibrio' coli* but his isolate is now lost.[6] For 27 years other workers[7-11] attempted to reproduce his results in conventional, specific pathogen-free and gnotobiotic pigs with varying degrees of success, building up a detailed picture of the clinical signs associated with *C. coli* infection. The link with swine dysentery was

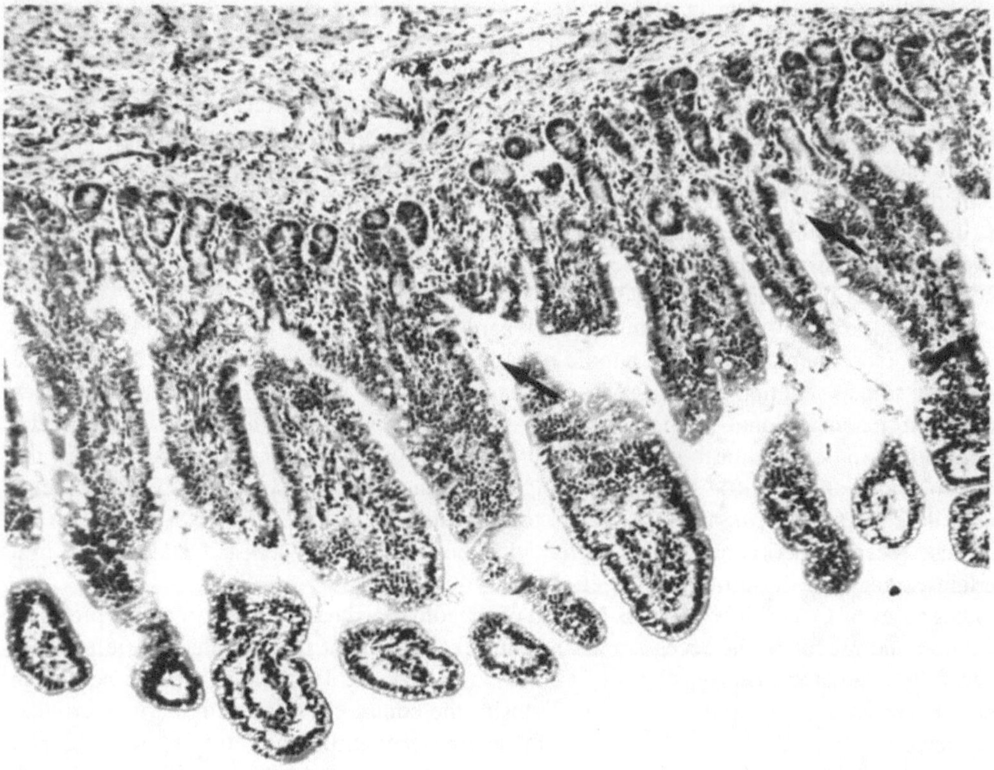

FIGURE 6. Ileum from a hysterectomy-derived, colostrum-deprived piglet killed 12 days postinfection with *C. coli*. Note the presence of cell debris between the villi (arrow) and the hypercellularity of the villi. (H.E.; magnification × 120.)

broken in 1971 by Taylor and Alexander[35] who showed that a large anaerobic spirochete was the cause of swine dysentery.

Reevaluation of the earlier literature,[36] coupled with the results of cultural and pathological studies of *C. coli* in pigs with enteritis, suggested that the organism was common in the small intestinal lesions found in sucking pigs with diarrhea and was also present in the colonic mucosa in certain situations.[36] The lesions with which it was associated included thickening of the small intestinal wall, enlargement of the mesenteric lymph nodes, and the presence of fluid contents containing mucus. The mucosal surface was covered with excess mucus and mildly inflamed, with some shortening of the villi. Inflammatory cells were frequently present in the crypts of the small intestine and the lymphoid areas of the terminal ileum were prominent and reactive. The crypts of the large intestinal mucosa were dilated and filled with organisms in cases where this organ was affected (Figure 5).

As in cattle and sheep, other pathogens such as coccidia, cryptosporidia, viruses, and enteropathogenic bacteria were also present in most of the lesions studied or in the gastrointestinal tracts in which they were found.

From the experimental studies described in the period from 1944 to 1968[6-11] and those described by Olubunmi and Taylor,[37] the following picture of *C. coli* infection in the pig can be constructed.

The syndrome is most obvious in hysterectomy-derived, colostrum-deprived piglets,[37] but some elements of it can be seen in weaned HDCD pigs and in conventional sucking and weaned pigs. In gnotobiotic pigs[11] the texture of the feces precluded an accurate assessment of the clinical signs and attention was, in any case, focused on the large intestinal changes so that small intestinal changes were poorly described. Mild colitis was recorded.

Following oral infection in piglets, there is a rise in rectal temperature within 48 to 72 hr and changes in fecal consistency become apparent within 3 days of infection. The feces become soft and, in some cases, diarrheic and mucoid. Within the mucus, flecks of blood may be seen. Looseness of the feces and the presence of excess mucus persist for at least 10 days after infection. Affected piglets do not die but appear depressed. At post-mortem examination the features described above are evident (Figure 6). Campylobacters cannot be seen adjacent to the mucosa and do not appear to lie within cells of the intestinal epithelium in silver-stained sections prepared from pigs killed from 4 to 10 days postinfection. Curved silver-stained rods may, however, be seen in the crypts of the colonic mucosa. Agglutinating antibody to the inocular strain develops.

Piglets sucking a sow from an infected herd also developed diarrhea and soft feces following experimental infection, but did not appear to lose so much condition. Nonimmune pigs lost some condition and passed some excess mucus, while only mild fecal changes were seen in conventional weaned pigs from a herd infected with *C. coli*. The rises in rectal temperature and serum antibody level occurred in all pigs following infection, as did the pathological changes. Little long-term change in productivity was seen in these experiments and this feature of *C. coli* infection needs to be examined further.

These recent experimental results agree with those of Doyle,[5,6] Deas,[8] Davis,[9] and to a certain extent with those of Andress et al.[11]

Diagnoses of *C. coli* infections as a cause of mucoid diarrhea in piglets or of the presence of blood and mucus in the feces are at present rare. Treatment for them has, therefore, not been fully evaluated. Many isolates of *C. coli* from pigs are sensitive in vitro to neomycin and this seems to be a satisfactory drug. Tylosin, the commonly used veterinary macrolide, may have a therapeutic effect on some farms as may ampicillin and tetracyclines if *C. coli* is thought to be involved in an enteric syndrome. In some cases, treatment with agents which suppress other elements of the flora, such as olaquindox, may expose disease in which *C. coli* is involved.

V. CONCLUSIONS

The range of species of catalase-positive *Campylobacters* found in the enteric tracts and feces of cattle, sheep, and pigs is considerable and many isolates do not fit into accepted species definitions. This wide range probably reflects species specificity, such as *C. coli* in pigs, and also the exposure of these animal species to feed and water contamination by infected feces from rodent and avian species. All the organisms so far tested in appropriate host species appear to produce broadly similar clinical and pathological syndromes, and biochemical characters may be only of laboratory relevance in animal disease associated with catalase-positive campylobacters. There is, however, a profound difference between the enteric lesions associated with catalase-negative campylobacters and those associated with catalase-positive organisms.

In view of the relatively mild nature of the syndromes associated with catalase-positive campylobacters in cattle, sheep, and pigs, further experimental studies are required to assess fully their effect on animal productivity. These studies should include further study of organisms such as *C. fecalis* in sheep and *C. jejuni* in conventional pigs. The role of campylobacters in the development of lesions initiated by other microorganisms should also be studied further. Perhaps, in addition, the tendency to regard campylobacters as normal inhabitants of the animal intestine should be replaced by the view that many clinically normal animals are recovered carriers of certain strains of campylobacter. The fact that different serotypes of campylobacter exist within a herd has been shown by Robinson,[19] and that different species can also be present in the same alimentary tract has been shown by Al-Mashat and Taylor.[14] These facts should be borne in mind by diagnostic laboratories.

These low-level infections do not appear to protect against colonization and the pathological consequences of fresh infections with a different serotype of the same species,[37] but they may affect the severity of the clinical signs produced. So, too, may antimicrobial treatment.

In short, infection with catalase-positive campylobacters in the enteric tracts of cattle, sheep, and pigs is a complex field which is at present only poorly understood.

REFERENCES

1. **Jones, F. S. and Little, R. B.**, The aetiology of infectious diarrhoea (Winter Scours) in cattle, *J. Exp. Med.*, 53, 835, 1931.
2. **Jones, F. S. and Little, R. B.**, Vibrionic enteritis in calves. *J. Exp. Med.*, 53, 845, 1931.
3. **Jones, F. S., Little, R. B., and Orcutt, M.**, A continuation of the study of the aetiology of infectious diarrhoea (Winter Scours) in cattle, *J. Am. Vet. Med. Assoc.*, 81, 610, 1932.
4. **Jones, F. S., Orcutt, M., and Little, R. B.**, Vibrios (*Vibrio jejuni* N. sp.) associated with intestinal disorders of cows and calves, *J. Exp Med.*, 53, 853, 1931.
5. **Doyle, L. P.**, A vibrio associated with swine dysentery, *Am. J. Vet. Res.*, 5, 3, 1944.
6. **Doyle, L. P.**, The etiology of swine dysentery, *Am. J. Vet. Res.*, 9, 50, 1948.
7. **Roberts, D. S.**, Studies on vibrionic dysentery in swine, *Aust. Vet. J.*, 32, 114, 1956.
8. **Deas, D. W.**, Observations on swine dysentery and associated vibrios, *Vet. Rec.*, 72, 65, 1960.
9. **Davis, J. W.**, Studies on swine dysentery, *J. Am. Vet. Med. Assoc.*, 138, 471, 1961.
10. **Warner, S. D.**, Studies on the Pathogenesis of *Vibrio coli* Infections in Swine, Ph.D. thesis, University of Minnesota, St. Paul, 1965.
11. **Andress, C. E., Barnum, D. A., and Thomson, R. G.**, Pathogenicity of *Vibrio coli* for swine. I. Experimental infection of gnotobiotic pigs with *Vibrio coli*, *Can. J. Comp. Med.*, 32, 522, 1968.
12. **Neill, S. D., Ellis, W. A., and O'Brien, J. J.**, Designation of aerotolerant *Campylobacter*-like organisms from porcine and bovine abortions to the genus *Campylobacter*, *Res. Vet. Sci.*, 27, 180, 1979.
13. **Ellis, W. A.**, personal communication, 1982.
14. **Al-Mashat, R. R. and Taylor, D. J.**, *Campylobacter* spp. in enteric lesions in cattle, *Vet. Rec.*, 107, 31, 1980.
15. **Al-Mashat, R. R. and Taylor, D. J.**, Production of diarrhoea and dysentery in experimental calves by feeding pure cultures of *Campylobacter fetus* subsp. *jejuni*, *Vet. Rec.*, 107, 459, 1980.
16. **Al-Mashat, R. R. and Taylor, D. J.**, Production of enteritis in calves by the oral inoculation of pure cultures of *Campylobacter fecalis*, *Vet. Rec.*, 109, 97, 1981.
17. **Al-Mashat, R. R. and Taylor, D. J.**, Production of enteritis in calves by the oral inoculation of pure cultures of *Campylobacter fetus* subsp. *intestinalis*. *Vet. Rec.*, 112, 54, 1983.
18. **Firehammer, B. D. and Myers, L. L.**, *Campylobacter fetus* subsp. *jejuni*. Its possible significance in enteric disease of calves and lambs, *Am. J. Vet. Res.*, 42, 918, 1981.
19. **Robinson, D. A.**, Campylobacter infection in milking herds, in *Campylobacter: Epidemiology, Pathogenesis, and Biochemistry*, Newell, D. G., Ed., MTP Press, Lancaster, U.K., 1982, 274.
20. **El Azhary, M. A. S. Y.**, An assay of isolation and differential identification of some animal vibrios and of elucidation of their pathological significance, *Med. Veeartensenijschool Rijksuniversiteit Gent*, 12, 1, 1968.
21. **Florent, A.**, Les deux vibrioses genitales de la bete Bovine: La vibriose venerienne due a *V. foetus veneralis*, et la vibriose d'origine intestinale duc a *V. foetus intestinalis*, in Proc. 16th Int. Vet. Congr., Madrid, 1959, 953.
22. **Allsup, T. N. and Hunter, D.**, The isolation of vibrios from diseased and healthy calves, *Vet. Rec.*, 93, 389, 1973.
23. **Allsup, T. N., Matthews, K. P., Hogg, S. D., and Hunter, D.**, Vibrios in diseased and healthy calves, *Vet. Rec.*, 90, 14, 1972.
24. **Firehammer, B. D.**, The isolation of vibrios from ovine faeces, *Cornell Vet.*, 55, 482, 1965.
25. **Smibert, R. M.**, The Campylobacters, in *Bergey's Manual of Determinative Bacteriology*, 8th ed., Buchanan, R. E. and Gibbons, N. E., Eds., Williams – Wilkins, Baltimore, 1974, 207.
26. **Smibert, R. M.**, The Genus *Campylobacter*, Ann. Rev. Microbiol., 32, 673, 1978.
27. **Shaw, I. G. and Ansfield, M.**, Fetopathogenicity of *Campylobacter jejuni* in sheep, in *Campylobacter: Epidemiology, Pathogenesis, and Biochemistry*, Newell, D. G., Ed., MTP Press, Lancaster, U.K., 1982, 177.

28. **Prescott, J. F. and Bruin-Mosch, C. W.,** Carriage of *Campylobacter jejuni* in healthy and diarrhoeic animals, *Am. J. Vet. Res.*, 42, 164, 1981.
29. **Sticht-Groh, V.,** Campylobacter in healthy slaughter pigs: a possible source of infection for man, *Vet. Rec.*, 110, 104, 1982.
30. **Skerman, V. B. D., McGowan, V., and Sneath, P. H. A.,** Approved list of bacterial names, *Int. J. Syst. Bacteriol.*, 30, 720, 1980.
31. **Skirrow, M. B. and Benjamin, J.,** '1001' Campylobacters: cultural characteristics of intestinal campylobacters from man and animals, *J. Hyg. (Cambridge)*, 85, 427, 1980.
32. **Gebhart, C. J. and Ward, G. E.,** Isolation and characterization of various campylobacter species from swine intestinal tissues, Proc. 7th Cong. Int. Pig Vet. Soc., Mexico City, 1982, 59.
33. **Kurtz, H. J. and Chang, K.,** Demonstration of a new campylobacter species in lesions of proliferative enteritis in swine, Proc. 7th Cong. Int. Pig Vet. Soc., Mexico City, 1982, 60.
34. **Whiting R. A., Doyle, L. P., and Spray, R. S.,** Swine dysentery, *Purdue Univ. Agr. Exp. Stn. Bull.*, 257, 1, 1921.
35. **Taylor, D. J. and Alexander, T. J. L.,** The production of dysentery in swine by feeding cultures containing a spirochaete, *Br. Vet. J.*, 127, lviii, 1971.
36. **Taylor, D. J. and Olubunmi, P. A.,** A re-examination of the role of *C. coli* in enteric disease of the pig, *Vet. Rec.*, 109, 112, 1981.
37. **Olubunmi, P. A. and Taylor, D. J.,** Production of enteritis in pigs by the oral inoculation of pure cultures of *Campylobacter coli*, *Vet. Rec.*, 111, 197, 1982.

Chapter 17

CAMPYLOBACTER SPUTORUM SUBSP. *MUCOSALIS*

G. H. K. Lawson and A. C. Rowland

TABLE OF CONTENTS

I. Historical 207

II. Morphology 207

III. Growth and Colonial Morphology 212
A. Atmosphere 212
B. Media 212
C. Colonial Morphology 212

IV. Physiological, Biochemical Characters, and Methods Used in Their Determination 213
A. Identification Tests 213
1. Catalase Test 213
2. Hydrogen Sulfide Production, Nitrate and Nitrite Reduction 213
3. Tests of Ability to Grow in the Presence of Sodium, Glycine, and Sodium Deoxycholate 213
B. Other Characteristics of the Organism 214

V. Antigens of *C. Sputorum* subsp. *mucosalis* 214

VI. Sensitivity to Antibiotics 215

VII. Isolation Media and Techniques 216
A. Dilution Technique 216
1. Mucosa 216
2. Chyme or Feces 216
B. Selective Media (Novobiocin, Brilliant Green, Trimethoprim Agar) 216
C. Filtration 216
D. Identification 217

VIII. Maintenance of Cultures 217

IX. Other Unclassified Catalase-Negative Porcine Vibrios 217

X. Taxonomy 217

XI. Association with Disease 218

XII. Clinical Symptoms 218

XIII. Pathological Changes 220

XIV. Other Related Conditions 221

XV. Antigenic Types in Association with Disease 221

XVI. Porcine Serology 221

XVII. Crohn's Disease 222

XVIII. Experimental Infection 222

XIX. Infection of Cell Cultures 222

XX. Other Proliferative Enteropathies 223

References 223

I. HISTORICAL

Vibrio sputorum was the name first given to a vibrioid bacterium recovered by Prévot[1] from a human bronchitic sputum sample, although it seems likely that others had previously isolated similar bacteria. Later it was realized that such bacteria could be recovered from the oral cavity of many healthy individuals where the organism is now considered to comprise some 5% of the flora of the gingival crevice.[2] Unlike many of the pathogenic vibrios, these organisms were negative in the catalase test. A similar bacterium, originally causing confusion with *Vibrio fetus*, was isolated from the bovine genital tract and named *Vibrio bubulus*.[3] As these organisms had many common features, they were later drawn together in a single species, *Campylobacter sputorum,* and each distinguished as separate subspecies, *sputorum* and *bubulus*, respectively. A third distinct organism was later isolated from pigs and because of its similarity to the other catalase-negative vibrios was assigned to the species, this time with the subspecies name of *mucosalis*.[4] It is possible that other workers may have isolated this organism earlier, but the descriptions of and the atmospheric conditions used for the isolation of these catalase-negative organisms suggests that these were not *mucosalis*.[5,6]

II. MORPHOLOGY

Campylobacter sputorum subspecies *mucosalis* is a Gram-negative, short, irregularly curved rod measuring 0.25 to 0.3 μm in width and 0.95 to 2.8 μm in length in preparations examined by light microscopy. Rod forms, S-shapes, seagulls, and spiral forms are present and may vary in proportion in cultures of different strains. The organisms do not stain strongly and many of the bacteria are indistinct. In older cultures, coccoid and filamentous forms become more prominent but never conceal the essential curved bacillary nature of the organism.[7]

All cultures are motile, this being easily demonstrated in the water of condensation of slope cultures or in the fluid phase of biphasic cultures *(vide infra);* phase microscopy is the easiest method of demonstrating the motility which is characteristically highly vigorous with rapid reversal of the direction of movement, interspersed on occasion with periods of inactivity.

Ultrastructural studies confirm the variability of form of the organism and thin sections indicate a maximum diameter of 0.36 μm with organisms tapering over a third of their length to either a blunt or a relatively pointed end (Figure 1). Most organisms show an

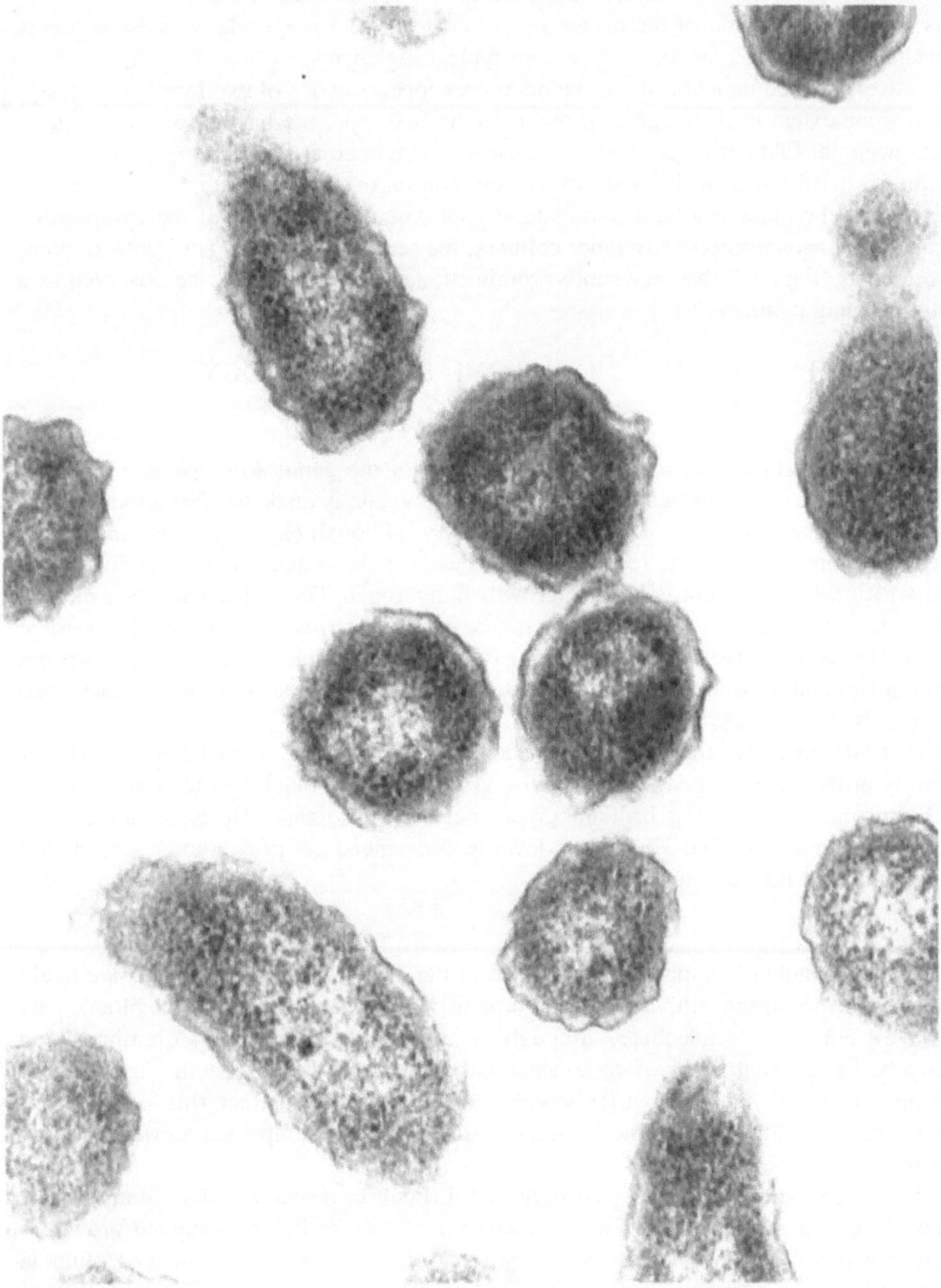

FIGURE 1. Ultrathin section of *C. sputorum* subsp. *mucosalis*. (Magnification × 92,000.)

outer, irregularly formed, double, electron-dense, wavy layer separated from the internal structure of the bacterium by an electron-lucent zone. The cytoplasmic membrane is poorly visible in intact organisms and cannot readily be distinguished at the edge of the dense granular electron-dense ribosomal cytoplasm. An area of the cytoplasm is commonly less electron-dense and presumably corresponds to the area of the nucleoid. In this less dense zone, it is often possible to visualize small, circular, electron-dense bodies. In many preparations, a small proportion of the bacteria present appear to have a relatively smooth coat without the marked ridges of the coat seen in typical organisms.

In negatively stained intact bacteria, the outer coat appears in one of two forms, apparently related to some extent to the stage of growth. In the first type, the bacteria show a smooth coat with irregular folds or ridges and may possess a single polar flagellum (Figure 2). The flagellum, 13 to 18 nm in width and unsheathed, can on occasion be seen to pass through a depression in the outer coat and terminate at approximately the level of the cytoplasmic membrane in a flagellar knob.[8,9] In older cultures, the second type, organisms with smooth, nonridged coats (Figure 3) become more prominent; such forms can also be observed as a minority in young cultures.

III. GROWTH AND COLONIAL MORPHOLOGY

A. Atmosphere

These organisms differ from many other members of the genus *Campylobacter* in that hydrogen is essential for satisfactory growth unless provision is made to offer an alternative electron donor. Small quantities of air are stimulatory, although exposure of cultures to air at atmospheric pressure at 37°C results in rapid death of the organism. A satisfactory gas mixture which will consistently support growth is hydrogen 77%, nitrogen 10%, oxygen 3% and carbon dioxide 10%. Such an atmosphere can be obtained by evacuating anaerobic jars to −650 mmHg, replacing the evacuated gas by hydrogen, reducing the new atmosphere to −76 mmHg, and replacing with carbon dioxide. Jars gassed in such a way are incubated without the use of a catalyst.

Individual strains differ in their tolerance to oxygen and while some laboratory strains consistently grow if the oxygen level is below 11%, others are much less tolerant. The use of inhibitory media containing brilliant green increases the lethal effects of higher than optimal levels of oxygen. The 3% oxygen level recommended has proven to be satisfactory with all strains and media employed.[10]

B. Media

Most media capable of supporting the growth of the more fastidious pathogens are likely to prove satisfactory in the cultivation of the organism on solid medium. All the blood agars tested have been perfecly adequate, although our own preference is for Columbia blood agar. Batches of nutrient bases of certain media have been found to become incapable of supporting growth of *mucosalis* after storage. It is not clear whether this is a general phenomenon, but it should be borne in mind where isolation attempts are carried out only infrequently.

Growth in fluid-phase media is often light and difficult to detect visually. Many simple peptones require enrichment with 5% inactivated horse serum before adequate growth is obtained. Where yield from fluid culture is important, our preference is to utilize a Columbia blood agar slope with a tryptose phosphate broth overlay (diphasic medium). Fluid cultures require to be incubated in the appropriate gaseous atmosphere with caps or plugs which allow free interchange of gas.

C. Colonial Morphology

The colonial appearance is modified by the moisture content of the medium. On undried

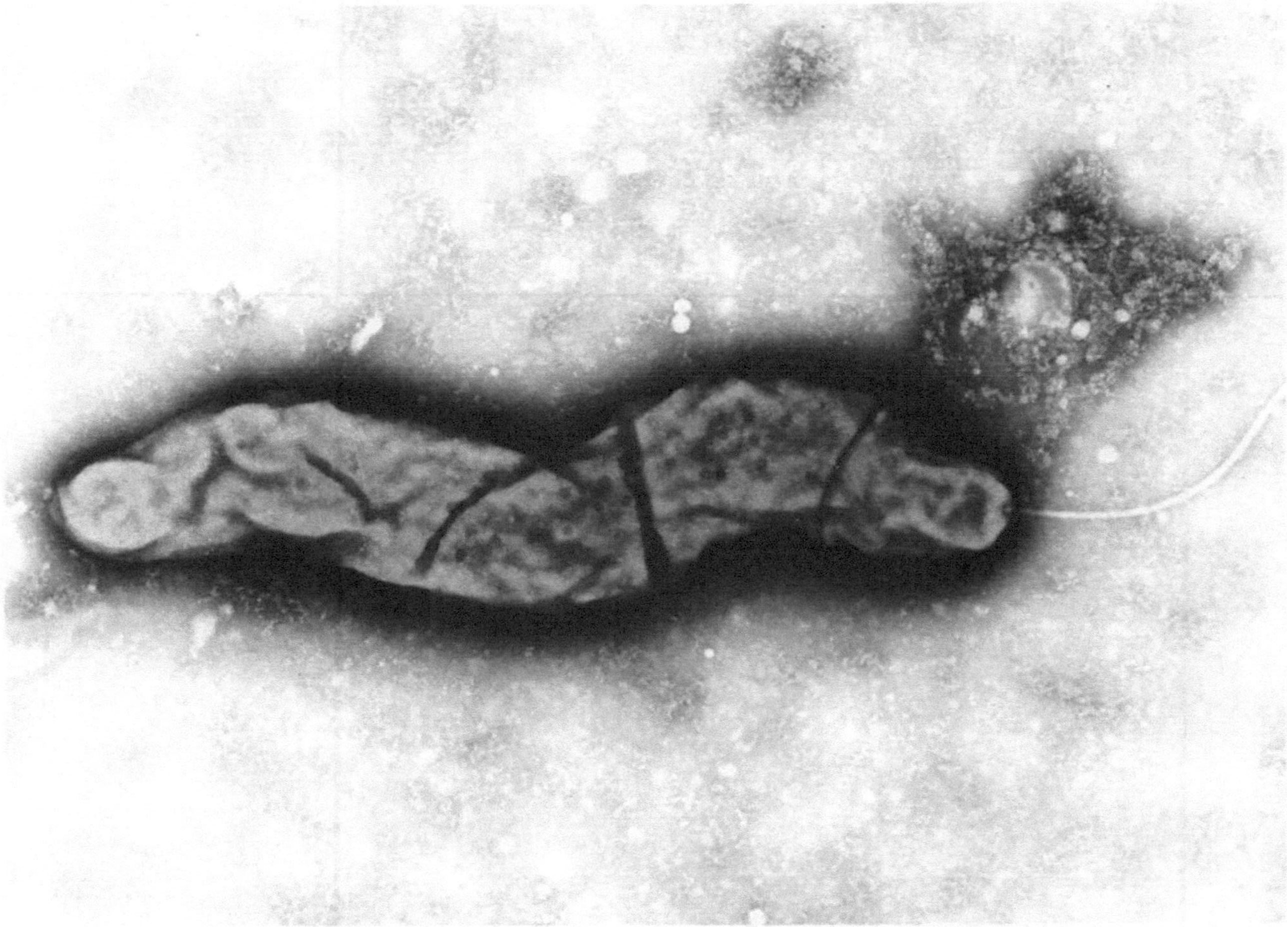

FIGURE 2. Negatively stained preparation of *C. sputorum* subsp. *mucosalis*. Note ridging of surface and single polar flagellum inserted in a pit. (Magnification × 55,000.)

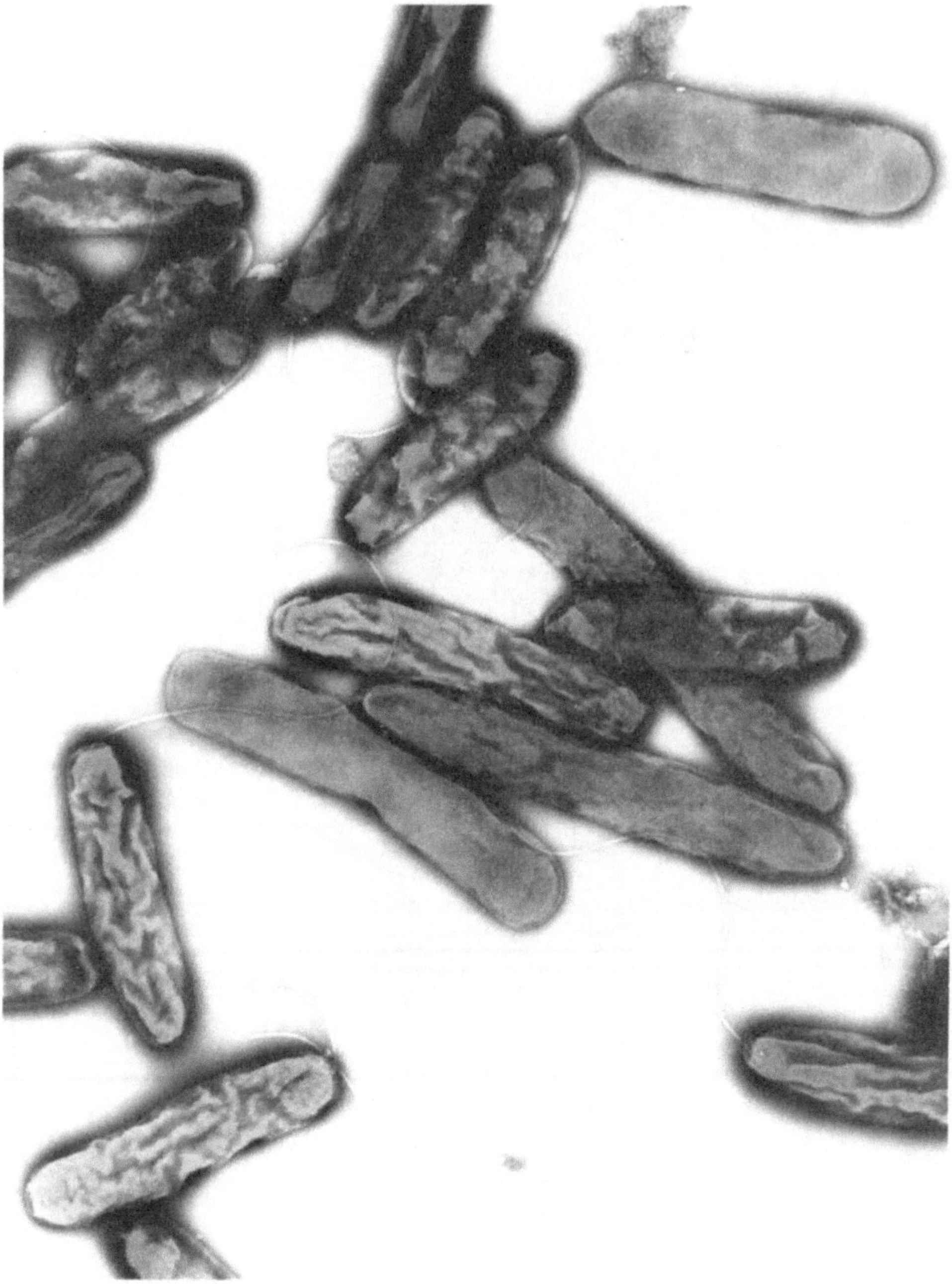

FIGURE 3. Negatively stained preparation of *C. sputorum* subsp. *mucosalis* showing rod forms with ridged and smooth coats. (Magnification × 28,125.)

plates, colonies are translucent with a tendency to overrun the medium, while on dried plates, colonies do not exceed 1.5 mm in diameter after 48 hr incubation, are circular and raised with a flat surface, and shiny gray. They can often be differentiated from catalase-positive organisms in primary culture by the more convex nature of colonies of the latter and by the presence of an indistinct yellow color to the *mucosalis* colonies. This pigment difference is more readily discernible when a loopful of growth is removed and smeared on white paper — pigment in *mucosalis* is dirty yellow, while the porcine catalase-positive organisms show a pink-tan color. Some care is required in making this assessment as pigments tend to darken with the age of the culture; it is, however, of great use in presumptive enumeration and selection of colonies for further study.

On some batches of horse blood agar, the organism forms ill-defined hemolytic zones, which are never complete, under individual colonies; such lysis is irregular in its appearance and most plates show no hemolysis.

IV. PHYSIOLOGICAL, BIOCHEMICAL CHARACTERS, AND METHODS USED IN THEIR DETERMINATION

These tests fall into two main groups, those that are likely to be of use in identifying the organism and those characters which are of interest in relation to either the taxonomy of the bacterium or to our appreciation of its metabolism. Some details of the methods used in the former will be included, while in the latter instances, only the results will be considered.

A. Identification Tests

The limited number of tests which define members of the genus is generally omitted in routine identification procedures, but it may be noted that the organisms do not ferment or oxidize carbohydrates or hydrolyze urea or gelatin. They do not require serum for growth, are methyl red and Voges Proskauer negative, do not produce lipase, and, as stated previously, are microaerophilic. They are also oxidase-positive but the reaction is delayed.[7]

1. Catalase Test

The slide test is often unreliable in differentiating weakly catalase-positive vibrios and the test should be carried out at 48 hr on cultures incubated on nutrient agar slopes containing 10% inactivated horse serum using 3% v/v H_2O_2.

2. Hydrogen Sulfide Production, Nitrate and Nitrite Reduction

These tests are carried out in triple-sugar iron agar, blood agar, or nutrient agar containing 0.05% cysteine with lead acetate strips to detect hydrogen sulfide production and nitrate broth with 5% serum for assessment of nitrate and nitrite reduction.[11]

3. Tests of Ability to Grow in the Presence of Sodium, Glycine, and Sodium Deoxycholate

The delicate turbidity denoting growth in fluid media makes the assessment of growth in the presence of inhibitory substances in broths unsatisfactory. For this and other practical considerations there are advantages in carrying out these tests on the surface of nutrient agar using a replicate plating system. With such a method it is possible to divide the three members of the species into their subspecies both readily and reproducibly (Table 1). Blood agar base No. 2 (Oxoid CM271) has proved a satisfactory nutrient base for carrying out the tests incorporating glycine 1.0 and 1.5 g/100 mℓ, NaCl 1.5 and 3.0 g/100 mℓ, and deoxycholate 0.05 and 0.2 g/100 mℓ. Master plates of blood agar are inoculated heavily with nine strains, each strain being inoculated on a circular area of approximately 1 cm diameter. The master plates are incubated for 24 hr at 37°C and then used to charge velour pads with inocula.[12] These pads are then used to inoculate up to nine plates serially, terminating with

Table 1
CHARACTERS ALLOWING LABORATORY DIFFERENTIATION OF THE SUBSPECIES *C. SPUTORUM*[7]

					Glycine %		NaCl %		Sodium deoxy-cholate %	
	Catalase	Oxidase	H_2S/TSI	H_2S	1.0	1.5	1.5	3.0	0.05	0.2
Subsp. *sputorum*	−	+D[a]	+ or −	+	+	+ or −	+	−	+ or −	−
Subsp. *bubulus*	−	+	+	+	+	+	+	+	+	−
Subsp. *mucosalis*	−	+D	+	+	−	−	+	−	+	−

[a] D: delayed result.

an additional nutrient plate without inhibitory substance to confirm the efficacy of inoculation. Ability to grow in the presence of the substance is assessed at 24 and 48 hr.

B. Other Characteristics of the Organism

The G+C% of the DNA from the type strain (NCTC 11,000) is 33.9%, indicating that the allocation of such organisms to the genus *Campylobacter* is appropriate despite the differences between this and other members of the species from the catalase-positive members of the genus.

Growth on the surface of solid media in which nitrogen completely replaces hydrogen in the atmosphere is minimal. The organism appears to contain an unusual c-type cytochrome (c-553), which may be reduced in cell extracts by either hydrogen or formate, and the latter substance in fluid media at 2.5 mg/mℓ will allow the growth of the organism in the absence of hydrogen.[13]

The organism is tolerant of bile (6%) but will not grow on MacConkey agar and is resistant to the action of brilliant green. Strains grow at 42°C, not at 45°C, and irregularly at 25°C; they do not fluoresce under UV light, but when grown on media containing neutral red the growth is irridescent and this is associated with fluorescence under UV light.[7,10]

Whole cell extracts of *C. sputorum* subsp. *(sputorum, bubulus*, and *mucosalis)* show similar fatty acid profiles and these are distinct from those of other members of the genus.[14]

V. ANTIGENS OF *C. SPUTORUM* SUBSP. *MUCOSALIS*

On the basis of heat-labile surface antigens, strains of the subspecies may be divided into three main groups, called serotype A, B, and C, respectively. Antisera prepared in rabbits against a member of the group using whole cells will react to high titer with all members of the group. Despite such cross reaction, which takes place close to the titer obtained with the homologous strain, major differences exist between many of the serotype A strains that have been examined.[15,16] There are no cross reactions between the surface heat-labile antigens of serotype A, B, and C nor have cross reactions been found with human strains of subsp. *sputorum* or bovine strains of *bubulus*. Using cross-absorption tests, the heat-labile surface antigens of a few strains of serotype A have been examined in some detail and the known antigens of these strains are set out in Table 2; there is no reason to believe that such a scheme exhausts the identifiable antigens which may be present in these strains.

Serotypes A and B show a common heat-stable antigen assessed by both agglutination and complement fixation tests. Extraction of phenol-soluble antigens by the method of Mitscherlich[17] and their use in a CFT indicates the presence of distinct antigens in both serotype A and B strains.[18]

Table 2
SEROLOGICAL CHARACTERISTICS OF SOME *CAMPYLOBACTER SPUTORUM* SUBSP. *MUCOSALIS* STRAINS[16-18]

	Heat-stable antigens (CFT)			
Serotype	Phenol/water extraction	Heat only	Surface heat-labile (agglutination)	Strain
A	A	X	I a b c e f g j	253/72
			II a c d e f i j	302/72
			III a b d e f h j	124/73 B4
			IV a b c d f g i	722/75
			V a b c d e g h i	140/76 VF220
B	B	X	No cross reaction with A or C	982/76
C	*[a]	*	No cross reaction with A or B	512/77

[a] * not examined.

Table 3
MINIMUM INHIBITORY CONCENTRATIONS OF ANTIBIOTICS (MIC) AND ZONE SIZE IN DISC TESTS WITH STRAINS OF *C. SPUTORUM* SUBSP. *MUCOSALIS*

		MIC μg/mℓ			Zone size	
	No. of strains	Range	Mean	SD[a]	Disc amount (μg)	Range (mm)
Ampicillin	22	0.12—16	1.40	3.37	10	16—40
Penicillin G	22	0.25—>10	6.65	5.00	10	9—28
Tetracycline	22	0.25—>4	1.45	2.19	30	0—41
Sulfonamide	22	>256	>256			
Streptomycin	22	0.50—>32	4.30	13.00	10	0—12
Lincomycin	27	2.50—>50	20.70	24.80	10	0—19
Dimetridiazole	26	0.10—200	77.70	92.60	50	0—40
Virginiamycin	10	5.00—>100	85.50	49.40	*[b]	*
Tylosine	11	2.50—100	29.70	23.10	15	0—25
Tiamulin	28	0.50—20	7.80	5.49	5	0—34

[a] SD: standard deviation.
[b] *, not examined.

VI. SENSITIVITY TO ANTIBIOTICS

C. sputorum subsp. *mucosalis* strains are unaffected by normal concentrations of trimethoprim, novobiocin, and bacitracin, most other antibiotics showing some activity. The range of minimum inhibitory concentrations (MIC) of a number of commonly used antibiotics is presented in Table 3 along with the normal zone size. Zone sizes are not greatly affected by the density of inoculum, although, as in other sensitivity tests, some crude standardization should be carried out. Many strains show markedly reduced sensitivity to antibiotics in comparison with the levels of susceptibility of "sensitive" strains. In some cases, this difference is so great as to indicate that it is likely to be acquired resistance, while in other cases either the differences are smaller or the drugs are not those normally expected to have been used in the treatment of diseases of the pig.

VII. ISOLATION MEDIA AND TECHNIQUES

The isolation of *C. sputorum* subsp. *mucosalis* from clinical cases selected for necropsy can often be achieved from the intestinal lesion without special media by the simple use of a dilution technique. Isolation of small numbers of bacteria or organisms from heavily contaminated material require the use of selective media. Although numerous attempts have been made to improve the selective medium originally developed,[4] this has not proved easy and the only improvement obtained to date has been with the inclusion of trimethoprim lactate at 5 μg/mℓ and the use of lysed horse blood in the original novobiocin-brilliant green medium.

A. Dilution Technique

1. Mucosa

The mucosal surface of the intestine is exposed and laid flat. The surface is then washed repeatedly with sterile phosphate buffered saline (PBS) until all traces of visible chyme are removed. The mucosa is then removed by scraping with a knife and diluted in a small quantity of reinforced clostridial (RCM — Oxoid CM149) broth. The mucosa is then homogenized at 13,000 r/min for 30 sec and made up to a 1/20 dilution. Further 1/20 dilutions in broth are prepared and 0.1-mℓ amounts spread over the surface of Columbia blood agar and selective media. It is possible that the RCM broth could be replaced by PBS and this is certainly true for the chyme of some animals known to be infected; at the present time, it has not been possible to examine a range of cases to determine if this is widely true.

2. Chyme or Feces

Dilutions of chyme or feces are prepared as above in PBS and 0.1 mℓ-plated on selective media and Columbia blood agar.

B. Selective Media (Novobiocin, Brilliant Green, Trimethoprim Agar)

Stock solutions of brilliant green are titrated against a known strain of *mucosalis* to find that level of dye which does not reduce the viable count of *mucosalis* grown originally in diphasic media and then plated on the basal media containing brilliant green alone. Strains of *mucosalis* vary slightly in their sensitivity to brilliant green solution and for critical work it may be best to acquire a known "sensitive" strain or in experimental work standardize the media employing the strain to be used in the experiments.

Brilliant green solution, novobiocin (5 μg/mℓ), trimethoprim (5 μg/mℓ) and lysed horse blood (5%) are added to the basal medium at 56°C. Even with the prior titration of brilliant green, it will be found that some batches are more inhibitory than expected and it is advisable to test each batch for the quantitative recovery of *mucosalis*. Each worker must decide his own standards depending on the occasion; in our work, batches of media were generally accepted if they do not reduce the viable count of NCTC 11,000, originally grown overnight in diphasic media then plated on selective medium, by more than 10^1 in comparison with noninhibitory media. In order to try and avoid certain problems encountered with selective media, the base used in this laboratory has recently been changed from Oxoid CM271 to Oxoid CM231 with added 1% yeast extract (Oxoid L21).

It should be noted that selective media developed for the isolation of *C. jejuni* are unsuitable for the recovery of *C. sputorum* subsp. *mucosalis*.

C. Filtration

Although there are theoretical advantages to be achieved by the use of filtration techniques, in general the practical advantages are probably outweighed in routine use by the technical problems of filtration. The one situation where filtration may be of value is in the examination of oral swabs; these are obtained by thoroughly swabbing the gingival margins, transferring the swab to saline, agitating, filtering through 1.2-μm pore diameter membrane filter, then plating the last drops on selective and nonselective media.

D. Identification

Colonies growing on selective media may take longer to develop than on noninhibitory media, plates are not finally rejected until after 5 days of incubation. Selective media also reduce the differences in colonial appearance that are readily observed when *mucosalis* and the catalase-positive organisms are grown on nonselective medium. Suspect colonies are subcultured to blood agar and examined for colonial pigment, morphology in Gram stain, slide catalase test, and slide agglutination tests using undiluted serotype A and serotype B "OH" antisera. Provided reasonable care is taken, such procedures are sufficient to identify *mucosalis*; in a series of isolates examined in detail, both biochemically and serologically, the slide agglutination test has never identified organisms which later turned out to be either catalase-positive campylobacters or other organisms. On a few occasions, some confusion could be caused by unstable (autoagglutinating) catalase-positive organisms but these could always be easily identified on other grounds including morphology.

VIII. MAINTENANCE OF CULTURES

Cultures maintained in microaerophilic conditions at 37°C should be subcultured at 10-day intervals. *Mucosalis* colonies on the surface of agar plates will remain viable for a number of days at either +5°C or at room temperature, and this method may be employed for short-term transport of cultures.

Where longer transit times are to be expected cultures should be inoculated into a suitable transport medium,* incubated overnight with loose cap in microaerophilic (hydrogen) conditions, the cap sealed without opening and the culture despatched.

In the long term, cultures may be preserved at −70°C in 5% serum broth with 17% glycerine. This can be easily achieved by suspending a heavy loopful of a 24- to 48-hr culture grown on blood agar in 1 mℓ of serum glycerin broth in a small glass tube, which is then immediately placed in the −70°C chest.

IX. OTHER UNCLASSIFIED CATALASE-NEGATIVE PORCINE VIBRIOS

Other catalase-negative vibrios may be isolated on occasion, either from the mouths or from rectal feces of some normal or diseased pigs.[15] Some of these organisms differ in their biochemical properties from *mucosalis;* glycine, salt-tolerant, and nonnitrite reducing organisms have been recovered from these sites; other organisms have also been recovered which are biochemically similar to *mucosalis* but not belonging to either of the serotypes A or B which have been associated with adenomatosis. The relationship of these bacteria to the adenomatosis organisms is not certain, and until such time as this becomes more clear we prefer to retain the name *mucosalis* for those organisms which may be isolated in large numbers from the lesions of adenomatosis, or for biochemically and serologically similar bacteria.

X. TAXONOMY

The marked enhancement of growth observed when strains of *C. sputorum* subsp. *mu-*

* Transport medium: nutrient broth (Oxid CM1), 1.3g; proteose peptone (Oxoid L46), 1.5g; sodium thioglycollate 0.05g, agar (oxoid L12) 0.5g, and distilled water, 100 mℓ, with final pH 7.0.

cosalis are grown in the presence of hydrogen is not demonstrated by many of the catalase-positive species of the genus. It is, however, a feature of strains of subsp. *sputorum* and also of the human oral catalase-positive vibrios.[19] *Vibrio succinogenes,*[20] an organism of uncertain status not included in the current taxonomy,[21] shows similar metabolic activity and this is also shared by some anaerobic bacteria, for example *Bacteroides corrodens.*[22] Such features have encouraged some authors to consider that subsp. *sputorum* should, therefore, be considered as anaerobic bacteria not requiring conditions of complete anaerobiosis.[23]

The members of the species *sputorum* have widely differing DNA base compositions, some of which are compatible with the genus *Campylobacter* in which G + C% should lie within 30 to 35%. Thus, *bubulus* has a G + C% of 30.1[24] and *mucosalis* 33.9; this is in contrast to subsp. *sputorum* in which the G + C% is 48 to 50%.[19] While enhancement of growth by hydrogen has been demonstrated for human oral vibrios and subsp. *mucosalis*, and both share similar metabolic properties, stimulation by hydrogen has not been reported for subsp. *bubulus.*

Clearly, there are major differences between these organisms and the other members of the genus, but it may be questioned whether the present time is appropriate for reallocation of the members of the species.

XI. ASSOCIATION WITH DISEASE

C. sputorum subsp. *mucosalis* can be recovered from the diseased intestinal mucosa of many cases of porcine intestinal adenomatosis (PIA), principally a disease of the weaned pig.[4,15] The organism is present in large numbers, up to 10^8/g of tissue, in the diseased mucosa; the bacteria are largely confined in the intestine to the abnormal tissue, being absent or present in restricted numbers in the nonadenomatous intestine of affected animals. Smaller numbers of organisms may be recovered from the chyme, while recovery from feces using the available selective media is irregular. Bacteria may be isolated from the mesenteric lymph nodes draining affected tissue, but there is no evidence to suggest that substantial widespread systemic invasion takes place.

In an examination of a number of nonadenomatous pigs of comparable age to those examined with adenomatosis, it was not possible to recover *mucosalis.* The recovery of small numbers of *mucosalis* from the chyme may have been prevented by other intestinal bacteria, it is clear that *mucosalis* is absent from the healthy mucosa of such animals.[4]

While the organisms appear absent from the mucosa of most healthy pigs, they may be recovered, albeit in smaller numbers, from the intestinal mucosa of some suckling diarrheic piglets.[8] Additionally, the organisms may be recovered from the oral cavity of pigs and this colonization has been shown to persist for a period of some 8 weeks in both experimental and natural infections.[25-27]

The majority of isolates recovered from adenomatosis cases have proved to be serotype A, although on occasion serotype B organisms may be recovered from the same lesions; one herd has yielded only serotype B strains from PIA cases and this situation has remained constant over a number of years.[10]

Serotype A isolates have now been recovered from the lesions of PIA in numerous herds in Scotland and England,[13] also from Sweden,[28] U.S.,[29] and Denmark.[30]

XII. CLINICAL SYMPTOMS

The principal symptoms of PIA are those of anorexia and wasting. Occasional diarrheic episodes may take place but are not a prominent feature of the disease. These clinical signs persist for a period of some 6 weeks and are followed by a return of appetite and normal progress to slaughter, at which time in those animals affected in the postweaning period, lesions are absent.

In some herds, the disease may be present at a level that does not attract attention and it is only when a detailed study of the growth of individual animals is made that its presence may be suspected.[31] Animals with lesions of the alimentary tract may be detected either at routine slaughter[32] or in herds slaughtered for other reasons.[33]

XIII. PATHOLOGICAL CHANGES

The lesions of PIA are a thickening of the mucosa of the intestine affecting principally the lower ileum and/or the cecum and proximal colon. The thickened mucosa is primarily composed of an immature hyperplastic epithelium in which much of the normal epithelial architecture is lost. Normal glands are replaced by crowded undifferentiated gland cells with, at least in some stages, very little conventional inflammatory response in the lamina propria.[34]

These altered gland cells can be shown to contain vibrio-like bacteria in smears stained by brucella differential stain,[35] in sections stained by silver techniques,[36] or in thin sections examined with the electron microscope.[34] Fluorescent techniques using conjungated convalescent pig serum or hyperimmune rabbit sera prepared against *mucosalis* indicate that antigen in these cells share common factors with *mucosalis*.[4,34]

Ultrastructural studies of the parasitized cells show a number of unusual features. The vibrio-like organisms appear to lie free within the apical cytoplasm of the cells and are not enclosed within a detectable membrane, the altered cells rarely show bacteria adjacent to or attached to the surface (Figures 4 and 5), and a number of the intracellular organisms can be seen apparently dividing in this situation. The intracellular vibrios are closely associated with the tissue changes and adjacent normal tissue fails to show the presence of intracellular bacteria.

XIV. OTHER RELATED CONDITIONS

A number of other conditions share the basic pathological changes of PIA upon which additional abnormalities are superimposed. Necrotic enteritis (NE), in which the altered mucosa undergoes a coagulative necrosis, and regional ileitis (RI), in which much of the mucosa is replaced by granulation tissue, both show intracellular vibrios in the undamaged surrounding epithelium and *mucosalis* may be recovered often along with comparable numbers of catalase-positive vibrios.[37]

A further clinical condition affects mainly young adult animals (>4 months of age) in which the proliferative intestinal lesion is complicated by substantial hemorrhage giving the disease its name, proliferative hemorrhagic enteropathy (PHE).[38] Although *mucosalis* may be isolated from such lesions, this is often not true, and when it can be demonstrated to be present the viable organisms recovered are few in comparison with the numbers that may be isolated from PIA.[39,40] One must consider the possibility that this clinical condition may involve other bacteria than *mucosalis*, although Roberts[8] has shown that the intracellular organisms in PHE are often degenerate and that the lesions show similarities to the recovery phase of PIA. Such damage may account for the difficulty experienced in the recovery of the bacteria. Additionally, more recently, phenol extraction of the intestinal mucosa has shown the presence of similar amounts of specific *mucosalis* CF antigen in PHE and PIA and that in PHE this relates specifically to the serotype of *mucosalis* isolated.[18]

XV. ANTIGENIC TYPES IN ASSOCIATION WITH DISEASE

In an examination of the antigenic types of the organism associated with PIA, NE, RI, or PHE, no particular antigenic type appeared to be associated with a particular clinical

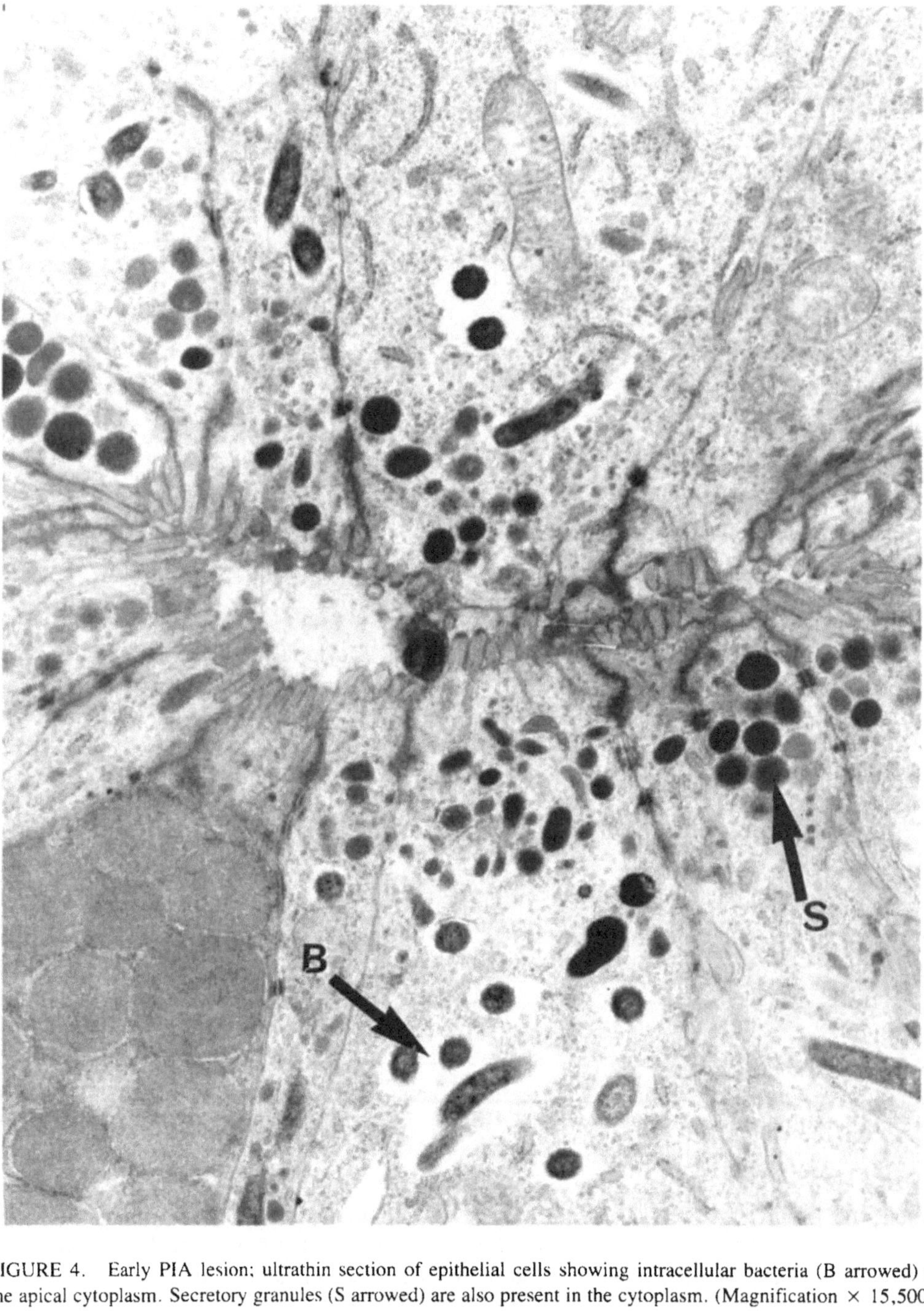

FIGURE 4. Early PIA lesion; ultrathin section of epithelial cells showing intracellular bacteria (B arrowed) in the apical cytoplasm. Secretory granules (S arrowed) are also present in the cytoplasm. (Magnification × 15,500.)

condition, although it must be emphasized that the selection of antigens for study has been arbitrary.[16] A sequential study was also made of strains isolated from clinical cases on two farms over a period of time. It was found that the antigens present in the strains changed from case to case with time.

XVI. PORCINE SEROLOGY

Investigation into the serological response to infection in pigs has been largely confined to the use of the agglutination test. Antigen for this test is normally grown on the surface

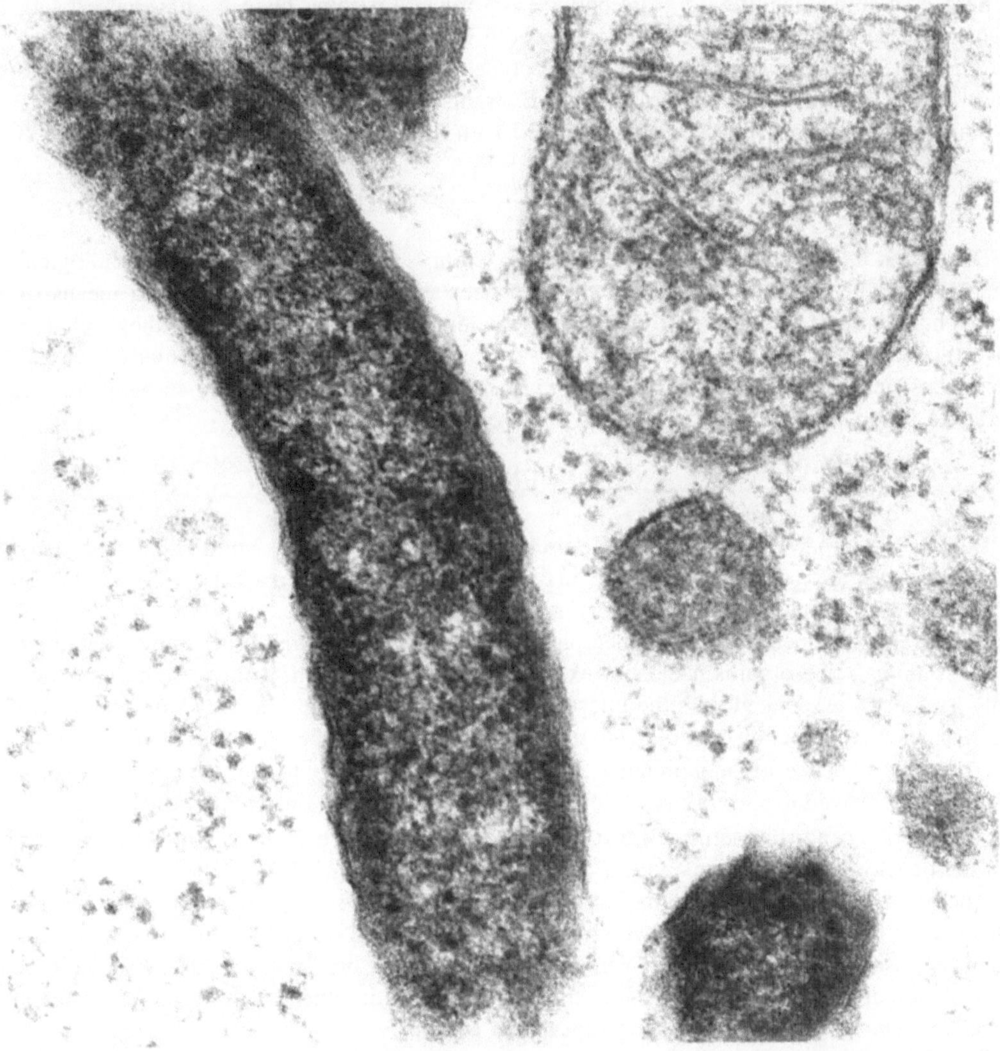

FIGURE 5. Ultrathin section of intracellular bacteria and mitochondrion in the cytoplasm of an intestinal epithelial cell from a case of PIA. Note absence of cell membrane surrounding the bacterial cell. (Magnification × 120,000.)

of solid media for 48 hr, washed off in 0.3% formol PBS, then washed and standardized for density.[15]

Most pigs affected with PIA have circulating agglutinating antibody during some of the period that lesions are present. In the early stages of disease, considerable differences are present in the serum agglutination titers depending on the particular antigenic type of serotype A strain chosen for antigen production. As pigs age, these differences become less and although some antigens appear more "sensitive" than others, activity will be present to all serotype A strains where infection involves one of these serotype A organisms.[41,42]

Arbitrarily selected sera obtained from healthy pigs at slaughter show a large number of animals with high titers. Investigation of one herd has shown that this seroconversion takes place at the age of 9 to 13 weeks and may not necessarily be associated with clinical symptoms of disease in all the animals. Seroconversion has also been shown to occur during the course of symptoms of PHE,[35] providing additional evidence that *mucosalis* is involved in the development of that condition despite the difficulty experienced in the recovery of the organism from the affected tissue.

Sera containing agglutinating antibody to serotype A strains have been shown, using fluorescent antibody techniques, to contain both IgG and IgM antibody reacting with *mucosalis*. Interpretation of reactions with serotype B strains may be more complex due to nonspecific agglutination of some serotype B strains by pig sera; this agglutination appears to principally involve a component associated with the albumin peak.[42]

XVII. CROHN'S DISEASE

Due to certain superficial similarities between Crohn's disease of man and the pathological changes present in PIA/RI, interest has often been expressed in the latter as a means of unraveling some of the problems posed by the nature of Crohn's disease. Matthews et al.[43] investigated the antibody response of Crohn's disease patients to a number of enteric bacterial pathogens, including *C. sputorum* subsp. *mucosalis*, but failed to find any serological evidence to suggest that this organism was involved in the condition.

XVIII. EXPERIMENTAL INFECTION

Despite the apparently convincing association found between *C. sputorum* subsp. *mucosalis* and members of the PIA complex in the field disease, the acquisition of experimentally based evidence to support the etiological association has not proved easy.

In conventional suckling and weaned piglets exposed to *mucosalis*, it became evident that in such animals the organism did not extensively colonize the alimentary chyme and was only irregularly associated with the intestinal mucosa, and then in relatively small numbers. Such exposure to infection did not result in the establishment of other than occasional irregular microscopic evidence of adenomatosis, an observation difficult to fully interpret.[44] Post-weaned pigs proved even more resistant to infection, and it was only by treatment with the anticholinergic drug, Benzetimide, that low level infection was established once again without the development of lesions within the period of experiment.[45]

After exposure to infection with *C. sputorum* subsp. *mucosalis*, neonatal pigs retain the organism in the oral cavity, from which it could be recovered irregularly for up to 8 weeks. Catalase-positive *Campylobacters* were also present in the oral cavity and, subsequent to the isolation of the serotype A-infecting strain from the mouth, other nonserotype A, catalase-negative *Campylobacters* made their appearance.[27]

On one occasion, exposure of neonatal pigs to adenomatous mucosa containing *mucosalis* and *mucosalis* culture appeared to result in the reproduction of PIA/NE.[46] This procedure has not proved reproducible but the original experiment has given an interesting insight into the histological and microbial development of the disease. *Mucosalis* could be recovered from the mucosa in moderate numbers in animals killed before the development of lesions, could be recovered in large numbers from the developed lesions, and then could not be isolated from the lesions which showed features of resolution.[8] A series of experiments involving gnotobiotic and conventional pigs infected with *mucosalis* or with *mucosalis* and other enteric microbial pathogens has not greatly furthered our knowledge of the particular relationship between *C. sputorum* subsp. *mucosalis* and the adenomatous intestinal cell.

XIX. INFECTION OF CELL CULTURES

Exposure of experimental animals to infection with *mucosalis* or to adenomatous mucosa apparently rarely results in the appearance of intestinal lesions of PIA in the exposed animals. Alternative methods have, therefore, to be developed to investigate the disease and the pathogenicity of *mucosalis*. Rajasekhar[9] has made a study of the relationship of *C. sputorum* subsp. *mucosalis* to mammalian and avian-derived cells in both primary and continuous cultures. *Mucosalis* attaches specifically with varying intensity to certain cells. While cells derived from many species had the capacity to bind *mucosalis*, certain cell cultures proved refractory. Following attachment, *mucosalis* enters the cells and shortly after this event the cells no longer permit attachment of this organism. A variety of techniques indicate that the bacteria, thereafter, either multiply intracellularly or produce large amounts of intracellular antigen depending on the cell type infected. This intracellular infection is associated with gross disturbance to the normal morphology and division of the cell, which in some cell lines results in early death.

These tissue culture studies, therefore, appear to indicate that *mucosalis* is capable of attachment to and ingestion by cells, persists intracellularly for prolonged periods, and brings about profound alteration of cell function without immediate cell death — features which are aspects of the cell parasitism in adenomatosis.

XX. OTHER PROLIFERATIVE ENTEROPATHIES

Of particular comparative interest is a histologically similar disease of hamsters, transmissible ileal hyperplasia or proliferative ileitis.[47] This disease also involves an intracellular vibrio-like organism which has been grown in tissue culture but not isolated in cell-free media. A further proliferative condition, murine colonic hyperplasia, may be induced by infection with certain strains of *Citrobacter freudii*.[48] In this latter disease, however, infection is of the mucosal surface and intracellular parasitism is not a feature.

Regional ileitis of lambs has also been described on occasion[49,50] and organisms with vibrio morphology have been found in the epithelial cells.[51] Preliminary cultural work has recovered both catalase-negative *Campylobacters* resembling *sputorum* and in an animal in which diphtheresis of the intestinal mucosa was present, *C. jejuni*.[52]

REFERENCES

1. **Prévot, A. R.,** in *Manual de Classification et de Détermination des Bacteries Anaérobies,* Masson & Co., Paris, 1940, 85.
2. **Socransky, S. S. and Manganiello, A. D.,** The oral microbiota of man from birth to senility, *J. Periodontal.,* 42, 485, 1971.
3. **Thouvenot, H. and Florent, A.,** *Vibrio bubulus* from the semen of the bull and the vagina of the cow, *Ann. Inst. Pasteur,* 86, 237, 1954.
4. **Lawson, G. H. K. and Rowland, A. C.,** Intestinal adenomatosis in the pig: a bacteriological study, *Res. Vet. Sci.,* 17, 331, 1974.
5. **Warner, S. D.,** Studies on the Pathogenesis of *Vibrio coli* Infection in Swine, Ph.D. thesis, University of Minnesota, Minneapolis, 1965.
6. **Söderlind, O.,** The isolation of *Vibrio coli* from pigs, *Vet. Rec.,* 77, 193, 1965.
7. **Lawson, G. H. K., Rowland, A. C., and Wooding, P.,** The characterization of *Campylobacter sputorum* subspecies *mucosalis* isolated from pigs, *Res. Vet. Sci.,* 18, 121, 1975.
8. **Roberts, L.,** A study of Porcine Intestinal Adenomatosis, Ph.D. thesis, University of Edinburgh, 1978.
9. **Rajasekhar, M.,** Growth Characteristics of *Campylobacter sputorum* Subsp. *mucosalis* in Cell Culture, Ph.D. thesis, University of Edinburgh, 1981.
10. **Lawson, G. H. K.,** unpublished data, 1981.
11. **Cowan, S. T.,** in *Manual for the Identification of Medical Bacteria,* 2nd ed., Cambridge University Press, Cambridge, 1974, 176.
12. **Lederberg, J. and Lederberg, E. M.,** Replica plating and indirect selection of bacterial mutants, *J. Bacteriol.,* 63, 399, 1952.
13. **Lawson, G. H. K., Leaver, J., Pettigrew, G., and Rowland, A. C.,** Some features of *Campylobacter sputorum* subsp. *mucosalis* subsp. nov., nom. rev. and their taxonomic significance, *Int. J. Syst. Bacteriol.,* 31, 385, 1981.

14. **Curtis, M. A.,** Cellular fatty acid profiles of Campylobacter, in *Campylobacter 1982*, Newell, D. G., Ed., MTP Press, Lancaster, U.K., 1982.
15. **Lawson, G. H. K., Rowland, A. C., and Roberts, L.,** Studies on *Campylobacter sputorum* subspecies *mucosalis*, *J. Med. Microbiol.*, 9, 163, 1976.
16. **Lawson, G. H. K., Rowland, A. C., and Roberts, L.,** The surface antigens of *Campylobacter sputorum* subspecies *mucosalis*, *Res. Vet. Sci.*, 23, 378, 1977.
17. **Mitscherlich, E. u. B. L.,** The differentiation of the Campylobacter strains on the basis of their antigen structure with the complement fixation test, *Dtsch. Tierarztl. Wschr.*, 65, 2, 1958; 65, 36, 1958.
18. **Mitscherlich, E. u. B. L. and Lawson, G. H. K.,** unpublished data, 1980.
19. **Van Palenstein Helderman, W. H. and Rosman, I.,** Hydrogen-dependent organisms from the human gingival crevice resembling *Vibrio succinogenes*, *J. Antonie van Leeuwenhoek*, 42, 107, 1976.
20. **Wolin, M. J., Wolin, E. A., and Jacobs, N. J.,** Cytochrome-producing anaerobic vibrio, *Vibrio succinogenes* S.P.N., *J. Bacteriol.*, 81, 911, 1961.
21. **Buchanan, R. E. and Gibbons, N. E., Eds.,** *Bergey's Manual of Determinative Bacteriology*, 8th ed., Williams & Wilkins, Baltimore, 1974.
22. **Smibert, R. M. and Holdeman, L. V.,** Clinical isolates of anaerobic Gram-negative rods with a formate-fumarate energy metabolism, *J. Clin. Microbiol.*, 3, 432, 1976.
23. **George, H. A., Hoffman, P. S., Smibert, R. M., and Kreig, N. R.,** Improved media for growth and aerotolerance of *Campylobacter fetus*, *J. Clin. Microbiol.*, 8, 36, 1978.
24. **Véron, M. and Chatelain, R.,** Taxonomic study of the genus *Campylobacter*, *Int. J. Systematic Bacteriol.*, 23, 122, 1973.
25. **Lawson, G. H. K., Rowland, A. C., and Roberts, L.,** Isolation of *Campylobacter sputorum* subsp. *mucosalis* from oral cavity of pigs, *Vet. Rec.*, 97, 308, 1975.
26. **Roberts, L.,** Natural infection of the oral cavity of young piglets with *Campylobacter sputorum* ssp. *mucosalis*, *Vet. Rec.*, 109, 17, 1981.
27. **Roberts, L., Lawson, G. H. K., and Rowland, A. C.,** Experimental infection of neonatal pigs with *Campylobacter sputorum* subspecies *mucosalis* with special reference to the oral cavity, *Vet. Microbiol.*, 5, 249, 1980.
28. **Gunnarsson, A., Hurvell, B., Jönsson, L., Martinsson, K., and Reiland, S.,** Regional ileitis of pigs, isolation of *Campylobacter* from affected ileal mucosa, *Acta Vet. Scand.*, 17, 267, 1976.
29. **Soto, J. A.,** *Campylobacter sputorum* Subsp. *mucosalis* Enteritis of Pigs, Ph.D. thesis, University of Minnesota, Minneapolis, 1979.
30. **Høgh, P. and Lawson G. H. K.,** unpublished data, 1979.
31. **Roberts, L., Lawson, G. H. K., Rowland, A. C., and Laing, A. H.,** Porcine intestinal adenomatosis and its detection in a closed pig herd, *Vet. Rec.*, 104, 366, 1979.
32. **Emsbo, P.,** Terminal or regional ileitis in swine, *Nord. Vet. Med.*, 3, 1, 1951.
33. **Rowland, A. C. and Hutchings, D. A.,** Necrotic enteritis and regional ileitis in pigs at slaughter, *Vet. Rec.*, 103, 338, 1978.
34. **Rowland, A. C. and Lawson, G. H. K.,** Intestinal adenomatosis in the pig: immunofluorescent and electron microscopic studies, *Res. Vet. Sci.*, 17, 323, 1974.
35. **Love, R. J., Love, D. N., and Edwards, M. J.,** Proliferative haemorrhagic enteropathy of pigs, *Vet. Rec.*, 100, 65, 1977.
36. **Young, B. J.,** A reliable method for demonstrating spirochaetes in tissue sections, *J. Med. Lab. Tech.*, 26, 248, 1969.
37. **Rowland, A. C. and Lawson, G. H. K.,** Porcine intestinal adenomatosis: a possible relationship with necrotic enteritis, regional ileitis and proliferative haemorrhagic enteropathy, *Vet. Rec.*, 97, 178, 1975.
38. **Rowland, A. C. and Rowntree, P. G. M.,** A haemorrhagic bowel syndrome associated with intestinal adenomatosis in the pig, *Vet. Rec.*, 91, 235, 1972.
39. **Love, D. N. and Love, R. J.,** Pathology of proliferative haemorrhagic enteropathy in pigs, *Vet. Pathol.*, 16, 41, 1979.
40. **Lawson, G. H. K., Rowland, A. C., Roberts, L., Fraser, G., and McCartney, E.,** Proliferative haemorrhagic enteropathy, *Res. Vet. Sci.*, 27, 46, 1979.
41. **Lawson, G. H. K., Roberts, L., and Rowland, A. C.,** Agglutinating activity to *Campylobacter sputorum* subspecies *mucosalis* in pig's sera, in *Proc. Int. Pig Vet. Sco.*, Copenhagen, 1980, 260.
42. **Lawson, G. H. K., Roberts, L., Rowland, A. C., McCartney, E., and Luckins, A. G.,** The presence of serum agglutinins to *Campylobacter sputorum* subspecies *mucosalis* in pigs, *Res. Vet. Sci.*, 32, 89, 1982.
43. **Matthews, N., Mayberry, J. F., Rhodes, J., Neale, L., Munro, J., Wensink, F., Lawson, G. H. K., Rowland, A. C., Berkhoff, G. A., and Barthold, S. W.,** Agglutinins to bacteria in Crohn's disease, *Gut*, 21, 376, 1980.
44. **Roberts, L., Lawson, G. H. K., and Rowland, A. C.,** The experimental infection of neonatal pigs with *Campylobacter sputorum* subspecies *mucosalis*, *Res. Vet. Sci.*, 28, 145, 1980.

45. **Roberts, L., Lawson, G. H. K., and Rowland, A. C.,** The experimental infection of pigs with *Campylobacter sputorum* subspecies *mucosalis*. Weaned pigs, with special reference to pharmacologically mediated hypomotility, *Res. Vet. Sci.*, 28, 148, 1980.
46. **Roberts, L., Rowland, A. C., and Lawson, G. H. K.,** Experimental reproduction of porcine intestinal adenomatosis and necrotic enteritis, *Vet. Rec.*, 100, 12, 1977.
47. **Frisk, C. S. and Wagner, J. E.,** Hamster enteritis: a review, *Lab. Anim.*, 11, 79, 1977.
48. **Barthold, S. W., Coleman, G. L., Bhatt, P. N., Osbaldiston, G. W., and Jonas, A. M.,** The aetiology of transmissible murine colonic hyperplasia, *Lab. Anim. Sci.*, 26, 889, 1976.
49. **Cross, R. F., Smith, C. K., and Parker, C. F.,** Terminal ileitis in lambs, *J. Am. Vet. Med. Assoc.*, 162, 564, 1973.
50. **Wensvoort, P.,** Chronic ileitis as a cause of "stretching" in lambs, *Tijdschr. Diergeneesk.*, 87, 841, 1962.
51. **Hoorens, J., Oyaert, W., Meyvisch, C., Vandenberghe, J., and Derijcke, J.,** Regional ileitis in a lamb, *Vlaams Diergeneesk. Tijdschr.* 46, 10, 1977.
52. **Vandenberghe, J. and Hoorens, J.,** Campylobacter species and regional enteritis in lambs, *Res. Vet. Sci.*, 29, 390, 1980.

INDEX

A

Abdominal cramps and pain
  enteritis experimental studies, 114
  traveller's diarrhea studies, 36—37
ABGT, see Antibovine globulin test
Abortion
  bovine, see Bovine abortion
  bovine genital campylobacteriosis studies, 182—184, 186—187
  endotoxin-caused, 139
  enteritis experimental studies, 115—116
  epidemiology studies, 145, 148, 151
  human infection studies, 23—25, 29
  ovine, see Ovine abortion
  porcine, 12
  septic, 151
  sporadic, see Sporadic abortion
  taxonomy studies, 2—4, 6, 11—12, 16
  vibrionic, 3—4, 11
Abscess, subcutaneous, 23
Absorbed antisera, specific, 62—63, 68, 71
Acid environment, *Campylobacter jejuni* sensitivity to, 169—170
Acid-extractable surface protein, 140
Acetidione, 47—48, 187
Acute endocarditis, 22—23
Acute appendicitis, 25
Acute enterocolitis, 37
Acute inflammation, 25—26
Acute pericarditis, 23
Acute purulent arthritis, 22
Acute purulent meningitis, 22
Acute ulcerative colitis, 25—27
Adenitis, 25
Adenomatosis, see also Porcine intestinal adenomatosis, 15—16, 217—223
Adenomatous intestinal cell, 222
Adventitious organisms, 114
Aerobic vibrios, 12
Aeromonas
  hydrophila, 34
  salmonicida, 136—137
Aerotolerant bacteria, 173
Aerotolerant *Campylobacter* sp.
  Campylobacter jejuni, 43, 173
  characteristics, key differential, 7—8, 12—13
  description and pathogenic significance, 12—13
  enteric infection studies, 194
  growth temperature, 12
  guanine + cytosine content, 13
  taxonomy studies, 2—3, 7—8, 12—13
Aflagellate mutants, 68
Agarose gel electrophoresis and visualization, plasmid DNA, 89—90
Age, incidence and, 34—35, 154—156
Age-specific ratios, blood-to-total isolates, 156
Agglutinating antibody, 186, 198—199, 204, 221—222
Agglutination tests, see also specific systems by name
  bovine genital campylobacteriosis studies, 186—187
  Campylobacter sputorum subsp. *mucosalis* studies, 214—215, 217, 221—222
  diagnostic serology studies, 106
  serological response studies, 98—102
  serotyping studies, 52—55, 62—74
Agglutinin antibodies, 36—37
Amikacin, 79—80, 82, 84
Aminoglycosides, 23, 29—30
  in vitro susceptibility studies, 79—80, 82, 84
Amoeba, infections caused by, 24—25
Amoxycillin, 78—79
Amphotericin B, 43, 47—48
Ampicillin, 78—79, 82—84, 204
Amplitude, cell-size determinations, 8, 11—12
Anaerobic bacteria, 218
Anaerobic growth, 7, 9, 11—12, 16, 185
Anaerobic vibrios, 14—15
Anamnestic booster reaction, 109
Anaphylactic shock, 123
Anhydrous environment, food storage in, 172—174
Animal feces
  Campylobacter sputorum subsp. *mucosalis* studies, 216
  enteric infection studies, 194—205
Animal infections
  bovine genital campylobacteriosis, 182—188
  Campylobacter sputorum subsp. *Mucosalis*-associated, 217—233
  carriers, 183, 187, 200, 202, 204
  causative agents, 183—186, 200—201, 203
  enteric, 194—205
  enteritis, 114—128
  reinfection, 188, 205
  spontaneous recovery from, 183
Animal isolates, serotypes, 57, 62, 68—74
Animal products, see also specific types by name
  epidemiology studies, 144—148, 157
  food studies, 164—179
Animal reservoirs, 144—149, 157
Animal studies
  abortion, 2—4, 6, 11—12, 16
  Campylobacter sputorum subsp. *mucosalis* studies, 208, 217—223
  direct animal contact, transmission following, 147—148
  enteric infections, 194—205
  enteritis, 114—128
  serotyping, 54, 57, 62—63, 68—74
  taxonomy, 2—4, 6, 9, 11—16
Anorexia, 218
Antibiotics, see also Antimicrobial agents; specific types by name
  bovine genital campylobacter studies, 186—188
  Campylobacter jejuni isolation studies, 40—42

enteric infection studies, 194, 200—201, 204—205
enteric microflora suppressed by, 128
epidemiology studies, 150
food studies, 177—178
human infection studies, 23, 27—30
MIC values, 78—84, 90, 186, 215
outer membrane studies, 134
resistance to, 40, 78—83, 88, 92, 215
susceptibility (sensitivity) to, see also under specific antibiotics by name, 78—84, 89, 186, 200
traveller's diarrhea studies, 37
Antibody, see also specific types by name
bovine genital campylobacteriosis studies, 186—188
Campylobacter sputorum subsp. *mucosalis* studies, 221—222
determination, techniques for, see also specific techniques by name, 98—100, 106
enteric infection, animal studies, 196, 198—199, 202, 204
enteritis experimental studies, 115, 119
epidemiology studies, 148, 150—152
immunoglobulin, see Immunoglobulin G antibody; Immunoglobulin M anitbody
local, production of, 187
monoclonal, 140
outer membrane studies, 139
serological studies, 99—103, 106—111
to homologous heated antigens, 63, 67—68
Antibody response
development, rapidity of, 101
duration of, 100—101
immunoglobulin class-specific, serology studies, 98—103, 106—111
pigs, to infections, 221—222
traveller's diarrhea studies, 36—37
Antibovine globulin test, 187
Anticholinergic drugs, 222
Antigen, see also specific types by name
bovine genital campylobacteriosis studies, 182—183, 186, 188
broadly reactive, 98—99, 103
Campylobacter spectorum subsp. *mucosalis* studies, 214—215, 219—221
epidemiology studies, 148
outer membrane studies, 134, 139—140
serological studies, 98—103, 106—111
serotyping studies, 52—58, 62—74
universal, 98
Antigenic factors
heat-labile, see Thermolabile antigenic factors
multiple, 72
Antimicrobial agents, see also Antibiotics; specific types by name
Campylobacter jejuni isolation studies, 41—43, 48
enteric infection studies, 194, 200—201, 204—205
human infection studies, 23
MIC values, 78—84, 90, 186, 215
plasmid studies, 88—95
resistance to, plasmid-mediated, 78, 81, 84, 88—95
susceptibility (sensitivity) to, see also under specific antimicrobial agents by name, 78—84, 89, 186,200
Campylobacter fetus subsp. *fetus*, 81—83
Campylobacter jejuni, 78—84, 89
disk sensitivity testing, 82—84
general discussion, 78—79,83—84
in vitro studies, 78—84
Antiphagocytic activity, 125, 139
Aniphagocytic surface protein, 139
Antisera
absorbed, specific, 62—63, 68, 71
monospecific, 55
monovalent, 68
polyvalent, 68
serotyping studies, 52—57, 62—74
Appendicitis, 25
Arthritis, 22, 25, 106
Articular forms, *Campylobacter fetus* subsp. *fetus* infections, 22
Artificial insemination centers, bulls infected in, 183, 186—187
A serogroup, 53, 62, 183—187, 214—215, 217—218, 221—222
Aspartate, 9, 11
Aspartate-fermenting *Campylobacter* sp., see Free-living aspartate-fermenting *Campylobacter* sp.
Asymptomatic infections, 34—37, 114, 117, 128, 148, 152
Attachment, *Campylobacter jejuni*, 122—124, 128, 139—140
Autoagglutination, 53, 63, 68, 98, 217
Avian infectious hepatitis, 6
Avian vibrionic hepatitis, 115

B

Bacitracin, 40, 43, 47, 81, 84, 186, 215
Bacteremia
enteric infections, animal studies, 198
enteritis experimental studies, 114—116, 125
epidemiology studies, 148—149, 151—152
frequency, 25
human infection studies, 22, 25
sporadic abortion studies, 183
taxonomy studies, 4
Bacteria
outer membrane, host-parasite relationships, role in, 134
typing, of, see Serotyping
Bacterial agglutination, see also Tube agglutination, 98
Bacterial diarrhea, see Diarrhea
Bacterial factors, see Virulence factors
Bactericidal activity, 78—79, 81, 139, 172

Bactericidal antibody, 99, 101
Bactericidal assay, serological response studies, 98—102
Bacteriocin, 13
Bacteriological diagnosis, *Campylobacter jejuni* isolation studies, 40—41
Bacteriophage, 13
Bacteriophage binding serotyping systems, 140
Bacteriophage-mediated transfer process, tetracycline resistance, 91, 95
Bacteriostatic activity, antimicrobial agents, 79, 81—82
Bacteroides
corrudens, 218
melaninogenicus, 15
Barium enema, 24
Bauer and Kirby technique, disk diffusion test, 83
Bedding, infected, cattle, infection transmitted by, 182—183, 186—187
Beef, contamination of, see also Cattle studies, 149, 157, 164—165, 167—168, 175—179
Bennington (Vt.) outbreak, epidemiology studies, 150
Benzetimide, 222
Beta-lactamase, see β-Lactamase
Bicozamycin, 82, 84
Bilateral salpingitis, 183
1% Bile, 14—15
Bimodal distribution, infections, 155
Biochemical properties, bovine genital campylobacteriosis causal agents, 185—186
Biotype *intermedius*, *Campylobacter fetus* subsp. *fetus*, 11
Biotypes and biotyping, 9—14, 148
1, 9—11, 72, 74, 184—185
2, 9—13, 72—74
3, 11, 72, 74
4, 72—74
limitations, systems, 128
serotype related to, 56, 70, 72—74
Bird studies, see also Poultry studies; Wild bird studies
enteritis, 114, 117, 122, 125
epidemiology, 151, 157
Blood
cultures, see Human blood cultures
in stools, see Stools, blood in
Blood agar media, 40—42, 47, 63, 68, 88, 177, 185, 210, 213, 216—217
Blood-to-total isolates, aga-specific ratios, 156
Bloody diarrhea
epidemiology studies, 151—152
human infection studies, 23—25
Bluecomb disease of turkeys, 6, 115, 122
Bovine abortion, 3—4, 11—12, 115, 139
Bovine enteritis, 3, 6
Bovine genital campylobacteriosis, 182—188
antigenic studies, 186
causative agents, 183—186
culture techniques, 185, 187
diagnosis, 187—188
enzootic infertility, see Enzootic infertility
historical background, 182
pathogenesis and chemical signs, 182—183
sporadic abortion, see Sporadic abortion
transmission and dissemination, 182, 186—187
treatment, 188
Bovine superoxide dismutase, 173
Braun's lipoprotein, 135, 138
Brilliant green, 10, 40, 187, 210, 214, 216
Broadly reactive antigen, 98—99, 103
Broth-mating method, 90
Brucella media, 41—42, 46—48, 166, 168—170, 173—174, 178
Brucellosis, 182
Brush borders, attachment to, 122, 129
B serogroup, 53, 62, 183—187, 214—215, 217—218, 222
Bulls, see Cattle, male
Butzler's medium Oxoid, 42—43, 46
Butzler's medium Virion, 43, 46
Butzler's selective medium, 44

C

Cake icing, contamination of, 1149—150
Campy-BAP medium, 43, 47—48
Campylobacter, general description, 2—3
Campylobacter coli
antibody to, 99
antiphagocytic surface protein, 139
biotype 3, 72, 74
biotyping, 13—14, 72, 74
Campylobacter jejuni differentiated from, 9—10, 47, 114, 128, 134, 201
Campylobacter jejuni strains BA37 and BA39 reclassified as, 95
characteristics, Key differential, 6—11
description and pathogenic significance, 6—10
distribution in, animals and men, 9
enteric infection studies, 201—204
enteritis experimental studies, 114, 116, 122—123, 128
epidemiology studies, 144—146
growth temperature, 6—10
guanine + cytosine content, 16
isolation, 202
nomenclature, 5—6
outer membrane studies, 134—135, 137, 139—140
serotyping studies, 13, 52—58, 62—74
individual schemes for, 55—56, 68—73
typable strains, 62, 68—72
strain discrimination methods, 13—14
surface, cell, 139—140
taxonomy, 3, 5—12, 201
Campylobacter concisus
characteristics, key differential, 14
description and pathogenic significance, 16
taxonomy studies, 3, 14—16
Campylobacter difficile, 24—25

Campylobacter fecalis
characteristics, key differential, 7—8, 12
colonial morphology, 195
description and pathogenic significance, 12
enteric infection studies, 194—195, 199—201, 204
enteritis experimental studies, 116, 122
epidemiology studies, 146
growth temperature, 12
guanine + cytosine content, 12
isolation, 199
taxonomy, 3, 6—8, 12, 15
Campylobacter fetus
antibiotics, sensitivity to, 186
antiphagocytic surface protein, 139
bovine genital campylobacteriosis studies, 182—188
Campylobacter jejuni isolation studies, 40, 47
characteristics, key differential, 6—11
cultures, 185
diagnostic serology studies, 106, 109
growth temperature, 6—9, 16, 185
infection caused by, 40
microcapsule, 139—140
nomenclature, 4—5
outer membrane studies, 135, 137, 139—140
serotyping and serotypes, 11, 53—54, 62, 68, 183—186
taxonomy, 3—12, 16, 182, 186
Campylobacter fetus subsp. *fetus*
antibiotics, sensitivity to, 186
antimicrobial agents, susceptibility to, 81—83
antiphagocytic surface protein, 139
bovine genital campylobacteriosis studies, 182—188
Campylobacter jejuni differentiated from, 82—83, 135
characteristics, key differential, 7—8, 11
colonial morphology, 195
description and pathogenic significance, 11
diagnostic serology studies, 109
enteric infection studies, 194—195, 199—201
enteritis experimental studies, 114, 116, 122—123, 125
erythromycin resistance in, 82, 94
growth temperature, 6, 16
guanine + cytosine content, 11
human infection caused by, types, prognosis, and treatment, 22—23
isolation, 44, 199—200
multiplication site, 23
outer membrane studies, 135, 137, 139
plasmid-mediated transfer of tetracycline resistance studies, 81, 88—92, 94
serological differences, 11
serotyping and serotypes, 53, 68, 183, 185—187
subsp. *venerealis* differentiated from, 11
taxonomy, 3—14, 16, 182, 186
tetracycline resistance in, 81, 88—92, 94
Campylobacter fetus subsp. *intestinalis*, 46, 68, 182, 186
Campylobacter fetus subsp. *jejuni*, 5
Campylobacter fetus subsp. *venerealis*
antibiotics, sensitivity to, 186
biotypes, 11, 184—185
bovine genital campylobacteriosis studies, 182—186
characteristics, key differential, 7—8, 11
description and pathogenic significance, 11
enteritis experimental studies, 114—115
guanine + cytosine content, 11
serological differences, 11
serotyping and serotypes, 11, 53—54, 183, 185—186
subsp. *fetus* differentiated from, 11
taxonomy, 3—14, 182, 186
Campylobacter hyointestinalis, 201
Campylobacteriosis, 29, 37, 114, 175, 182—188
bovine genital, see Bovine genital campylobacteriosis
pathogensis of, 138—139, 182—183
Campylobacter jejuni
acid environment, sensitivity to, 169—170
acid-extractable surface protein, 140
aerotolerance, 43, 173
ampicillin resistance in, 83, 90, 93
animal infections caused by, 114—128, 183, 185, 194—205
antimicrobial agents
resistance to, 88—95
susceptibility to, 78—84, 89
antiphagocytic surface protein, 139
attachment, 122—124, 128, 139—140
BA37 and BA39 strains, reclassification, 95
bacteriological diagnosis, 40—41
biotypes, 9—14, 72—74
biotyping, 13—14, 72—74
bovine genital campylobacteriosis studies, 183, 185
Campylobacter coli differentiated from, 9—10, 47, 114, 128, 134, 201
Campylobacter fetus subsp. *fetus* differentiated from, 82—83, 135
carbon dioxide requirement, 173
centrifugation of, 122—123, 125—126, 139—140
characteristics, key differential, 6—11
cold resistance, 167—169
colonial morphology, 195
colonization, 115—119, 122—124, 128
colony types, 45—46
dehydration affecting, 172—174
description and pathogenic significance, 6—10
disinfectant effects on, 173—175
distribution, in aimals and man, 9
enteric infection studies, 194—201, 204
enteritis studies, 114—128
erythromycin resistance in, 80—82, 84, 90, 93—94
erythromycin susceptibility in, 79—84
growth, in foods, 165—177
growth temperature, 6—10, 165—167, 172

guanine + cytosine content, 10
heat resistance, 168—169
human infections caused by, 22—30
  complications, 24—25, 29
  epidemiology see also Human infections, epidemiology, 144—157
  food-borne, 164—179
  pathology, 25—29
  traveller's diarrhea see also Traveller's diarrhea, 34—37
  treatment, 27—30
identification, 46—47
immunity to, 36
inactivation of, 166, 168—169, 175
incubation, 43—46, 128, 150, 165, 169, 196
infection studies
  animal, see *Campylobacter jejuni*, animal infections caused by
  asymptomatic, 36
  clinical aspects, 22—30
  diagnostic serology, 106—111
  differential diagnosis, 24—29
  enteric infections, 194—201, 204
  enteritis, 114—128
  epidemiology, see also Human infections, epidemiology, 144—157
  food-borne, 164—179
  human, see *Campylobacter jejuni*, human infections caused by
  incubation period, 25
  natural course, self-limiting nature, and spontaneous recovery, 27—30, 114
  prodomal state, 23, 114, 123
  serological responses, see also Serological responses, 98—103
  transmission, 30
  traveller's diarrhea studies, 34—37
  treatment of, 27—30, 114—115
inoculation, 42—43
invasion by, 123—128, 139—140
isolation
  Campylobacter sputorum subsp. *mucosalis* studies, 216, 223
  enteric infection studies, 195—199
  enteritis experimental studies, 114—119, 122, 128
  epidemiology studies, 144—146, 152—157
  food studies, 145—146, 177—179
  human feces studies, 40—48
  rate, 145—146, 152—157
matrix protein, 135, 139
minimal dose, 25
nomenclature, 5—6
outer membrane and surface structure, see also Outer membrane, 134—140
oxygen requirement and tolerance, 173
pathogenesis, 115, 117, 122—128
pathogenic nature, 114—116, 122, 128
penetration by, 123—128
pH effects on, 169, 176
plasmid-mediated transfer, tetracycline resistance see also Plasmid-mediated transfer, tetracycline resistance, 88—95
pMAK 175 plasmid, restriction endonuclease analysis, 92—94
porin, 135, 140
SD2 strain, nalidixic acid resistance, 89—90
serotyping studies, 13, 92—93
  individual schemes for, 55—56, 68—73
  slide agglutination method, 62—74
  thermostable antigen method, 52—58
  typable strains, 62, 68—72
sodium chloride effects on, 170—172
strain discrimination methods, 13—14
surface structure, 128, 134—140
survival, in food, 165—177
taxonomy, 3, 5—12, 16, 201
temperature effects on, 165—169, 172—173, 175—177
tetracycline-resistance in, see also Plasmid-mediated transfer, tetracycline resistance, 80—81, 84, 88—95
tetracycline susceptibility in, 79—84
thermal inactivation, 166, 168—169
traveller's diarrhea studies, 34—37
Campylobacter laridis, 12
Campylobacter-like organism, 149
Campylobacter Omp 1 protein, 135
Campylobacter sputorum
  DNA base composition, subspecies, 218
  subspecies differentiaion, 213—214, 218
Campylobacter sputorum subsp. *bubulus*
  bovine genital campylobacteriosis studies, 185
  Campylobacter sputorum subsp. *mucosalis* differentiated from, 208, 214, 218
  characteristics, key differential, 14
  description and pathogenic significance, 15
  enteritis experimental studies, 123
  guanine + cytosine content, 15
  taxonomy studies, 3, 14—15
Campylobacter sputorum subsp. *mucosalis*
  antibiotic sensitivity and resistance, 215
  antigens, 214—215, 219—221
  characteristics, key differential, 14
  cultures, 210—214, 216—217, 223
  description and pathogenic significance, 16
  differentiation, 213—214, 218
  diseases, association with, 218—223
  experimental infections, 222—223
  growth, 210, 213—214, 217—218, 223
  guanine + cytosine content, 16, 214, 218
  historical description, 208
  identification, 213—214, 217
  isolation media and techniques, 216—217
  morphology, 208—212, 217
  motility, 208
  oxygen tolerance, 210
  pathogenicity and pathology, 16, 219—221, 223
  physiological and biochemical characteristics, 213—214
  porcine serology, 221—222

serotypes, 214—215, 217—218, 221—222
studies, 208—223
subspecies *bubulus* and *sputorum* differentiated from, 208, 214, 218
taxonomy, 3, 14—16, 208, 217—218
Campylobacter sputorum subsp. *sputorum*
characteristics, key differential, 14
description and pathogenic significance, 15
differentiated from subsp. *mucosalis*, 208, 214, 218
taxonomy, 3, 14—15
Campy-Thio medium, 42, 44, 46, 48
C antigen, 53
Capnophilic nature, campylobacters, 44
Capsular antigen, see K antigen
Capsule, 52, 54, 62, 68
micro-, 125, 139—140
Carbenicillin, 78—79, 82
Carbon dioxide requirements, *Campylobacter jejuni*, 173
Cardiac forms, *Campylobacter fetus* subsp. *fetus* infections, 22
Carriers, animal, infections, 183, 187, 200, 202, 204
Cary-Blair medium, 40, 47
Case-to-infection ratio, traveller's diarrhea studies, 36
Catalase, oxygen tolerance enhanced by, 173
Catalase-negative campylobacters, see also specific types by name
bovine genital campylobacteriosis studies, 185
Campylobacter sputorum subsp. *mucosalis* studies, 208, 222—223
catalase-positive campylobacters distinguished from, 6, 15
characteristics, key differential, 14—15
description and pathogenic significance 14—16
differentiation, 12, 15
enteric infections, animal studies, 204
guanine + cytosine content, 15—16
historical background, 14—15
taxonomy studies, 2—3, 6, 12, 14—16
Catalase-negative porcine vibrios, 217
Catalase-negative vibrios, 208
Catalase-positive campylobacters, see also specific types by name
bovine genital campylobacteriosis studies, 185
Campylobacter sputorum subsp. *mucosalis* studies, 208, 213, 217—218, 222
catalase-negative campylobacters distinguished from, 6, 15
causative agent in infections, 114
characteristics, key differential, 6—13
description and pathogenic significance, 6—13
differentiation, 6—8
enteric infections caused by, in animals, see also Enteric infections, 194—205
growth temperature, 3—4, 6—12, 16
guanine + cytosine content, 2, 10—13
historical background, 3—6, 16
morphology, 2, 4, 7, 10—12
taxonomy studies, 2—14, 16
Catalase-positive vibrios, 213, 218
Catalase test, 3, 14, 47, 208, 213, 217
Cat studies
enteritis, 115, 117
epidemiology, 146, 148, 157
Cattle studies, see also Beef, contamination of; headings under Bovine
abortion, see Bovine abortion
bovine genital campylobacteriosis, see also Bovine genital campylobacteriosis, 182—188
endotoxic shock, 139
enteric infections, 194—200, 204—205
enteritis, see also Bovine enteritis, 114—116, 122—123, 125
enzootic infertility in, see Enzootic infertility
enzootic sterility in, see Enzootic sterility
epidemiology, 144—146, 148—149, 157
female (heifer), bovine genital campylobacteriosis studies, 182—188
food-borne infections, 164—165
male (bull), bovine genital campylobacteriosis studies, 182—188
serotyping, 62, 69, 72, 74
taxonomy, 2—4, 6, 11—12, 14—15
Causative agents, see also specific types by name
animal infections, 183—186, 200—201, 203
human infections, 114—115, 149, 152—154, 164, 172
Cefaclor, 82
Cefamandole, 80, 82
Cefazolin, 42, 78—79, 82
Cefoperazone, 48, 80
Cefotaxime, 78—80, 82, 84
Cefotoxamine, 82
Cefotoxin, 80, 82
Cefuroxime, 80
Cell size (dimensions), 8, 10—12
Cellular fatty acid composition, campylobacters, 8—9, 137, 185, 214
Cellular protein profile, 9
Centrifugation, *Campylobacter jejuni*, enteritis studies, 122—123, 125—126, 138
Cephalexin, 78, 82
Cephaloridine, 78, 82
Cephalosporin, 29—30
in vitro susceptibility studies, 78—80, 82, 84
Cephalosporin C, 78, 82
Cephalothin, 7, 42—43, 47, 168
disk test, 83
resistance to, 9, 11—12, 40, 78, 201
sensitivity to, 7, 9, 12, 78, 82—83
Cervix, cow, infection in, 183
Characteristics, key differential
catalase-negative campylobacters, 14—15
catalase-positive campylobacters, 6
Charcoal-yeast extract agar test, 14
Chemotherapeutic agents, 83
Chemotherapy, 27, 37, 83
Chicken studies, see also Poultry studies
enteritis, 115, 117, 122, 125

epidemiology, 144—145, 149, 157
food-borne infections, 164—165, 167—169, 175—176, 178—179
intestinal carriage, 144
Childhood, transmission during, 152
Chloramphenicol, 29, 29—30, 79, 82, 84, 151, 186
Chlorine, in water, 144, 150—151, 173—175
Cholecystitis, 25
Chyme, 216
CI *Campylobacter* species, 3, 8, 13
Circular colored reaction zones, DIG-ELISA, 106—107
Citrobacter freundii, 223
Claims, contamination of, 150
Classification, see Taxonomy
Climate, effect of, 154
Clindamycin, 79, 81—82, 84
Clinical aspects (features; signs)
bovine genital campylobacteriosis, 182—183
enteric infection studies, 194—195, 202—203
enzootic infertility, 182—183
human infections, 22—30
porcine intestinal adenomatosis studies, 218—219
sporadic abortion, 183
traveller's diarrhea, 36—37
Clinical studies, serotyping, 57—58
Cloxacillin, 78, 84
Coagglutination, 53
Coccal transformation, 7, 9—12, 46—47
Coccidia, 195, 203
Coccoid forms, 46—47, 134, 208
Cold resistance, *Campylobacter jejuni*, 167—169
Colistin, 29, 47, 79, 81, 83
Colitis, see also specific types by name, 40, 116—117, 194, 203
histology and differential diagnosis, 24—29
Colonial characteristics, bovine genital campylobacteriosis causative agents, 185
Colonial morphology
Campylobacter sputorum subsp. *mucosalis*, 210—213, 217
enteric infection studies, 195
Colonization, *Campylobacter jejuni*, 115—119, 122—124, 128
site of, 122
Colony types, *Campylobacter jejuni*, 45—46
Colorectal inflammation, 24
Colostrum deprivation, 116, 203
Columbia agar, 42, 48, 88, 210, 216
Commensals, 144—145, 184, 186
Communicable Disease Surveillance Center, epidemiologic data from, 25, 152, 156—157
Complement fixation
bovine genital compylobacteriosis studies, 186
Campylobacter sputorum subsp. *mucosalis* studies, 214—215
diagnostic serology studies, 106, 109—111
diffusion-in-gel enzyme-linked immunosorbent assay compared with, 109—111
serological response studies, 98—102
serotyping studies, 54
titer, low, 109
Complement-fixing antibodies, 36—37, 99, 101—103, 148
Complement-fixing antigen, 219
Complications, human infections, 24—25, 29
Conjugative transfer, plasmids, 89—91, 95
Contact infections, cattle, 182—183, 186—187
Convalescent serum, human, 98—101, 150
Core lipopolysaccharide, 136—137
Corona virus, 195
Corynebacterium pyogenes, 183
Cotrimoxazole, in vitro susceptibility studies, 79, 82—84
Countries, see Developed countries; Developing countries
Cows, see Cattle
Crohn's disease, 24, 26, 222
Cross-absorption tests, 55
Cross-contamination, foods, 164
Cross-titration tests, 55
Crypt abscesses, 26—29, 116
Cryptosporidia, 195, 203
Crystal violet stain, 41, 46
C serogroup, 53, 62, 214—215
Cycloheximide, 42—43, 47—48, 178
Cyst, ovarian, infected, 23
Cysteine-containing media, 4, 7, 11, 13, 183, 185—186, 213
Cytochrome C-553, 214
Cytotoxic enzyme, 123
Cytotoxic factor, 123
Cytotoxic phenomena, 123, 125, 127—128
Cytotoxins, 123, 128

D

Dairy products, epidemiology studies, see also Milk, 148
Defective mutants, 11
Dehydration
foods, effects of, 172—174
human infections causing, 29—30
Deoxycholate, 213—214
Deoxyribonuclease, 63, 91
Deoxyribonucleic acid
base compositions, 185, 218
guanine + cytosine content, 10, 185, 214, 218
hydrolysis test, 14
plasmid, 88— 90, 92, 94
relatedness, NARTC strains and *Campylobacter jejuni* biotypes 1 and 2, 11—12
Developed countries, incidence in, 147, 152—154
Developing countries, incidence in, 34—37, 147, 154, 157
Diagnosis
bacteriological, *Campylobacter jejuni* isolation studies, 40—41
bovine genital campylobacteriosis, 187—188
differential, see Differential diagnosis

Diagnostic serology, *Campylobacter jejuni* infections, diffusion-in-gel enzyme-linked immunosorbent assay technique, 106—111
Diarrhea
antimicrobial susceptibility studies, 78, 83
bloody, see Bloody diarrhea
Campylobacter jejuni, 23—25, 29—30, 34—37, 40—42, 46
causative agents, 115, 152—154
causative mechanism, 123—125
enteric infections, animal studies, 194—195, 199—204
enteritis experimental studies, 114—118, 122—125, 128
epidemiology studies, 146, 148, 150—154, 156—157
porcine intestinal adenomatosis studies, 218
serotyping studies, 57—58, 62
serum antibody response, 36—37
taxonomy studies, 2—4, 6, 11
traveller's *Campylobacter jejuni* in, 34—37
treatment of, 29—30, 114—115, 150
watery, see Watery diarrhea
winter, 194—195
Differential characteristics, key
catalase-negative campylobacters, 14—15
catalase-positive campylobacters, 6—13
Differential diagnosis
human infection studies, 24—29
traveller's diarrhea studies, 34, 36—37
Diffusion-in-gel enzyme-linked immunosorbent assay
circular colored reaction zones, 106—107
complement fixation test compared with, 109—111
diagnostic serology studies, 106—111
serological response studies, 99
DIG-ELISA, see Diffusion-in-gel enzyme-linked immunosorbent assay
Dilution techniques, *Campylobacter sputorum* subsp. *mucosalis* isolation, 216
Dimetridiazole, 215
Diphasic media, 210, 216
Diphtheresis, 223
Direct agglutination, diagnostic serology studies, 106
Direct examination, stools, 40—41
Direct inoculation, 42
Dirty yellow colonies, 14—16, 213
Disease-to-infection ratio, see Infection-to-illness ratio
Disinfectants, see also specific types by name, 173—175
Disk sensitivity testing, 82—84, 215
Dissemination, bovine genital campylobacteriosis, 186—187
DNA, see Deoxyribonucleic acid
DNase, see Deoxyribonuclease
Dog studies
enteritis, 114—115, 117, 128
epidemiology, 146, 148, 151
Domestic animal studies, see also specific animals by name
enteritis, 114—115, 117, 128
epidemiology, 144—148
serotyping, 57
taxonomy, 6, 9
Double-diffusion technique, serotyping, 54
Doxycycline, 29, 79, 81—82
Drinking water, see Water systems, municipal
Drying, foods, effects of, 172—174
Duck studies, epidemiology, 144—146
Dysentery, 3, 36—37, 115—117, 194—195, 201—202
swine, see Swine dysentery
winter, see Winter dysentery

E

ELISA, see Enzyme-linked immunosorbent assay
Elongated organisms, 125, 127
Embryo, cow, bovine genital campylobacteriosis studies, 183
Endocarditis
acute, see Acute endocarditis
human infection studies, 22—23
Endometritis, 183
Endotoxicity, 138—139
Endotoxin, 123, 138—139
Endotoxin-induced anaphylactic shock, 123
Energy sources, 2
Enriched brucella medium, 46, 48
Enrichment media, 40—41, 48
Enrichment techniques, 145, 164
Entamoeba histolytica, 25, 34, 51
Enteric infections, catalase-positive campylobacter-caused, animal studies, 194—205
catalase-negative campylobacter-caused infections differing from, 204
cattle, 194—200, 204—205
clinical aspects, 194—195, 202—203
experimental infections, 196—201, 203—204
general discussion, 194, 204—205
pathology and pathogenesis, 194—205
pigs, 194, 201—205
sheep, 194, 200—201, 204—205
species specificity, 204
treatment, 204—205
Enteric pathogens, see Enteropathogens
Enteritis
antimicrobial susceptibility studies, 78, 81, 84
bovine, see Bovine enteritis
Campylobacter jejuni isolation studies, 40—41
catalase-positive campylobacters, animal studies, see also Enteric infections, 194—205
clinical aspect studies, 24—25, 29
diagnostic serology studies, 106, 109—110
epidemiology studies, 145, 147—152, 154
experimental studies, animals, 114—128
general discussion, 114, 128
models, 115—121, 128

pathogenesis, 115, 117, 122—128
pathogenicity, 114—116, 122, 125, 128
food studies, 164, 173, 178
gastro-, see Gastroenteritis
necrotic, see Necrotic enteritis
poultry as vehicle for, 149
serotyping studies, 52, 55, 57—58
world health problem, status as, 128
Enterocolitis, 24, 27, 37
Enteropathogens, see also specific types by name
Campylobacter sputorum subsp. *mucosalis* studies, 222
enteric infections, animal studies, 194—195, 199, 201, 203
enteritis experimental studies, 114—116, 122, 125
epidemiology studies, 149, 151, 154
outer membrane studies, 134
serotyping studies, 52, 62
traveller's diarrhea studies, 34
Enterotoxins, 123, 128
Environmental reservoirs, see Reservoirs
Enzootic infertility, 182—188
causative agents, 183—186
diagnosis, 187—188
historical background, 182
pathogenesis and clinical signs, 182—183
transmission and dissemination, 182, 187—188
treatment, 188
Enzootic sterility, 2, 4, 11, 182
Enzyme-linked immunosorbent assay, see also Diffusion-in-gel enzyme-linked immunosorbent assay
diagnostic serology studies, 106
enteritis experimental studies, 117, 119
serological response studies, 98—102
Epidemiology
diagnostic serology studies, 106, 109
enteritis experimental studies, 115—116, 128
human *Camyplobacter jejuni* infections, see also Human infections, epidemiology, 144—157
serotyping studies, 52, 55, 57, 62, 73
taxonomy studies, 2
traveller's diarrhea, 34—36
Error rate analysis, disk sensitivity testing, 83—84
Erythema nodosum, 25
Erythrocyte
serotyping studies, 54—55
traveller's diarrhea studies, 36—37
Erythromycin, 24, 29—30, 114—115, 188
resistance to, 29, 80—82, 84, 94, 115
sensitivity to, 79—84, 115, 186
Erythromycin stearate
Escherichia coli
antimicrobial susceptibility studies, 81
bovine genital campylobacteriosis studies, 188
enteric infections, animal studies, 195, 199
enteritis experimental studies, 114
epidemiology studies, 147
outer membrane studies, 135, 137—138
plasmid studies, 88—89, 91, 95
serotyping studies, 52
traveller's diarrhea studies, 34
ETEC, see Enterotoxigenic *Escherichia coli*
Ethyl alcohol, food treatment with, 175
Examination, direct, stools, 40—41
Experimental infection
Campylobacter sputorum subsp. *mucosalis* studies, 222—223
enteric, 114—128, 196—201, 203—204
serological response following, 100—101
Extracted antigens, 52—54

F

False resistance and susceptibility, disk sensitivity testing, 83
Farm animals, see Domestic animals; specific animals by name
Fatty acid composition, campylobacters, 8--9, 137, 185, 214
FBP medium, see Iron/metabisulfite medium
Febrile forms, *Campylobacter fetus* subsp. *fetus* infections, 22
Fecal cultures, see Human fecal cultures
Fecal erythrocytes, 36—37
Fecal leukocytes, 27, 36—37
Feces, see Animal feces; Human fecal cultures; Human feces
Female sex, risk to, 156
Ferret studies, enteritis, 117
Ferrous metabisulfite, see Iron/metabisulfite
Ferrous sulfate, 42—43, 48, 173
Fetopathogenic effects, 116
Fetus
cow, bovine genital campylobacteriosis studies, 182
human, epidemiology studies, 151
Fever
human infection studies, 22
traveller's diarrhea studies, 36—37
Filter-mating method, 90
Filtration, isolation by, 40—41, 187, 195, 216—217
Fimbriae, 52
Flagella, 2, 14, 52, 54, 62, 68, 123—124, 134—135, 138—140, 185, 210—211
Flagellar antigen, see H antigen
Flagellar antigenic factors, 68
Flagellin, 134—135
Flucloxacillin, 78
Fluid enrichment media, 40—41, 48
Fluorescent techniques, procine intestinal adenomatosis studies, 219, 222
Food-borne infections, see also specific foods by name
Campylobacter jejuni studies, 145—145, 164—179
cross-contamination, 164
epidemiology studies, 144—150, 152, 157
general discussion, 164

isolation procedures, 145—146, 177—179
occurrence, 164—165
organism survival and growth, factors affecting, 165—177
raw foods, 175—177
secondary food contamination, 150
source of infection, 106, 108
storage effects, see Storage, food
Formation, food treatment with, 175
N-Formimidoyl thienamycin, 78
Free-living aspartate-fermenting *Campylobacter* sp., 3, 8, 13
Freezing, effects of, see also Refrigeration, 144, 147, 167—168
Fumarate, 7, 9
Furazolidore, 29, 79, 81, 83—84

G

Gastroenteritis, 3, 6, 23—24, 70, 73, 114, 151
Gastrointestinal complaints and symptoms, 36, 101—103, 151, 164
G + C ratio, see Guanine + cytosine ratio
Genital campylobacteriosis, bovine, see Bovine genital campylobacteriosis
Gentamicin, 30, 79—80, 82, 84, 151, 186
Geographic occurrence, tetracycline-resistant plasmids, similarities among examples, 95
Giardia lamblia, 25, 34
Gluteraldehyde, food treatment with, 175
10% Glycerol, 168
1% Glycine, 4, 7, 11, 14—15, 184—186
Glycine tolerance, 213
Glycoprotein, 125—126
Glycoprotein antigen, 99—100, 103
Glycoprotein antigen fraction, 140
Glycoprotein microcapsular antigen, 139
Gnotobiotic animal studies, 117, 122, 128, 201—203, 222
Gram-negative cell, outer membrane, function, 134
Gram stain, 41
Granulomatous colitis, 24
Ground beef, see Hamburger
Group 2 *Campylobacter* strains, 12—13
Growth
anaerobic, see Anaerobic growth
Campylobacter cells, prior to plasmid DNA isolation, 88
Campylobacter jejuni, in foods, 165—177
Campylobacter sputorum subsp. *mucosalis*
conditions, general, fecal cultures, 43—44
microaerophilic conditions affecting, 41, 106
Growth rate, factors affecting, 128, 166—167, 169
Growth supplements, 43—44, 48
Growth temperature
Campylobacter jejuni, 6—10, 165—167, 172
catalase-positive campylobacters, taxonomy studies, 3—4, 6—12, 16
optimum, 165, 172, 185
Guanine + cytosine content
catalase-negative campylobacters, 15—16
catalase-positive campylobacters, 2, 10—13
DNA, 10, 185, 214, 218
tetracycline-resistant plasmids, 92
vibrios, 182
Guillain-Barré syndrome, 25

H

Halophilic organisms, 63
Hamburger (ground beef; minced beef), contamination of, 149, 164—165, 167—168, 175—179
Hamster studies
enteritis, 117
epidemiology, 146
transmissible ileal hyperplasia, proliferative ileitis, 223
H antigen (flagellar antigen), 52, 62, 68, 188
Heated cell method, serotyping studies, 52—54
Heat-labile antigen, see Thermolabile antigen
Heat-labile antigenic factors, see Thermolabile antigenic factors
Heat resistance, *Campylobacter jejuni*, 168—169
Heat stability, dependence on, serotyping studies, 52
Heat-stable antigen, see Thermostable antigen
HeLa cells, enteritis experimental studies, 122—127
Hemagglutination, see also Indirect hemagglutination; Passive hemagglutination
diagnostic serology studies, 106
enteritis experimental studies, 122
serological response studies, 98—101
Hemorrhagic enteropathy, 16
Hepatitis, 6, 115, 122
Herpes simplex virus, 151
Heterologous suspensions, serological response studies, 98, 100, 102
Heterologous titers and reactions, slide agglutination serotyping studies, 63—67
High temperature-short time pasteurization process, effectiveness of, 168
Hippurate
negativity, 10—12, 55—57, 73—74
positivity, 10, 56, 74
Hippurate hydrolysis test, 5, 7, 9—12, 17, 55—57, 72
Histology, 24—29
HLA B27 antigen, 25
Homologous heated antigens, antibodies to, 63, 67—68
Homologous suspensions, serological response studies, 98—100, 102
Homologous titers and reactions, slide agglutination serotyping studies, 62—68
Homosexuals, proctitis in, 25, 151
Host-parasite relationships, 134
Host response, enteric infections, animal studies, 199
Human antibody response, see Antibody response

Human blood cultures
age-specific ratios, blood-to-total isolates, 156
diagnostic serology studies, 110
enteritis experimental studies, 114
epidemiology studies, 156
human infection studies, 22—25
taxonomy studies, 3—4
Human convalescent serum, see Convalescent serum, human
Human fecal cultures
Campylobacter jejuni isolation studies, 40—48
diagnostic serology studies, 109—110
enteritis experimental studies, 114—115
epidemiology studies, 152—154, 156—157
filtration of, 40—41
growth conditions, general, 43—44
human infection studies, 23, 25
incubation, 43—46
plates, examination of, 45—46
storage, isolates, 46
taxonomy studies, 3, 15
temperature, optimal, 44
traveller's diarrhea studies, 34—36
Human feces, *Campylobacter jejuni* isolated from, 40—48
bacteriological diagnosis, 40—41
direct examination, stools, 40—41
general discussion, 40
identification, 46—47
incubation, 43—46
inoculation, 42—43
primary isolation, 41
sample collection, transport, and storage, 40
Human infection studies
animal source, 114
asymptomatic, see Asymptomatic infections
Campylobacter fetus subsp. *fetus*-caused, 22—23
Campylobacter jejuni-caused, 22—30, 34—37
causative agents, 114—115, 149, 152—154, 164, 172
clinical aspects (features), see also Clinical aspects (features), 22—30
complications, 24—25, 29
differential diagnosis, see also Differential diagnosis, 24—29, 34, 36—37
epidemiology, 144—157
general discussion, 144, 157
incidence and prevalence, 144, 152—157
reservoirs, see also Reservoirs, 144—147, 157
transmission, modes of, 144, 147—152, 157
experimental, serological response following, 100—101
experimental enteritis studies, 114—115, 122, 128
food borne, see Food-borne infections
general discussion, 22
milk borne, see Milk-borne infections
natural course and spontaneous recovery, 27—30
pathology, 25—29
reinfection, different serotypes in, 58, 106, 111
serological response, see also Serological responses, 98—103
simultaneous, more than one serotype in, 103
site of, 24—25
transmission, 30
treatment, see also Treatment, 23, 27—30
water-borne, see Water-borne infections
Human isolates, serotyping studies, 56—57, 62 68—74
Human reservoirs, 147
Human studies
Campylobacter jejuni isolation from feces, 40—48
diagnostic serology, 106—111
epidemiology, see also Human infections, epidemiology, 144—157
experimental, enteritis, 114—115, 122, 128
infections, see also Human infections, 22—30, 34—37
serological responses, 98—103
serotyping, 56—58, 62, 68—74
taxonomy, 2—4, 6, 9, 11, 14—16
traveller's diarrhea, 34—37
Humidity, effects on food, 173—174
Hydrogen peroxide, 13, 188
Hydrogen requirement, 16
Hydrogen sulfide test
bovine genital campylobacteriosis studies, 182—186
Campylobacter sputorum subsp. *mucosalis* studies, 213
enteric infections, animal studies, 201
taxonomy studies, 4, 6—7, 10—15
Hyperplasia, 116—117, 125, 197, 223
Hypogammaglobulinemia, 25

I

Icing, cake, contamination of, 149—150
Identification tests, 213—214
Ileitis, 16, 24, 223
Ileum, involvement of, 195—200, 202—203, 219
Illness-to-infection ratio, see Infection-to-illness ratio
Imidozole derivatives, 82
Immune response, 128, 134, 183
Immune status, 122, 128
Immune systems, 134
Immunity studies, 36, 98, 106, 111, 194
Immunizing agents, 188
Immunocompromised hosts, 11, 44, 114
Immunofluorescence, see also Indirect immunofluorescence
bovine genital campylobacteriosis studies, 187
diagnostic serology studies, 106
serological response studies, 98—99, 101
Immunogenicity, 140
Immunogenic response, 68
Immunoglobulin class-specific antibody response, 98—103, 106—111

Immunoglobulin G antibody
bovine genital campylobacteriosis studies, 187—188
diagnostic serology studies, 108—110
enteritis experimental studies, 114
epidemiology studies, 150—151
outer membrane studies, 139
porcine serology studies, 222
serological response studies, 98—102
Immunoglobulin M antibody
bovine genital campylobacteriosis studies, 187—188
diagnostic serology, studies, 108—111
enteritis experimental studies, 114
porcine serology studies, 222
serological response studies, 98—102
Immunological response, see Immune response
Immunosuppression, 128
Inactivation, *Campylobacter jejuni*, 166, 168—169, 175
Inagglutinability, 68, 139
Inanimate reservoirs, 147
Incidence and prevalence
age and sex, 34—35, 154—156
location, 34—37, 147, 152—154
occupation, 148, 157
season, 34—35, 144, 147, 157
Incubation, *Campylobacter jejuni*, 43—46, 128, 150, 169, 196
methods, see also specific methods by name, 44—45
temperature, optimal, 44, 165
Incubation time (period)
human infections, 25
traveller's diarrhea, 34
India, epidemiology studies, 155
Indirect hemagglutination, diagnostic serology studies, 106
Indirect immunofluorescence, serological response studies, 98, 103
Indirect inoculation, 42
Infected bedding, cattle, infection transmitted by, 182—183, 186—187
Infected ovarian cyst, 23
Infection, see specific types by name
animal, see Animal infections
human, see Human infections
Infection-to-illness ratio, 154, 157
Infectious hepatitis, 6, 122
Infectious infertility, 4
Infectious proctitis, 25
Infectious sterility, 182
Infective colitis, differential diagnosis, 24—29
Infertility, 4, 115
Inflammation, in infection, 25—27, 199, 201, 203
Inflammatory bowel disease, differential diagnosis, 24—29
Inflammatory phenomena, bovine genital campylobacteriosis, 183
Inflammatory response, in enteritis, 116, 122
Inhibition zone, disk sensitivity testing, 83
Inoculation
direct and indirect, 42
media, see also specific media by name, 42—43
Interspecies transfer, tetracycline resistance, 90—91
Intestinal adenomatosis, 15—16
Intestinal carriage, infections, in chickens, 144
Intestinal mucosa, *Campylobacter sputorum* subsp. *mucosalis* studies, 216, 218—219, 222—223
Intracellular vibrio-like organism, 223
Intracellular vibrios, 219—221
Intususception, 25
In utero acquirement, infection, 151
Invasion, *Campylobacter jejuni*, 123—128, 139—140
In vitro susceptibility, to antimicrobial agents, 78—84
Iodophor, food treatment with, 175
Iron/metabisulfite growth supplement, 43—44, 48
Iron/metabisulfite medium, 7, 10—11, 13
Iron salts, 173
Isolation, primary, 41, 185
Isolation rate, *Campylobacter jejuni*, see also *Campylobacter jejuni*, isolation, 145—146, 152—157
Isoprotein, 135, 140

K

Kanamycin, 80, 82, 186
K antigen (capsular antigen), 52, 62, 68, 125, 188
Kennel population studies, epidemiology, see also Cat studies; Dog studies, 146
Key differential characteristics
catalase-negative campylobacters, 14—15
catalase-positive campylobacters, 6—13
KI medium, see Kligler's iron medium
Kitten studies, see also Cat studies, 146, 148, 157
Kligler's iron medium, 6, 12

L

Laboratory animal studies, see also specific animals by name, 148
β-Lactamase, 78, 93
Lactic acid, 169—170
Lamina propria involvement, 25—29, 116, 119, 122, 125, 197
Laparotomy, 24—28
Lau serotypes, 55—58
Lead acetate medium, 13
Leptospira isolation media, 12
Leukocytes, 27, 36—37, 114, 116, 125
Lincomycin, 81, 83—84, 215
Lincosamide antibiotics, in vitro susceptibility studies, 79, 81—84
Lipopolysaccharide
core, 136—137
enteritis experimental studies, 123
outer membrane studies, 134, 136—140

serotyping studies, 52, 54
Lipoprotein, Braun's, 135, 138
Liquid enrichment medium, see also Fluid enrichment media, 48
Listeria sp., 183
Local antibody, production of, 187
Location, incidence and, 34—37, 147, 152—154
LPS, see Lypopolysaccharide
Lymphocytes, in enteric infections, 198—199
Lymphoid system, enteric infection, animal studies, 196—197, 199, 203
Lyophilation, 46

M

Macrophages, in enteric infections, 198—199
Macrolides, 79, 81—84, 204
Male sex, risk to, 156
Mammals, small, see Small mammal studies; specific types by name
Man, see Human
Mastitis, 12, 116, 145, 149
Mating procedures, plasmid studies, 90—91
Matrix protein, 135, 139
Meat studies, see also specific meats by name
  epidemiology, 145, 148—149, 152, 157
  food-borne infections, 164—169, 173, 175—178
Mecillinam, 78
Membrane, outer, see Outer membrane
Meningism, 25
Meningitic form, *Campylobacter fetus* subsp. *fetus* infections, 22—23
Meningitis, 22—23, 25, 30, 151
Meningoencephalitis, 22
Mentzing's epidemiologic data, 155
Mesenteric adenitis, 25
Metabolism, 2
Metronidazole, 14, 79, 82
Meyers method, plasmid isolation, 88
Mezlocillin, 78
MIC, see Minimum inhibitory concentrations
Mice, see Mouse studies
Microaerophilic bacteria, 13, 43
Microaerophilic conditions, growth and, 41, 106, 217
Microaerophilic nature, campylobacters, 43—44, 83, 88, 213
Microaerophilic nitrogen-fixing bacterium, 13
Microaerophilic organisms, 173, 185
Microaerophilic vibrios, 2—4, 12—13, 15
Microcapsular antigen, 139
Microcapsule, 125, 139—140
Microvilli involvement, 119—120, 123—124
Milk-borne infections
  diagnostic serology studies, 106
  enteritis experimental studies, 116, 128
  epidemiology studies, 144—145, 149, 152, 157
  food studies, 164—169, 173, 175—178
  outer membrane studies, 134
  pasteurization, see Pasteurization
  raw milk, see Raw milk
  serological response studies, 101—103
  serotyping studies, 73
Milk filters, isolation from, 149
Millipore® filtration, 187
Minimum inhibitory concentrations, antibiotics, 78—84, 90, 186, 215
Minocycline, 79, 81—82
Molecular weight determination, plasmids, 89
Monkey studies, 115, 148
Monoclonal antibody, 140
Monospecific antisera, 55
Monovalent antisera, 68
Morphology
  bovine genital campylobacteriosis causative agents, 185
  Campylobacter jejuni isolation studies, 41, 46—47
  Campylobacter sputorum subsp. *mucosalis* studies, 208—212, 217
  catalase-positive campylobacters, 2, 4, 7, 10—12
  colonial, see Colonial morphology
  human infection studies, 23
  taxonomy studies, 2, 4, 7, 10—12
Motility, 41, 47, 68, 122—123, 128, 208
Mouse lethality, endotoxin-caused, 139
Mouse studies
  enteritis, 114—122, 125
  epidemiology, 146
Moxalactam, 79—80
Mucoid diarrhea, 202, 204
Mucosa, involvement of, 195—200, 202—204, 216, 218—219, 222—223
Mucus
  depletion, human infection studies, 25—27
  in stools, see Stools, mucus in
Mucus agglutination test, 187
Mueller-Hinton blood agar plate, 63, 68
Mueller-Hinton broth, 41—42, 63
Multiple antigenic factors, 72
Multiple infections, 114, 116
Multiple serotypes, 68
Multiple thermolabile antigens, 62
Multiply-reacting isolates, 55
Municipal water systems, contamination of, see also Water-borne infections, 150—151
  disinfectants, see also chlorine, 173—175
Murine campylobacter enteritis, 117
Murine colonic hyperplasia, 223
Mutants
  aflagellate, 68
  defective, 11
  nalidixic acid-resistant, *Campylobacter jejuni*, 89

N

Nalidixic acid
  disk test, 83
  resistance to, 9, 12, 82—83, 89—90, 201
  sensitivity to, 7, 9, 12, 79, 82—83

Nalidixic acid-resistant strains, *Campylobacter jejuni*, 89—90
Nalidixic acid-resistant thermophilic campylobacters
biotype 4, 72—74
characteristics, key differential, 7—8, 11—12
description and pathogenic significance, 11—12
distribution, in animals and man, 9
DNA relatedness to *Campylobacter jejuni* biotypes 1 and 2, 11—12
food studies, 170—174, 176—177
growth temperature, 11
guanine + cystosine content, 12
serotyping studies, 69—70, 72—74
taxonomy studies, 3, 7—9, 11—12
Nalidixic acid tolerance tests, 47
NARTC, see Nalidixic acid-resistant thermophilic campylobacters
Natural course, diseases, 27—30, 114
Nausea, see also Vomiting
human infection studies, 23—24
traveller's diarrhea studies, 36
NE, see Necrotic enteritis
Necrotic enteritis, 219—220, 222
Neisseria
gonarrheae, 63, 151
meningitidis, 63
sp., 137—138
Nematodes, 195
Neomycin, 80, 82, 186, 188, 200, 204
Neonates, epidemiology studies, 151—152
Neosensitabs, 83—84
Neutrophils, in enteric infections, 197, 199
Nifuroxazide, 81, 83—84
Nitrate
anaerobic growth in presence of, 9
negativity, 5—6, 12
positivity, 5—6
reduction of, 2, 185, 213
Nitrite, reduction of, 6, 15, 213
Nitrofuran compounds, in vitro susceptibility studies, 79, 81, 83—84
Nitrofurantoin, 81, 83
Nitrogen fixation, 13
Nitrogen-fixing *Campylobacter* sp. CI, 3, 8, 13
Nomenclature
serotypes, 55
taxonomy, 2, 4—6, 14, 182, 201
Nonhuman isolates, serotyping studies, 57, 62, 68—74
Noninhibitory media, 216—217
Nonlocalized sepsis, 23
Norepinephrine, 173
Norwalk agent, 34
Nosocomial transmission, 157
Novobiocin, 40, 42, 47, 82, 84, 186—187, 215—216
resistance to, 186, 215
Nuclease, 63
Nursery, outbreak in, 152
Nutrient Broth No. 2, 48

O

O antigen, 52—54, 186, 188
Occupation, incidence and, 148, 157
Olaquindox, 204
Omp 1 protein, 135
O polysaccharide, 136—137
Opportunist pathogens, 23, 44
Oral enteric infections, experimental, animal studies, 200—201, 203—204
Oral swabs, cultures, 217—218
O serogroup, types 1 and 2, 182
Outbreaks, investigation of,
enteric disease, animal studies, 195
epidemiology studies, 144—145, 149—152, 155
food studies, 164, 173
serological response studies, 99, 101—103
serotyping studies, 57, 72—73
Outer membrane and surface structure, *Campylobacter jejuni*, 128, 134—140
characterization of surface structure, 128
function, 134
general discussion, 134, 140
isolation of, 134, 138
lipopolysaccharide in, 134, 136—140
release of, 137—138
structure and composition, 134—137, 140
surface antigens, 139—140
virulence properties, 137—140
Outer membrane fraction, 138—139
Outer membrane proteins, 134—137, 139—140
Ovarian cyst, infected, 23
Oviducts, cow, infection in, 183
Ovine abortion, 3—4, 6, 16, 115—116, 145, 148, 182, 200
Oxidase positivity, 2, 185, 213
Oxidase test, 47
Oxoid SR 84 supplement, 43
Oxygen, effects of, 2, 6, 15, 43, 173, 210

P

Passive hemagglutination, 54, 101
Pasteurization, effects of, 145, 149, 168
Pathogenesis and pathology
bovine genital campylobacteriosis, 182—183
campylobacteriosis, 138—139, 182—183
Campylobacter jejuni infections, 115, 117, 122—128
Campylobacter sputorum subsp. *mucosalis* studies, 219—221, 223
enteric infections, 194—205
enzootic infertility, 182—183
human infections, 25—29
sporadic abortion, 183
Pathogenic nature, campylobacters, 114—116, 122, 125, 128, 144, 223
Pathogenic significance
catalase-negative campylobacters, 14—16
catalase-positive campylobacters, 6—13

Pathogenic vibrios, see also specific types by name, 185, 208
Pathology, see Pathogenesis and pathology
Patient response, individual, serological response studies, 100—101
Penetration, *Campylobacter jejuni*, 123—128
Penicillin, 186—187
in vitro susceptibility studies, 78—79, 82, 84
resistance to, 23, 29—30, 78, 82, 84
Penicillin G, 29, 78—79, 84, 215
Penicillin V, 78
Penos, bull, involvement in bovine genital campylobacteriosis, 183, 188
Pen serotypes, 55, 52—58
Peptidoglycan, 137
Peptidoglycan-associated protein, 135
Peptone iron media, 12, 14—15
Pericarditis, acute, 23
Perinatal transmission, 151—152
Peritonitis, 25
Peroxide, 43
Person-to-person transmission, 147, 151—152
Pet studies, see also Cat studies, Dog studies, 147, 157
pH effects, *Campylobacter jejuni*, 169, 176
Phage, see Bacteriophage
Phage-typing methods, 13
PHB, see Polyhydroxybutyric acid
PHE, see Proliferative hemorrhagic enteropathy
Phenol-soluble antigens, 214—215, 219
Phosphate-buffered saline, 63, 216
PIA, see Porcine intestinal adenomatosis
Pigment production, 13, 15
Pig studies, see also Pork, contamination of
abortion, 12
adenomatosis, see Porcine intestinal adenomatosis
Campylobacter sputorum subsp. *mucosalis* studies, 217—223
enteric infections, 194, 201—205
enteritis, 114, 116—117, 122—123
epidemiology, 145—146
experimental infections, 222
food-borne infections, 164—166, 172
necrotic enteritis, 219—220, 222
proliferative hemorrhagic enteropathy, 219—220, 222
regional ileitis, 219—220, 222
serological response to infection, 221—222
serotyping, 57, 69, 71—74
swine dysentery, see Swine dysentery
taxonomy, 3, 6, 12, 15—16
Pili, 52, 122
Placenta, cow, bovine genital campylobacteriosis studies, 183
Placentitis, 183
Plant studies, 13
Plasmid-associated erythromycin resistance, 94
Plasmid DNA, 88—90, 92, 94
Plasmid mediated antibiotic resistance, 78, 81, 84
Plasmid-mediated transfer, tetracycline resistance, 81, 84, 88—95
ampicillin resistance, 90, 93
buoyant density, plasmids, 92, 95
conjugative transfer, 89—91, 95
erythromycin resistance, 90, 93—94
general discussion, 88, 95
quanine + cytosine content, plasmids, 92
interspecies transfer, 90—91
isolation procedures, 88—89
molecular weight, plasmids, 88—92, 95
physical characterization of plasmids, 91—95
restriction endonuclease digestion analysis, 88, 92—95
similarity of examples, different geographical locations, 95
Plate examination, fecal cultures, 45—46
Plate-mating procedure, 90—91
Pleomorphism, 12
Plesiomonas shigelloides, 34
Pleurisy, purulent, 23
pMAK 175 plasmid, restriction endonuclease analysis, 92—94
Polyhydroxybutyric acid, 2
Polymixin, 84, 187
Polymixin B, 40, 47—48, 81, 178, 186
Polmixin E, see also Colistin, 81
Polymorphs, 23, 25—27, 29, 116, 119
Polypeptide antibiotics, in vitro susceptibility studies, 79, 81, 83—84
O Polysaccharide, 136—137
Polyvalent antisera, 68
Porcine abortion, 12
Porcine intestinal adenomatosis, 15—16, 218—223
Porcine serology, 221—222
Pore protein, 140
Porin, 135, 140
Pork, contamination of, 152, 164—166
Portnoy and White technique, plasmid isolation 88—90
Poultry studies, see also Chicken studies; Turkey studies
commercial poultry, isolation of *Campylobacter* sp. from, 145, 164—165
epidemiology, 144—145, 148—149, 151
enteritis, 114, 117, 122, 128
food-borne infections, 164—169, 173, 175—176, 178—179
major source of infections, 149, 164—165
Pregnancy
cows, effect of, on infections, 183, 188
humans, epidemiology studies, 151
Prepuce, bull, involvement in bovine genital campylobacteriosis, 183—184, 186—188
Preston medium, 43, 48
Prevalence, see Incidence and prevalence
Primary isolation, 41, 185
Primate studies, 114—115, 128, 148
Proctitis, 24—25, 151
Prodomal state, *Campylobacter jejuni* infection, 23, 114, 123
Proliferative colitis, 117
Proliferative enteropathies, 223

Proliferative hemorrhagic enteropathy, 219—220, 222
Proliferative ileitis, 223
Proliferative intestinal adenopathy, 201
Protective antibodies, 188
Protein profile, 9
Protocolitis, 24
Pseudomonas aeruginosa, 88
Pseudomonas aeruginosa bacteriocins, 13
Puppy studies, see also Dog studies, 146, 148, 157
Purulent arthritis, acute, 22
Purulent meningitis, acute, 22
Purulent pleurisy, 23
Pus, human infection studies, 22—24
Pustule, 23

Q

Quackenbush method, plasma isolation, 88

R

Rabbit studies, enteritis, 117, 123, 125
Rat studies, 117, 146
Raw foods, see also Food-borne infections, 175—177
Raw milk, see also Milk-borne infections, 149, 164—165, 168, 175—176, 178—179
RCM broth, 216
Reactive arthritis, 25, 106
Rectal histology, 24—29
Rectal swabs, 40, 42, 150
Red meat see Beef; Meat; Pork
Refrigeration, effects of, see also Freezing, 164, 167, 169, 172—173, 175
Refrigeration temperature, effects of, 167, 172—173, 175
Regional ileitis, 219—220, 222
Regression line equations, disk sensitivity testing, 83
Rehydration, 29—30
Reinfection, 58, 106, 111, 188, 205
Reiter's syndrome, 25
Related vibrios, 4, 52, 62
Replicate plating, *Campylobacter sputorum* subsp. *mucosalis* studies, 213—214
Reptile studies, epidemiology, 146
Reservoirs
  animal, 144—149, 157
  enteritis studies, 114
  epidemiology studies, 144—149, 157
  human, 147
  inanimate, 147
  outer membrane, studies, 134
Restistance, antibiotic, see Antibiotics, resistance to
Restriction endocnuclease digestion analysis, tetracycline-resistant plasmids, 88, 92—95
RI, see Regionalileitis
Rifampicin, 43, 48
Rifampin, 82, 84
River water, epidemiology studies, see also Water, 147, 150
Rodent studies, see also specific rodents by name, 114, 117—123, 125, 146
Rosaramicin, 84
Rosco neosensitabs, 83—84
Rotavirus, 34, 195
R-plasmids, 78
Rural areas, epidemiology, 148, 150—151

S

Salmonella
  species
    bovine genital campylobacteriosis studies, 183
    enteric infection studies, 195
    enteritis experimenal studies, 114, 125
    epidemiology studies, 146, 148, 150, 152—154, 157
    human infection studies, 24—25
    outer membrane studies, 135, 137
    serotyping studies, 52, 62, 68
    traveller's diarrhea studies, 34
  typhimurium, 135, 137
Salmonella 0.5 antigen, 52
Salmonella serotyping scheme, 52
Salmonellosis, 195
Salpingitis, bilateral, 183
Salt tolerance, 15
Salt treatment, intestines, effect of, 145
Saprophytic vibrios, 14, 185
Sausage, contamination of, 145, 165, 172
Schwartzmen reaction, 139
SDS, see Sodium dodecyl sulfate
SDS-PAGE, see Sodium dodecyl sulfate-polyacrylamide gel electrophoresis
Season, incidence and, 34—35, 144, 147, 157
Seawater, epidemiology studies, see also Water, 147
Secondary food contamination, 150
Secondary transmission, 150, 152
Selective media, 40—44, 47—48, 82, 128, 186, 216—217
Selective pressures, effects of, 128
Selenite reduction test, bovine genital campylobacteriosis studies, 182—185
Self-limiting nature, diseases, see also Natural course, diseases,
Semen, cattle, infected, 187
Sensitivity testing, disk method, 82—84
Sepsis, nonlocalized, 23
Septic abortion, 151
Septicemia, 11, 22—23, 25, 30, 40, 78, 84, 117, 151
Seroconversion, 222
Serodeterminants, 140
Serogroups, 68—74, 139
  A, 11, 53, 62, 183—187, 214—215, 217—218, 221—222

B, 53, 62, 183—187, 214—215, 217—218, 222
C, 53, 62, 214—215
10 common, 69, 72
O_1 and O_2, 182
subserogroups, 71—72
Serological differences, *Campylobacter fetus* subsp. *venerealis* and subsp. *fetus*, 11
Serological responses
assay techniques, see also specific techniques by name, 98—100
Campylobacter jejuni infections, 98—103, 116
general discussion, 98, 103
individual patient response, 100—101
outbreaks, 99, 101—103
pigs, to infections, 221—222
specific, 116
Serological screening, 68
Serologic surveys, 148
Serology
diagnostic, *Campylobacter jejuni* infections, see also Diagnostic serology, 106—111
porcine, 221—222
response to infection studies, 98—103
traveller's diarrhea, 37
Serotype-specific antigens and antibodies, 99, 103, 106
Serotyping and serotypes
A, see A serogroup
applications, 56—58
B, see B serogroup
bacterial agglutination, 98
bacteriophage binding systems, 140
biotyping related to, 56, 70, 72—74
bovine genital campylobacteriosis studies, 182—187
C, see C serogroup
Campylobacter jejuni and C. *coli* studies, 13, 52—58, 62—74, 92—93
Campylobacter sputorum subsp. *mucosalis* studies, 214—215, 217—218, 221—222
enteric diseases, animal studies, 195, 204—205
epidemiology studies, 144, 147—151, 154, 157
general discussion, 52—53, 58, 62
historical development, 53—54, 62
Lau serotypes, 55—58
limitations, systems, 128
multiple serotypes, 68
nomenclature, 55
outbreaks, investigation of, 57, 72—73
outer membrane studies, 134, 137, 139—140
Pen serotypes, 55, 57—58
prevalence, serotypes, in human isolates, 56—57
reinfection, different serotypes in, 58, 106, 111, 205
results, reproducibility of, 68
serogroups, see Serogroups
serological response studies, 98—99, 101, 103
simultaneous infections, more than one serotype in, 103
slide agglutination method, see also Slide agglutination, serotyping system, 62—74, 98, 100—101, 217
specificity, 99, 103, 106, 137, 140
subserogroups, 71—72
thermolabile antigen method, see also Thermolabile antigen serotyping system, 13, 52—54, 62—63, 68, 106, 140, 214—215
thermostable antigen method, see also Thermostable antigen serotyping system, 13, 52—58, 106, 139—140, 214—215
tube agglutination, 53, 63—65, 98
typable strains, 62, 68—72
Serum antibody response, see Antibody response
Sex, incidence and, 34—35, 154, 156
Sexual transmission, 151
Sheep studies
abortion, see Ovine abortion
bovine genital campylobacteriosis, 183
food-borne, infections, 165
enteric infections, 194, 200—201, 204—205
enteritis, 115—116, 123
epidemiology, 145—146, 148, 157
regional ileitis of lambs, 223
taxonomy, 3—4, 6, 11, 15
Shellfish, contamination of, 150
Shigella sp.
enteritis experimental studies, 114, 125
epidemiology studies, 151—154, 156
human infection studies, 24—25
serotyping studies, 52
traveller's diarrhea studies, 34—37
Shigellosis, 34, 36—37
Sigmoidoscopy, 24
Silver-stained bacteria, 198, 204
SIM medium, 6—7, 12
Simultaneous infections, more than one serotype in, 103
Skirrow's medium, 41—43, 47, 82, 118, 178
Slide agglutination
Campylobacter sputorum subsp. *mucosalis* studies, 217
heterologous titers and reactions, 63—67
homologous titers and reactions, 62—68
live cultures, 62—63, 66—67
methodology, 62—68
serological response studies, 98, 100—101
serotyping system, 53, 62—74
Small mammal studies, see also specific types by name, 117—123, 125, 128
Sodium, tolerance to, 213
Sodium bisulfite, 173
Sodium chloride, in foods, effects of, 170—172
3.5% Sodium chloride, 14—15
Sodium deoxycholate, 213—214
Sodium dodecyl sulfate lysis procedure, plasmid isolation, 89
Sodium dodecyl sulfate-polyacrylamide gel electrophoresis, outer membrane studies, 134—139
Sodium metabisulfite, 42—43, 48, 178
Sodium pyruvate, 42—43, 48, 173, 178
Sodium selenite reduction test, see Selenite reduction test
Soil-borne infections, 147

Somatic antigens, 63, 68
Sonicate, 98—99, 103
South India, epidemiology studies, 155
Species and subspecies
catalase-negative, see also Catalase-negative campylobacters, 2—3, 14—16
catalase-positive, see also Catalase-positive campylobacters, 2—14, 16
types, see also specific types by name, 2—3
Species specificity, 128, 204
Spirillaceae, 2
Spirillum-like organisms, 12
Spirillum sp. 2
Spirochetal organisms, 146
Spirochetes, 202—203
Spontaneous recovery, from disease, see also Natural course, diseases, 27, 183
Sporadic abortion, 2, 4, 6, 11
bovine genital campylobacteriosis studies, 182—184, 186—187
causative agents, 184, 186
historical background, 182
pathogenesis and clinical signs, 183
transmission and dissemination, 187
S-shaped organisms, 119—122, 125, 185, 208
Staining, fecal cultures, 41, 46—47
Sterility, 2, 4, 11, 182
Steroids, 24, 183
Stools, see also Animal feces; Human fecal cultures; Human feces
blood in,
enteric infection studies, 195—196, 199—201, 204
enteritis experimental studies, 114, 116, 119, 125
human infection studies, 23—25
traveller's diarrhea studies, 36
Campylobacter jejuni isolation from, 40—48
human, see Human fecal cultures; Human feces
isolation of *Campylobacter jejuni* from, 40—48
mucus in
enteric infection studies, 195—196, 199—201, 204
enteritis experimental studies, 114, 116—117
traveller's diarrhea studies, 36
specimens, direct examination of, 40—41
Storage
anhydrous environment, effects of, foods, 172—174
food, infection and, see also Freezing; Refrigeration, 165—177
isolates, 46
temperature, foods, 167, 172—173, 175—177
Storage media, 46, 48
Strain discrimination methods, *Campylobacter jejuni-C. coli* group, 13—14
Stream water, epidemiology studies, see also Water, 147, 151
Streptomycin, 80, 82, 186—188, 215
Streptomycin-resistant strains, 186
Stuart medium, 40
Subactue diffuse mucopurulent endometritis, 183
Subcutaneous abscess, 23
Subserogroups, 71—72
Subspecies, see Species and subspecies
Sulfadiazine, 82
Sulfamethoxozole, 79, 82
Sulfonamide, 79, 82, 84, 215
Summer months, infection peaks during, 144, 147, 157
Superoxide anion, 43, 173
Superoxide dismutase, 173
Suppuration, other sites, in *Campylobacter fetus* subsp. *fetus* infection, 23
Surface antigens, 68, 98—99, 106, 139—140
Surface-associated proteins, 139—140
Surface-exposed outer membrane proteins, 135—136, 139—140
Surface structure, *Campylobacter jejuni*, see also Outer membrane and surface structure, *Campylobacter jejuni*, 128, 134—140
Surface water, contamination of, see also Water, 150—151
Surveillance artifacts, 144, 155
Survival, *Campylobacter jejuni*, in foods, 165—177
Suspending medium, 48
Swarming, 7, 9—12
Sweden, outbreak in, epidemiology studies, 150, 155
Swine, see Pigs
Swine dysentery, 3, 115—116, 194, 201—203
Systemic infection and spread of infection, 114, 119, 125

T

Tarrozi's liver broth, 46
Taxonomy, 2—16, 186
bovine genital campylobacteriosis studies, 182
Campylobacter sputorum subsp. *mucosalis* studies, 3, 14—16, 208, 217—218
catalase-negative campylobacters, see also Catalase-negative campylobacters, 2—3, 6, 12, 14—16
catalase-positive campylobacters, see also Catalase-positive campylobacters, 2—14, 16
general description, campylobacters, 2—3
general discussion, 2, 16
nomenclature, 2, 4—6, 14, 182, 201
Temperature
food-borne infections, 165—169, 172—173, 175—177
growth, see Growth temperature
incubation, optimal, 44
refrigeration, foods, 167, 172—173, 175
storage, food, 167, 172—173, 175—177
Temperature tolerance tests, 47
Tetracycline, 23, 29—39, 188, 204
resistance to, 29, 80—81, 84, 88—95, 215
bacteriophage-mediated, 91, 95

plasmid-mediated transfer, see also Plasmid-mediated transfer, tetracycline resistance, 88—95
susceptibility to, 79—84, 186, 215
Tetracycline-resistant plasmids
buoyant density, 92, 95
conjugative transfer, 89—91, 95
guanine + cytosine content, 92, 95
molecular weight, 88—92, 95
restriction endonuclease analysis, 92—95
Therapy, see Chemotherapy; Treatment
Thermal inactivation, *Campylobacter jejuni*, 166, 168—169
Thermolabile antigen, serotyping system, 13, 52—54, 63, 68, 106, 140, 214—215
multiple, 62
Thermolabile antigenic factors, slide agglutination serotyping studies, 62—74
role of, 68
Thermophilic campylobacters, 55, 134—137, 139
growth rate, 128
nalidixic acid-resistant, see Nalidixic acid-resistant thermophylic campylobacters
Thermostable antigens, serotyping system, 13, 52—58, 62—63, 68, 106, 139—140, 186, 214—215
procedure, 54—55
Thiamphenicol, 79, 82
Thienamycin, 84
Thioglycollate media, 40, 42, 46—48, 185
Tiamulin, 215
Ticarcillin, 82
Tinidazole, 82
TMP, see Trimethoprim
Tobramycin, 79—80, 84, 117
Todd-Hewitt broth, 41
Torbal jar, 45
Toxins, see also specific types by name, 123, 128
Transfer
bacteriophage-mediated, tetracycline resistance, 91, 95
conjugative, see Conjugative transfer
plasmid-mediated, see Plasmid-mediated transfer
Transmissible ileal hyperplasia, 223
Transmission, modes of
bovine genital campylobacteriosis, 182, 186—187
childhood, 152
contaminated food ingestion, see also Food, 148—150
contaminated water, see also Water-borne infections, 150—151
direct animal contact, 147—148
epidemiology studies, 144, 147—152, 157
human infection studies, 30
milk, see also Milk-borne infections, 149
nosocomial, 157
perinatal, 151—152
person-to-person, 147, 151—152
secondary, 150, 152
sexual, 151
undetermined, 148
vehicles, see Vehicles
Transport media, 40, 46—47, 217
Traveller's diarrhea, 34—37
diagnostic serology studies, 109
epidemiology studies, 34—36, 150, 154, 157
Treatment
antimicrobial agents, susceptibility to, see also Antibiotics; Antimicrobial agents, 78, 81, 83—84
bovine genital campylobacteriosis, 188
Campylobacter fetus subsp. *fetus* infections, 23
Campylobacter jejuni infections, 27—30, 114—115
chemotherapy, see Chemotherapy
diarrhea, 29—30, 114—115, 150
enteric infections, animals, 204—205
food, to eliminate *Campylobacter* sp., 164—177
Treponema hypodysenteriae, 201
Trimethoprim, 29, 42, 47—48, 178, 215—216
in vitro susceptibility studies, 79—82, 84
Trimethoprim agar, 216
Trimethylamine *N*-oxide, 7, 11
Triphenyl tetrazolium chloride, 8, 10, 14, 47, 201
Triple sugar iron medium, 6—7, 12—14, 201, 213
Triton® X-100 lysis procedure, plasmid isolation, 88—89
TSI medium, see Triple sugar ion medium
T-soy broth, 41
TTC, see Triphenyl tetrazolium chloride
Tube agglutination
serological response studies, 98, 100—102
serotyping studies, 53, 63—65, 98
Turkey studies
bluecomb disease, 6, 115, 122
epidemiology, 145, 149, 157
food-borne infections, 165, 173, 175
Tylosine, 204, 215
Typable strains, *Campylobacter jejuni* and *C. coli*, serotyping studies, 62, 68—72
Typhoid fever, 25

U

Ulcerative colitis, 24—27
Ultrastructural studies, *Campylobacter sputorum* subsp. *mucosalis*, 208—210, 219—221
Universal antigen, 98
Urban areas, epidemiology, 148
Urea, hydrolysis of, 13
Urinary tract infections, 25
Uterus, cow, infection in, 183

V

Vaccine and vaccination, in bovine genital campylobacteriosis, 186, 188
Vaginal infections
bovine genital camplyobacteriosis, 183, 186—188
human, epidemiology studies, 151—152

Vaginal mucus agglutination test, 187
Vago stain, 41, 46
Vancomycin, 42, 47, 81, 83—84, 178
Vehicles, transmission, epidemiology studies, 149—150, 152, 155, 157
Venereal bovine campylobacteriosis, see also Enzootic infertility, 182—183
Vi antigen, 52
Vibrio
  bubulus, 14—15, 185, 187, 208
  cholerae, 2, 34
  coli, 3—5, 116, 203
  fecalis, 12
  fetus, 4—5, 14, 62, 156, 182, 188, 208
  fetus var. *intestinalis*, 4—5, 182
  fetus var. *venerealis*, 4—5, 182
  jejuni, 3—5, 194
  sputorum, 14—15, 208
  sputorum var. *bubulus*, 15
  sputorum var. *sputorum*, 14—15
  succinogenes, 218
Vibrio-like bacteria, 219—221
Vibrio-like organism, intracellular, 223
Vibrionic abortion, 3—4, 11
Vibrionic hepatitis, 115
Vibrios
  aerobic, 12
  anaerobic,14—15
  bovine genital campylobacteriosis, studies, 182, 185
  Campylobacter sputorum subsp. *mucosalis* studies, 208, 213, 217—218, 221
  enteric infection studies, 194, 196, 203
  guanine + cytosine content, 182
  intracellular, 219—21
  microaerophilic, 2—4, 12—13, 15
  pathogenic, 185, 208
  related, 4, 52, 62
  saprophytic, 14, 185
  taxonomy studies, 2—4, 12—15
Virginiamycin, 215
Virulence factors, 122, 125, 128
Virulence properties, see also Attachment, Invasion, 137—140
Viruses, 195, 203
Voles, enteritis studies, 117
Vomiting, see also Nausea
  enteritis experimental studies, 117
  human infection studies, 23—24, 30
  traveller's diarrhea studies, 36

W

Water-borne infections
  diagnostic serology studies, 106, 109—110
  epidemiology studies, 144, 147, 150—151, 157
  outer membrane studies, 134
  serotyping studies, 73
Waterman's enrichment broth, 48
Water systems, municipal, contamination of, see also Water-borne infections, 150—151, 173—175
Watery diarrhea
  experimental enteritis studies, 114—117
  human infection studies, 24, 30
  traveller's diarrhea studies, 36
Wavelength, cell size determinations, 8, 11—12
Wild animal studies, epidemiology, 144, 146—147
Wild bird studies, see also Bird studies, 122, 144
Winter diarrhea, 194—195
Winter dysentery, 3, 194—195
Winter scours, 115

Y

Yeast extract-aspartate nutrient broth, 46, 48
Yersina, 125
YNAAB, see Yeast extract-aspartate nutrient broth

Z

Zone size, disc tests, *Campylobacter sputorum* subsp. *mucosalis* studies, 215
Zoo animals, epidemiology studies, 146—147